Contested Waterscapes in the Mekong Region

Hydropower, Livelihoods and Governance

EDITED BY

François Molle, Tira Foran and Mira Käkönen

earthscan
from Routledge

First published by Earthscan in the UK and USA in 2009

For a full list of publications please contact:
Earthscan
2 Park Square, Milton Park, Abingdon, Oxfordshire OX14 4RN
711 Third Avenue, New York, NY 10017

First issued in paperback 2014

Earthscan is an imprint of the Taylor & Francis Group, an informa business

ISBN: 978-1-84407-707-6 (hbk)
ISBN: 978-1-138-02118-1 (pbk)

Typeset by JS Typesetting Ltd, Porthcawl, Mid Glamorgan
Cover design by Susanne Harris

A catalogue record for this book is available from the British Library

Library of Congress Cataloging-in-Publication Data

Contested waterscapes in the Mekong Region : hydropower, livelihoods, and governance / edited by François Molle, Tira Foran, and Mira Käkönen.
p. cm.
Includes bibliographical references and index.
ISBN 978-1-84407-707-6 (hardback)
1. Water resources development--Mekong River Watershed. 2. Water-supply--Mekong River Watershed. 3. Hydroelectric power plants--Political aspects--Mekong River Watershed. 4. Water rights--Mekong River Watershed. 5. Watershed management--Mekong River Watershed. I. Molle, François. II. Foran, Tira. III. Kakonen, Mira.
TC513.M45C67 2009
333.91'150959--dc22

2008051424

Contents

List of Figures, Tables and Boxes

Figures

Tables

Boxes

List of Contributors

Robert Arthur is a fisheries management scientist at the WorldFish Centre in Cambodia, researching fisheries as complex dynamic social-ecological systems, and the role of science and research in fisheries management, with a focus on approaches such as adaptive management and co-management. He has been working in the region since 1999 and has recently been coordinating WorldFish support to the Wetlands Alliance.

Ram C. Bastakoti is an M-POWER coordinator based at Chiang Mai University, Thailand, with experience in agricultural and resource economics. He is involved in the study of socio-economic aspects of common pool resource management with a focus on water resources. He is currently researching the institutional dynamics of irrigation management, comparing cases from Nepal and Thailand.

David Biggs is professor of history at the University of California at Riverside, US. A specialist in Southeast Asian environmental topics, his past research has focused on environmental politics and history in the waterscapes of the Mekong Delta and his current research examines environmental dimensions of war in Vietnam and Southeast Asia. His essays have appeared in the *Journal of Southeast Asian Studies* (2003), *Environmental History* (2005) and *Technology and Culture* (2008), and his book *Between the Rivers and Tides: An Environmental History of Nation-Building in the Mekong Delta* (2009) is in preparation by University of Washington Press.

David J. H. Blake worked for over 12 years in northeast Thailand and Laos in relation to rural development, sustainable agriculture and smallholder aquaculture extension. Most recently, he worked as a technical adviser with the World Conservation Union (IUCN) to the Mekong Wetlands Biodiversity Conservation and Sustainable Use Programme's Thailand 'demonstration site' in the lower Nam Songkhram River Basin. He is currently completing his PhD at the School of Development Studies, University of East Anglia, UK.

John Dore is programme director with M-POWER, as well as a political economy and governance writer with particular interest in regionalism, transboundary and multicultural deliberative engagement and negotiation processes. He is an adviser to the AusAID Mekong Region Water and Infrastructure Unit based at the Australian Embassy in Laos.

Thierry Facon is a senior water management officer at the Regional Office for Asia and the Pacific of the United Nations Food and Agriculture Organization (FAO). He has been working in the region since 1999 on the development of tools, methodologies and capacity-building for the modernization of irrigation systems, and on the evaluation of irrigation performance, irrigation-sector reform, integrated water resources management, and intersectoral strategic planning in the water sector.

Philippe Floch is a researcher at the University of Natural Resources and Applied Life Sciences in Vienna. His research interests include river basin management, the governance of water resources planning and the design and implementation of irrigation water policies.

Tira Foran was born in Thailand and has been based there since 2001. He is a Research Fellow at Chiang Mai University – Unit for Social and Environmental Research. His background is in contentious politics, natural resource management and anthropology. His recent work has focused on transitioning to more sustainable energy systems: policy instruments for energy efficiency and renewable energy, estimation of commercially achievable alternatives to centralized electric power systems, and improved public participation.

Richard Friend is an independent consultant based in Bangkok, Thailand, and is the research leader of the fisheries and livelihoods research theme of M-POWER. He has a background in social anthropology and development studies and has been working in the Mekong region for over 15 years on issues related to livelihoods and policy analysis, with an interest in water resource management and fisheries.

Jelson Garcia is based in Bangkok, Thailand, and is a coordinator for the Mekong/Southeast Asia region of the Bank Information Centre, an independent non-profit NGO with headquarters in Washington DC. He has a background in anthropology and comparative sociology. He works with civil society organizations that advocate for the protection of human rights, participation, transparency and public accountability in the governance and operations of the World Bank, Asian Development Bank and International Monetary Fund. He is an M-POWER research fellow studying state–civil society relations in the context of watershed management in Yunnan, China.

Po Garden is now the project director of Internews' Earth Journalism Network. He has interests in climate change, science and communication and local government. At the time of work on this book he was a researcher at the Unit for Social and Environmental Research (USER), Chiang Mai University, Thailand.

Philip Hirsch is professor of human geography and director of the Australian Mekong Resource Centre at the University of Sydney, Australia. He has written widely on natural resource management, development and the politics of environment in the Mekong region. His current research is on new contexts of agrarian change in the region, and water governance in the context of irrigation development and catchment management.

Chu Thai Hoanh is a senior water resources specialist at the International Water Management Institute with experience in research and planning in agriculture and water management. His specific areas of focus include hydraulic modelling, crop modelling, land-use optimization, balancing rice supply and demand, and climate change scenario analysis in South and Southeast Asia, with an emphasis on the Mekong River Basin.

Patchamuthu Illangovan has been the World Bank's Country Manager for Laos since 2005. He has many years of experience as an environmental specialist in the Asia-Pacific region, particularly in the fields of environmental policy and institutions, water resources management and urban environmental management. He also served as the World Bank's Team Leader for the Nam Theun 2 Environment and Social Program between 2001 and 2006. He joined the World Bank in 1993; prior to that he worked both in the public and private sectors in his native Sri Lanka.

Mira Käkönen is a researcher with the Water and Development Research Group, Helsinki University of Technology, Finland. She has experience in natural resource management-related projects in the Mekong region and her research interests include water governance from the perspective of power and knowledge relations and changes in expertise.

Shawn Kelley is visiting research fellow at Chulalongkorn University's Social Research Institute in Bangkok, Thailand, and an M-POWER research fellow focusing on transboundary hydropower development. He has 15 years of experience in the Mekong region, specializing in finance and investment, energy security, and media and human rights, with a current emphasis on political risk and foreign investment in Southeast Asia.

Marko Keskinen is a researcher with the Water and Development Research Group, Helsinki University of Technology, Finland. His research concentrates on the interactions between water and society, as well as on interdisciplinary approaches to water management and impact assessment, with emphasis on the Mekong region.

Jorma Koponen is a senior modeller at the Environmental Impact Assessment Centre of Finland (EIA Ltd). He has long-term experience in development, validation, use and practical application of hydrological, hydrodynamic, water quality and ecosystem simulation models. He has been working in the Mekong Basin for several years.

Matti Kummu is a researcher at the Water and Development Research Group, Helsinki University of Technology, Finland. His research concentrates on human impact upon hydrology and sediment transport. He has several years of experience of working in the Mekong Basin, including intensive historical analysis of the hydrology of the Tonle Sap ecosystem and the Angkor complex.

Shannon Lawrence is Lao programme director with International Rivers, an NGO with headquarters in Berkeley, California. Her work involves documenting the impacts of hydropower projects in Laos, advocating for livelihood restoration for affected communities, and promoting better planning and accountability in the sector.

Kate Lazarus is a freelance consultant on water resources management and governance in the Mekong region. She is a member of the M-POWER steering committee and is currently co-leader of its dialogues research theme. Kate has previously worked for Oxfam America, IUCN and Amnesty International.

Louis Lebel is founding director of the Unit for Social and Environmental Research, affiliated with the Faculty of Social Science at Chiang Mai University, Thailand. He has interests and research publications in ecology, epidemiology, resilience and environmental governance. He has lived and worked in Thailand for most of the past 17 years.

Darrin Magee is assistant professor of Asian environmental studies at Hobart and William Smith Colleges in Geneva, New York. He is a geographer who specializes in water and energy in China, as well as in broader questions about the geographies of resources and waste. His current research interests include large dams and other water infrastructure, transboundary rivers and natural resource governance.

Kanokwan Manorom is assistant professor and currently a director of the Greater Mekong Sub-Region Social Research Centre at the Faculty of Liberal Arts, Ubon Ratchathani University, Thailand. She has long-term experience in social and political research and social impact assessment on water development and management in the northeast of Thailand, especially the dams located in the Mun River: the Pak Mun and Hua Na projects. She also worked with the Mekong River Commission (MRC) as a social specialist in the Integrated Basin Flow Management (IBFM) Programme during 2006 to 2007.

Carl Middleton is the Mekong programme coordinator with International Rivers, an NGO with headquarters in Berkeley, California. His work focuses on monitoring the dam-building industry in the Mekong region and on promoting better planning processes and technologies for sustainable and equitable use of water and energy resources.

Fiona Miller is a Future Generation fellow in the Department of Resource Management and Geography, University of Melbourne, Australia. She has a PhD in human geography from the University of Sydney for her research on the political ecology of risk and society–water relations in the Mekong Delta, Vietnam. Fiona has considerable experience conducting research on the social dimensions of environmental change in the Asia Pacific region – notably, Vietnam as well as Australia. She specializes in research on social vulnerability, water resources management, political ecology, sustainable livelihoods, resilience and community participation

François Molle is a senior researcher at the Institut de Recherche pour le Développement, France, and holds a joint appointment with the International Water Management Institute. He has 25 years experience in irrigation and water management in South America, Africa, Asia and the Middle East region. He now focuses his research on institutional and political aspects of river basin management and on water policy; he is co-editor of the *Water Alternatives* journal.

Fongsamuth Phengphaengsy is programme officer for Agriculture and Irrigation at the Mekong River Commission Secretariat. He has experience in irrigation and water management, with special focus on water-use efficiency, water productivity, irrigation scheduling modelling and irrigation system performance.

Buapun Promphakping is Associate Professor in social development at the Faculty of Humanities and Social Sciences, Khon Kaen University, Thailand. His research interest is mainly focused on poverty, rural development, people participation, gender and wellbeing. He was recently a member of ESRC Research Group called 'Well-being in Developing Countries' and led a research project under this Group in the Northeast of Thailand.

Jeffrey E. Richey is a professor at the University of Washington, US. His research focuses on the biogeochemistry and hydrology of river basins, from Hood Canal and Puget Sound to the large tropical basins of the Amazon, Mekong and Zambezi.

Juha Sarkkula is a senior research scientist at the Finnish Environment Institute, Finland. He has long-term experience in hydrological, hydrodynamic, sediment and water-quality modelling projects for water resources management and environmental impact assessment. He has considerable experience in working in the Mekong Basin.

Suong Seng is executive director of the Centre for Development-Oriented Research in Agriculture and Livelihood Systems, Cambodia. He is an M-POWER research fellow (2008 to 2009) conducting research on flood issues. His research interests include farming systems and natural resource management, livelihood systems and disaster risk management and mitigation.

Bach Tan Sinh is director of the Department of Science and Technology Human Resource Policy and Organization, National Institute for Science and Technology Policy and Strategy Studies under the Ministry of Science and Technology in Vietnam. He is currently leading the M-POWER research theme on floods and flood-related disasters. He is also the policy engagement coordinator of the Sustainable Mekong Research Network.

Try Thuon was formerly M-POWER research fellow during 2006 to 2007. He is the focal person for M-POWER research work in Cambodia and is based at the Centre for Study and Development in Agriculture in Phnom Penh. His research interests include the politics of natural resource management, river-based livelihood systems, political ecology, water resource, and irrigation management.

Duong Van Truc is project manager of the Centre for Environmental Engineering, Cadastre and Engineering Survey Company under the Ministry of Natural Resources and Environment, Vietnam. Duong has more than six years of experience with environmental planning and management. He is a member of the research group of the National University of Ho Chi Minh City, exploring solutions for urban flood management.

Le Anh Tuan is a senior lecturer on environmental and water resources engineering at Can Tho University, Vietnam. He has experience in teaching and research in agro-hydrology, water management and environmental engineering in Vietnam and southern Laos. Specific areas of research interest include constructed wetlands, rural water supply and sanitation, flood management and climate change.

Olli Varis is a professor at the Water Resources Laboratory of Helsinki University of Technology, Finland. He has broad and interdisciplinary experience of water and environmental issues, and has worked in development research and consultation. He is a frequently used expert by various international organizations and the author of over 300 scientific papers.

Preface: About M-POWER

This book is a product of M-POWER, which stands for the Mekong Programme on Water, Environment and Resilience. The network brings together people committed to improving local, national and regional governance in Cambodia, China, Laos, Burma/Myanmar, Thailand and Vietnam.

The ultimate goals of M-POWER are improved livelihood security and human and ecosystem health in the Mekong region. We contribute to this by focusing on improving water governance.

Our action research, practical policy support and facilitation involve pursuing fair and effective governance, which takes account of possible rewards, voluntary and involuntary risks, and rights and responsibilities of all authorities and stakeholders. We are committed to ensuring that water-related negotiations and decision-making, which almost always have political dimensions, are more fully informed and transparent.

M-POWER's action research programme is organized around comparative and regional studies and cross-cutting governance themes. Synthetic activities are guided by research leaders that build up multi-country and multi-organization teams.

M-POWER is primarily supported by the efforts and resources of the partner organizations who choose to cooperate in this transnational effort to improve water governance. Substantial financial support for 2006 to 2010 comes from the Consultative Group on International Agricultural Research (CGIAR) Challenge Programme on Water and Food via resources from Echel-Eau (Government of France) and the International Fund for Agricultural Development (IFAD).

M-POWER is coordinated by the Unit for Social and Environmental Research (USER), Chiang Mai University, Thailand. For more information, see www.mpowernet.org.

This volume is the second of a three-volume multi-authored book series on water governance in the Mekong region produced as part of the Mekong Programme on Water, Environment and Resilience. The first volume, *Democratizing Water Governance in the Mekong Region*, was published in 2007 by Mekong Press, Chiang Mai, Thailand. The series editors are Rajesh Daniel and Louis Lebel.

Acknowledgements

This volume has been conceptualized, shaped and written by partners in the Mekong Programme on Water, Environment and Resilience (M-POWER) water governance network, with authors from across the Mekong region. The editors acknowledge and thank all of the writers for their efforts, and the many who provided constructive feedback on drafts.

We are grateful to the many individuals who contributed their time and effort to reviewing and providing comments on individual chapters. We especially wish to thank Chris Barlow, Bryan Bruns, Simon Bush, Stéphanie Duvail, Wolf Hartman, Okudaira Hiroshi, Aviva Imhof, Kevin Yuk-shing Li, Jesse Manuta, Elena Nikitina, Andrew Noble, Jonathan Rigg, Theresa Wong and Emily Yeh.

Special thanks go to Geeta Bastakoti at the Unit for Social and Environmental Research (USER) for assistance with the final stages of publishing this volume and to Alison Kuznets and Tim Hardwick at Earthscan for their support and patience.

List of Acronyms and Abbreviations

ADB	Asian Development Bank
AFD	Agence Française de Développement
AIT	Asian Institute of Technology
amsl	above mean sea level
AOP	Assembly of the Poor
ARD	Accelerated Rural Development Programme
ASEAN	Association of Southeast Asian Nations
BCE	before the Common Era (formerly known as BC)
BDP	*Basin Development Plan* (of the MRC)
BDP	basin development planning
BMA	Bangkok Metropolitan Authority
BOT	build–operate–transfer
BOOT	build–own–operate–transfer
CE	Common Era (formerly known as AD)
CEO	chief executive officer
CEPA	Culture and Environmental Preservation Association
CGIAR	Consultative Group on International Agricultural Research
CIA	cumulative impact assessment
CSIRO	Commonwealth Scientific and Industrial Research Organization
CSO	civil society organization
CSR	corporate social responsibility
CSS	Country Safeguard Systems
CWRC	Changjiang (Yangtze River) Water Resources Commission
3D	three dimensional
DC	direct current
DEDP	Department of Energy Development and Promotion (Thailand)
DHP	Department of Hydropower Planning (Myanmar)
DKBA	Democratic Karen Buddhist Army
DSF	Decision Support Framework (of the MRC)

DSHEP	Don Sahong Hydro Energy Project
DWR	Department of Water Resources (Thailand)
ECAFE	United Nations Economic Commission for Asia and the Far East
ECOSORN	Economic and Social Re-Launch of Northwest Provinces in Cambodia
EDF	Electricité de France
EdL	Electricité du Laos
E-Flows	Environmental Flows
EGAT	Electricity Generating Authority of Thailand
EGCO	Electricity Generating Public Company of Thailand
EIA	environmental impact assessment
EP	Equator Principles
EPF	Electric Power Forum
EU	European Union
EVN	Electricity of Vietnam
FAO	United Nations Food and Agriculture Organization
FIVAS	Association for International Water and Forest Studies
FWUC	farmers' water user community
GDP	gross domestic product
GEF	Global Environment Facility
GMS	Greater Mekong Sub-Region
GOL	Government of Laos
ha	hectare
HPID	Hydropower Implementation Department (Myanmar)
HU	Health Unlimited
IAG	International Advisory Group
IBP	*Indicative Basin Plan*
IBFM	integrated basin flow management
IFAD	International Fund for Agricultural Development
IFI	international financial institution
IGA	Intergovernmental Agreement
IKMP	Information and Knowledge Management Programme
IMA	independent monitoring agency
IMF	International Monetary Fund
IMT	irrigation management transfer
INGO	international non-governmental organization
IPP	independent power producer
IRP	integrated resources planning
ISFP	Initiative on Soaring Food Prices (FAO)
ITD	Ital-Thai Development
IUCN	International Union for Conservation of Nature (formerly World Conservation Union)
IWMI	International Water Management Institute

IWRM	integrated water resources management
JC	Joint Committee
JICA	Japan International Cooperation Agency
KCM	Khong-Chi-Mun Project
km	kilometre
km^2	square kilometre
KNU	Karen National Union
kWh	kilowatt hours
kV	kilovolt
KVHG	Kamnan and Village Headmen's Group
LMB	Lower Mekong Basin
LSRB	Lower Songkhram River Basin
m	metre
m^3/s	cubic metres per second
MAFF	Ministry of Agriculture, Forestry and Fisheries (Cambodia)
MARD	Ministry of Agriculture and Rural Development (Vietnam)
MASSCOTE	Mapping System and Service for Canal Operation Techniques
MDB	multilateral development bank
MEP	Chinese Ministry of Electric Power
MoNRE	Ministry of Natural Resources and the Environment (Vietnam and Thailand)
MoU	memorandum of understanding
MOWRAM	Ministry of Water Resources and Meteorology (Cambodia)
M-POWER	Mekong Programme on Water, Environment and Resilience
MRC	Mekong River Commission
MRCFP	Mekong River Commission Fisheries Programme
MRCS	Mekong River Commission Secretariat
MW	megawatt
MWBP	Mekong Wetlands Biodiversity Conservation and Sustainable Use Programme
MWRAS	Mekong Water Resources Assistance Strategy
n	total population sample size
NDRC	National Development and Reform Commission (China)
NEA	National Energy Authority (Thailand)
NEB	National Environmental Board (Thailand)
NESDB	National Economic and Social Development Board (Thailand)
NESDP	National Economic and Social Development Plans (Thailand)
NGO	non-governmental organization
NLF	National Liberation Front (Vietnam)
NMC	National Mekong Committee
NMCS	National Mekong Committee Secretariat
NPA	national protected area
NPIMP	National Pump Installation Management Project

NSO	National Statistics Office (Thailand)
NT2	Nam Theun 2 Dam
NTEC	Nam Theun 2 Electricity Consortium
NTFP	non-timber forest product
NTPC	Nam Theun 2 Power Company
O&M	operation and maintenance
OAA	other aquatic animals
OECD	Organisation for Economic Co-operation and Development
ONEP	Office of Natural Resources and Environmental Policy and Planning (Thailand)
ORDPB	Office of the Royal Development Project Board
PDP	*Power Development Plan*
PER	Project for Ecological Recovery
PIM	participatory irrigation management
PIMD	participatory irrigation management and development
PIP	*Project Implementation Plan* (of the WUP)
PNCPA	Procedures for Notification, Prior Consultation and Agreement
PoE	International Panel of Environmental and Social Experts
PPA	Power Purchase Agreement
PPC	Provincial People's Committee (Thailand)
PRC	People's Republic of China
RID	Royal Irrigation Department (Thailand)
RMK-BRJ	Raymond Morrison Knudsen-Brown Rootes Jones
RVN	Republic of Vietnam
SCIP	Stung Chinit Irrigation Scheme (Cambodia)
SEMFOP	*Social and Environmental Framework and Operational Plan*
Sida	Swedish International Development Agency
SIP	Small Irrigation Project(s)
SOE	state-owned enterprise
SPDC	State Peace and Development Council (Myanmar)
SRAL	Special Rehabilitation Assistance Loan
t/ha	tonnes per hectare
TAO	Tambon Administration Organization
TAR	Tibet Autonomous Region
TDRI	Thailand Development Research Institute
TERRA	Towards Ecological Recovery and Regional Alliance
THHP	Theun-Hinboun Hydropower Project
TISTR	Thailand Institute of Scientific and Technological Research
TNMC	Thai National Mekong Committee
TVA	Tennessee Valley Authority
UBU	Ubon Ratchathani University
UK	United Kingdom
UN	oUnited Nations

UNDP	United Nations Development Programme
UNESCO	United Nations Educational, Scientific and Cultural Organization
US	United States
USAID	United States Agency for International Development
USBR	United States Bureau of Reclamation
USER	Unit for Social and Environmental Research
USOM	United States Overseas Mission
VIC	Variable Infiltration Capacity
VLPC	Vietnam–Laos Joint Stock Company
WANI	Water and Nature Initiative
WCD	World Commission on Dams
WMPA	Watershed Management Protection Authority
WREA	Water Resources and Environment Agency
WUA	water user association
WUP	Water Utilization Programme
WWF	World Wide Fund for Nature

1

Introduction: Changing Waterscapes in the Mekong Region – Historical Background and Context

François Molle, Tira Foran and Philippe Floch

Introduction

The Mekong region fans out from the folds of the eastern Himalayas that give birth to its main arteries, including, from west to east, the Irrawaddy, the Nu-Salween, the Chao Phraya, the Lancang/Mekong and the Red rivers (see Figure 1.1). These rivers have constituted defining features of Southeast Asian cultures, religions, ways of life and substantive economies. Winding through deep gorges in their upper reaches, the region's rivers, together with their tributaries, have lent themselves to the construction of dams and hydropower generation plants. Entering large plains and ending in wide deltas, they have been diverted to support large-scale irrigation, while all along their course, they have long provided fish and other aquatic products to local dwellers, as well as means of transportation. In upper catchments, their tributaries have, for centuries, been tapped by highlanders for small-scale irrigation and other domestic uses.

As a result, agrarian landscapes have traditionally been divided between forested highlands, exploited directly or through swidden farming techniques; intermountain valleys with bottoms mostly under paddy cultivation; large plains and deltas devoted to rice cultivation under various guises; and uplands planted to both rice and field crops. With time and the closure of agricultural frontiers, water use and cultivation have intensified; greater levels of control over water have also been achieved through continuing investments in embankments, canals, drains, reservoirs, pumping stations and on-farm irrigation infrastructures. More recently, water came to be partly 'recaptured' and further domesticated by urban and industrial interests, supplying cities, diluting waste and generating hydroelectricity.

This book focuses on the dynamics of *waterscape transformation* in the Mekong region: by waterscape we mean here the surface and groundwater resources of an area of land and their interrelationships with other physical, climatic and biotic elements, as well as with human activities. Waterscapes are an expression of the interaction between humans and their environment and encompass all of the social, economic and political processes through which water in nature is conceived of and manipulated by societies. In other words, waterscapes are landscapes viewed through the lens of their water resources, taken as a defining element of both ecosystems and human life.

As the subtitle indicates, this volume puts particular emphasis on three dimensions of Mekong waterscape transformations: first, many current changes and challenges revolve around *hydropower*. For various reasons that include the Indochina wars and other political circumstances, the region is still characterized by a low density of large dams compared with other parts of the world. But current economic growth rates combined with high fossil fuel prices have spurred a rush towards hydropower generation that has the potential to completely remodel regional waterscapes. Second, *livelihoods* refer to the means of subsistence of rural, often impoverished, populations for whom a substantial part of their livelihoods is linked to the use and management of forest and wetland ecosystems, fisheries and the practice of rain-fed and irrigated agriculture. As such, they are directly threatened by large-scale transformations designed and decided in other spheres, often without their knowledge. The third issue of *governance* refers to the distribution of decision-making power. All transformative options that result in large-scale alterations of the hydrological regime, in terms of quantity, quality, timing or sediment load, tend to generate externalities that affect particular ecosystems and users. These externalities result from the nature of the hydrological regime, which interconnects individual or groups across river basins, and from its manipulation through hydraulic infrastructure and associated management rules. All interventions, whether implemented by the state (dams, flood control, irrigation schemes, inter-basin transfers, etc.) or resulting from combined small-scale decisions (e.g. individual well-drilling and construction of farm ponds), tend to generate costs, benefits and risks. Governance, thus, refers to the way in which decisions are made and power exercised, and to the spatial and social distribution of related benefits and externalities. The intent of this volume is to contribute to a better understanding of the transformation currently under way in the Mekong region, what is at stake, who benefits and who is at risk, and to improved water

Figure 1.1 *The main river basins of the Mekong region*

Source: adapted from Kummu (2008)

governance by reopening and investigating the political dimensions of decision-making over water resources in the Mekong region.

Brief history of water resources development in the Mekong region

Much scholarly work has described and analysed the history of water resources development in the Mekong region, notably the Mekong Basin itself, both in physical and institutional terms (see, for example, Bakker, 1999; Friesen, 1999; Thi Dieu, 1999; Browder, 2000; Hori, 2000; Le-Huu and Nguyen-Duc, 2003; Ratner, 2003; Hirsch and Jensen, 2006). This section only recaps the main historical benchmarks as a way of contextualizing the questions addressed in the chapters of this volume.

Early planning and the formation and demise of the Mekong Committee (1951 to 1975)

The initiation of 'modern' and coordinated efforts to 'harness' the Mekong River are generally associated with the establishment of the United Nations Economic Commission for Asia and the Far East (ECAFE), which was created in 1946 in an effort to promote post-war economic development in the region. But it was not until the seventh ECAFE session, held in 1951, that a call to study technical problems of river flood control would shape the Lower Mekong Basin's water developmental visions for at least the next 40 years. By 1952, ECAFE's Bureau of Flood Control had drawn up a working paper (ECAFE, 1952) that, far from dealing solely with flood control, also detailed a wider vision for water resources development in the Mekong Basin.

Apart from ECAFE's interest in promoting regional development, the US increasingly looked at Southeast Asia as a critical terrain in its efforts to contain the spread of communism after Mao's takeover of China in 1949, and saw economic development as one measure of its wider containment policy for the region. By 1955, the International Cooperation Administration (a precursor to the United States Agency for International Development, or USAID) commissioned the United States Bureau of Reclamation (USBR) to conduct a study that was published in 1956 (USBR, 1956)[1] but largely ignored by the riparian governments. ECAFE and its executive secretary produced their own study, which was presented at ECAFE's tenth anniversary meeting (ECAFE, 1957). The consultants hired by ECAFE laid out a preliminary development scheme that identified five primary dam projects on the Lower Mekong mainstream (Pa Mong, Khemerat, Khone Falls, Sambor and Tonle Sap), two more mainstream possibilities (near Luang Prabang and Thakhek) and a tributary site (Nam Theun River in Laos).

In October 1957, the Committee for Coordination and Investigations of the Lower Mekong Basin (in short, the Mekong Committee) was established with the mandate to 'promote, coordinate, supervise and control planning and investigations of water resources development projects in the lower Mekong Basin' (Article 4 of the statute). It was also given the authority to prepare and submit plans for coordinated research, study and investigations, make special financial and technical funding requests, and recommend to the four riparian governments criteria for sharing water resources – an authority that would officially sanction the role of the committee in 'harnessing' the Mekong River (Friesen, 1999). An executive agent was posted in 1959 and a permanent office created later.

A review study of the earlier USBR (1956) and ECAFE (1957) reports was entrusted to Lieutenant General Raymond A. Wheeler, a retired engineer from the US Army Corps of Engineering, who recommended three top-priority projects: the Pa Mong, Sambor and Tonle Sap dams. Wheeler was seized by what he called 'a majestic river' and was readily 'convinced of the great potential of the Lower Mekong for service to the riparian countries in the fields of navigation, hydropower generation, irrigation and other related water uses'. The Japanese, likewise, promoted the development of the Lower Mekong and surveyed 34 'promising' tributaries, among them the rivers of northeast Thailand, for which they envisioned a 'remarkable development of agriculture' if the Mekong waters could be diverted to this otherwise little fertile region (Hori, 2000). US geographer Gilbert White's (1962) report called for carefully designed tributary projects, but underlined the economic risk of over-enthusiasm and large 'monolithic concrete structures whose immediate return is inflation of national ego'. The development focus of the Mekong Committee shifted somewhat to tributary projects, with a total of eight dams constructed up to the early 1970s under its auspices, including the Nam Ngum Dam in Laos and several others in northeast Thailand.

Thailand, the closest Cold War ally to the US in mainland Southeast Asia, received substantial economic aid and advice from the US and the World Bank, with an emphasis on electrification, roads, reservoirs and canals (Muscat, 1990). Technical and financial support were instrumental in helping Thailand to construct several large power generation projects, including the 535MW multipurpose Bhumipol (Yunhee) Dam on the Ping River (a tributary of the Chao Phraya), commissioned in 1964, and the early stages of Thailand's electricity transmission network (Greacen, 2004). Under their advice, in 1968, Thailand established a state-owned electricity utility, the Electricity Generating Authority of Thailand (EGAT).

The 150MW Nam Ngum 1 Dam, the first large hydropower dam in Laos, was built with technical advice from the Mekong Committee and the World Bank in the late 1960s. Located in Vientiane Province, 90km north of the capital, the project was built as the US-backed Royal Army and the Vietnam-backed Pathet Lao Army fought for control of the country (Thi Dieu, 1999). To make way for the project, at least 800 families were resettled; yet none received any compensation

(Hirsch, 1998). Inaugurated in 1971 and foreshadowing what would become the predominant development strategy of Laos from the 1980s onwards, the Nam Ngum 1 Dam became a significant earner[2] for Laos and now sells 70 to 80 per cent of its power to Thailand.

The US was the largest non-riparian aid donor and provided 37 per cent of the total US$86 million contribution to the Mekong Committee in the first ten years of its existence (Friesen, 1999). In 1958, the US government signed an agreement with the Mekong Committee for the collection of basic scientific data for the whole of the Lower Mekong mainstream, and in 1961 agreed to fund the phase 1 pre-feasibility study of the Pa Mong mainstream project, together with a later feasibility study. The Pa Mong Dam was the cornerstone and poster child of US strategy in the region. The dam, as laid out in studies by USBR (1970), was of truly awesome dimensions: 98m in height, a storage capacity of over 100 billion cubic metres, could generate up to 4000MW, irrigate some 2 million hectares and inundate a total area of almost $4000km^2$ (see Figure 1.2). It would displace 250,000 people, a figure that was later revised upwards to 400,000. At the cost of US$1 billion, the Pa Mong would be the world's largest multipurpose dam at the time, an engineer's dream and a 'once in a lifetime' project for Lyle Mabbott, the Pa Mong project manager (Jenkins, 1968).

In 1970, the Mekong Committee published its first major basin-wide development plan: the *Indicative Basin Plan* (IBP) (Mekong Secretariat, 1970). The report built on the previous studies and was, by any standards, grandiose and comprehensive, listing some 180 possible projects on the tributaries and the mainstream: it defined short- and long-term (up to the year 2000) goals; it emphasized that large-scale irrigation was necessary to transform the agricultural sector; it saw hydroelectric power generation as a key to securing industrialization for greater prosperity; it found that flood control relied on dikes and dams on the mainstream; and it foresaw transportation from the mouth of the Mekong in Vietnam to upper Laos facilitated by a series of cascading dams equipped with navigation locks (see Figure 1.3). While the tributary projects were seen as attractive in dealing with the short-term developmental needs of riparian countries, it was the long-term development potential of the major dams that would comprehensively uplift the region (see Figure 1.3). The IBP report also proposed additional field investigations that would include fisheries, forestry, resettlement, wildlife, sedimentation, Mekong River crossings, navigation facilities, urban studies, archaeological studies and environmental studies (Friesen, 1999).

Four mainstream sites were to be completed by the 1980s (the Pa Mong in 1983, the Stung Treng and Sambor in 1985 and the Tonle Sap Barrage in 1987) at a total cost of US$10 billion. However, growing unrest and resistance by the Pathet Lao guerrillas in the region eventually derailed the Pa Mong Project, making it both too costly and too risky (Biggs, 2006). The Mekong Committee ultimately disbanded in 1975, when the Pathet Lao and the Khmer Rouge acceded to power, while Vietnam was about to reunify.

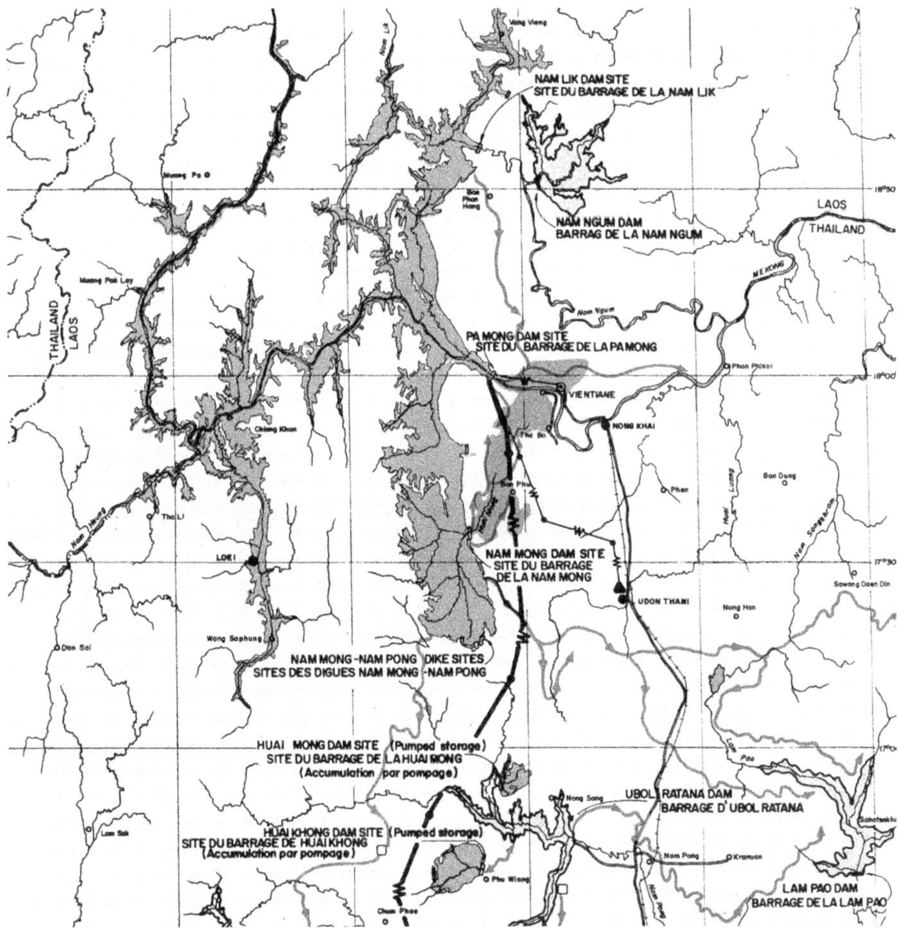

Figure 1.2 *The Pa Mong Dam Project (1970)*

Source: USBR (1970)

In Vietnam, throughout the Cold War period, Russia provided support in much the same way that the US and World Bank supported Thailand (Greacen and Palettu, 2007). Russian support for development of Vietnam's electricity sector was channelled through the state-owned monopoly, Electricity of Vietnam (EVN). Significant technical and financial support was provided for Vietnam's earlier hydropower projects, including the massive 1920MW Hoa Binh Dam that commenced in 1979, was completed in 1994, and remains mainland Southeast Asia's largest dam.

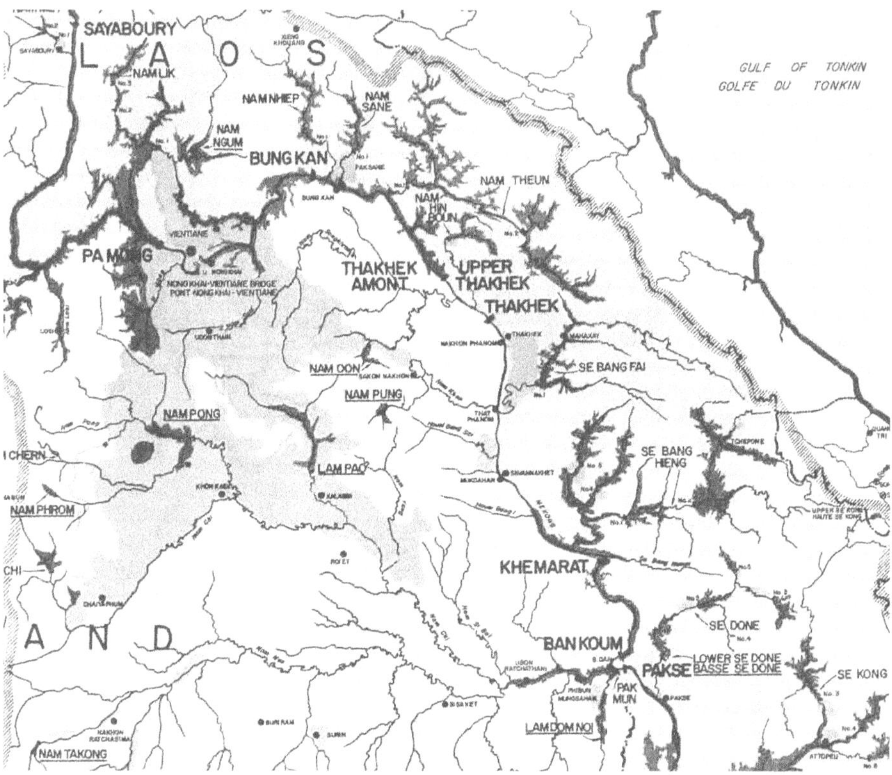

Figure 1.3 *Plans for hydropower development in Laos (1970)*

Note: Reservoir water bodies appear in the darkest shade of grey.

Source: Mekong Secretariat (1970)

The Interim Committee and the revision of a development vision (1975 to 1992)

The US withdrawal from Southeast Asia left a large hole in project funding and the United Nations Development Programme (UNDP), which had picked up funding where the now-defunct ECAFE had left off, reduced its contribution from US$5.6 million in 1973 to zero in 1976. Contributions to the committee from riparian nations also dropped significantly because of their own financial situation (Friesen, 1999). These changes altered the working base of the Mekong Committee dramatically. In 1978, after one year of negotiations, the three remaining country members of the original group negotiated new terms of cooperation and decided to form the Interim Committee for Coordination of Investigations of the Lower Mekong Basin, with a base in Bangkok.

Apart from the lack of financial resources, the withdrawal of Cambodia made the dream of developing the Mekong mainstream look more distant; subsequently, the focus of the committee shifted to smaller and national (tributary) projects. This, however, did not entail that the vision of comprehensive development of the mainstream had vanished. In 1980, an Interim Mekong Committee study reiterated that a mainstream cascade of dams should have priority once the committee was reunited to its initial four-member structure. In an effort to reframe and reassess the options for water resources development, a revised *Indicative Basin Plan* was published in 1987 (Interim Committee, 1988) as only 16 out of the 180 possible projects outlined in the *Indicative Basin Plan* had been implemented. Unlike the first basin plan, out of pragmatism and political realism, the 1987 plan refocused its attention on the development potential of each individual country, yet still proposing a cascade of eight dams on the mainstream as the best option for long-term development of the basin's water resources (Mitchell, 1998). The Pa Mong, still seen as the cornerstone of the overall development scheme, was now considered 'problematic' and its proposed height reduced from 250m above mean sea level (amsl) to 210m amsl in order to reduce the scale of resettlement. The Lower Pa Mong and Nam Theun 2 (NT2) projects were seen as 'enjoy[ing] very attractive economics ... [and] should, therefore, from an economic and technical point of view, be built as soon as possible'. Significantly, unlike the 1970 plan, the revision now saw the generation of hydropower as the largest benefit of developing a cascade on the Mekong mainstream, with other benefits, such as flood control, fisheries and navigation, insignificant in economic terms.

In the wake of the 1987 revised plan, in 1990 the Interim Committee commissioned another study of the potentialities for mainstream development: *Mekong Mainstream Development Possibilities: Summary Report* (Interim Committee, 1990). Although a more environmentally sensitive rhetoric was deployed and (minimal) changes to the cascade scheme were made, the report did not really change the overall configuration of the Mekong development project. Resettlement, however, was – at least in this report – considered a priority parameter of a project selection process guided by 'limitation imposed by resettlement requirements, conservation of the environment, minimum flow requirements for downstream interests, reduction of downstream effects caused by varying releases for power production in the case of peaking operations' (Interim Committee, 1990). The total number of people to be displaced by the development of the cascade, however, was still estimated at 330,000; and this was only a rough and preliminary estimate, which led to the conclusion that more studies on social and environmental impacts would be required on a project-by-project basis.

In the late 1980s, Laos started to parallel Thailand's effort at developing hydropower on the Mekong River's tributaries. Following a 1991 World Bank-endorsed feasibility study, which stated that it was the 'best option' for hydropower development in Laos, the NT2 Dam was more vigorously pushed forward; together with other plans at Ho Houay and Nam Theun-Hinboun, a total of 23 projects were targeted for construction up until 2010.

In 1991, the Mekong Secretariat commissioned the Compagnie Nationale du Rhône (France) and Acres International (Canada) to study (again) alternative ways of putting the Mekong River's resources to use. Unlike previous studies, the planners switched from the classical cascade of storage dams on the mainstream to a cascade of 'smaller' run-of-river projects (CNR and Acres International Limited, 1994).

The Mekong River Commission and its *Indicative Basin Plan* (1992 to present)

The geopolitical implications of the collapse of the Soviet Union in 1991 and the weakening of its satellite states since the mid 1980s set in motion substantial shifts in the Mekong region's political and economic landscape. In 1986, the governments of Laos and Vietnam, while remaining socialist states, initiated market-oriented economic reforms (*Doi moi tu duy* in Vietnam and the New Economic Mechanism in Laos). In Cambodia, the signing of the Paris Peace Accords paved the way for democratic elections in 1993 and the country's transition to a market-oriented economy. In Thailand, then Prime Minister Chatichai Choonhavan famously called for a transformation of the Mekong region 'from battlefields to marketplaces', heralding his government's policy shift from that of Cold War hostility towards the promotion of regional trade and investment and triggering renewed hope that the Mekong Committee could finally be reinstated with all its original member states.

As regional stability was restored step by step, Western bilateral aid agencies, the World Bank, and the Asian Development Bank (ADB) returned in earnest, offering aid and investment opportunities. Support for hydropower projects was high on their agendas (Ryder, 2004). By 1991, with funding from Sweden, Norway, the ADB and UNDP, the second largest hydropower dam in Laos, the 45MW Xeset 1 Dam, was completed, generating electricity for export to Thailand and domestic consumption. In 1995, a new arrangement between the four original members of the Mekong Committee was signed, and the four governments re-established their cooperative efforts under the new banner of the Mekong River Commission (MRC), despite its weakened mandate compared with that of the original Mekong Committee (Ratner, 2003; see Chapter 14 in this volume).

In 1992, the ADB launched the Greater Mekong Sub-Region (GMS) programme, endorsed by the region's governments, which set a path towards regional economic integration (ADB, 2007). Centred on establishing a market-based economy, the GMS programme, to date, has emphasized physical interconnectivity of the region, entailing the construction of major infrastructure projects such as transnational highways, railways, hydropower dams and regional transmission lines, and programmes that encourage cross-border trade and the integration of markets. The GMS programme has shaped many Western bilateral donors and the World Bank's aid strategies towards the Mekong region.

Undeniably, aggregate economic wealth has grown remarkably throughout the region. Far less, however, has been achieved in addressing the environmental and social issues that have accompanied this economic growth (Cornford and Matthews, 2007). The fact that much of the economic activity promoted under the GMS programme relies on the exploitation of the region's natural resources leads to a readily apparent contradiction within the programme's goals of widespread economic growth and helping to 'ensure sustainable development and conservation of natural resources' (ADB, 2007). Furthermore, much of economic growth has benefited urban areas rather than rural areas, leading to negative impacts on subsistence-based rural livelihoods and growing inequality (UNEP and TEI, 2007).

The ADB's GMS programme has replaced the earlier Mekong Committee as the principal framework for channelling economic development assistance into regional projects (Ratner, 2003). This allowed the ADB to focus unhindered on regional economic development, while leaving the potentially contentious management of the Mekong River to the MRC (see Chapter 14).

Current challenges and dynamics

The Mekong region's economic dynamism is associated with social, economic and environmental transformations that include deforestation and environmental degradation; growing commercialization of agriculture and increasingly multi-sectoral rural livelihoods; urbanization and industrialization; increased migration and the spread of diseases such as HIV; and population growth in the Mekong Basin that rose from 35 million in 1970 to 65 million at present (Parnwell and Bryant, 1996; Rigg, 1997; de Koninck, 2003).

Natural resources are under pressure and countries such as Laos or Cambodia are opening up to foreign investors interested in exploiting mines or expanding plantations of trees for either pulp or oil/biofuel production. A paramount current dynamic is the groundswell of hydropower projects in the region. Dams recently concluded or under construction include a cascade of dams in China's upper reaches in the Lancang River (the Upper Mekong), the NT2 Dam in Laos, and several others in the '3S' region[3] shared by Vietnam, Laos and Cambodia. The growing enthusiasm for hydropower is increasingly driven and exploited by private companies, financiers and government elites who largely bypass the traditional players such as the MRC, the ADB or the World Bank, with complex impacts upon political decision-making (Chapters 2 and 14). While electricity-dependent segments of society (particularly industry, but also urban elites) may benefit from hydropower plants, the manner in which many projects are currently being developed offers little comfort to those affected. Across the region, one finds no shortage of easy rhetoric about how export-oriented hydropower will help 'kick-start development', help 'eradicate poverty' or 'power progress', but far fewer examples of tangible links between investor-owned dams and rural electrification or improved livelihoods.

In parallel with the interest in hydropower, Mekong countries display ongoing interest in expanding irrigation and flood control infrastructure. Despite disappointing experiences with recent irrigation development or rehabilitation projects in Laos and Cambodia, the promise of improved productivity, food security and poverty alleviation puts irrigation expansion on the agenda of politicians and development banks. Thai politicians also mobilize such arguments when making renewed proposals for massive irrigation development (e.g. with the Water Grid Project), most particularly in northeast Thailand (see Chapter 10). Opportunities for rent-seeking from large construction contracts may also drive irrigation agencies and consulting firms. Globally, the World Bank has argued that it is necessary to boost investment in water infrastructure (Grey and Sadoff, 2007), while high rice prices in 2007 to 2008 have quickened new donor interest in expanding irrigation works in Cambodia.

In the Mekong region, the burst of investor interest in hydropower and the revival of donor interest in irrigation take place in a governance context where developers externalize costs; where authorities do not systematically screen and rank projects according to economic, environmental and social criteria; and where planners think in terms of supply-side, not demand-side, alternatives (Greacen and Palettu, 2007). In short, recent water resources development occurs in a context where evidence of coordinated, rigorously justified river basin development is not strong. Despite a process of democratization and the emergence or strengthening of civil society organizations (NGOs, academics and community-based organizations), megaproject triumphalism complemented by faith that socio-political and ecological impacts can be mitigated and transcended remains pervasive.

As the volume's opening chapters on hydropower assert, currents of modernist progress in the Mekong are being challenged by important counter-currents of critique and resistance. Such critique, when informed by credible knowledge (e.g. regarding irrigation design and implementation) offers a set of lessons about making development work (see Chapter 6). But, of course, in the gulf between lessons offered and lessons learned we find the full spectrum of politics. How political processes unfold varies among Mekong countries; but one important dynamic since the 1990s is that of national and transnational civil society advocacy. Obvious targets for such advocacy are the MRC (ostensibly set up to harmonize river basin development plans) and international development banks. But advocacy networks have also raised concerns about the downstream impacts of China's plans to build a cascade of hydropower dams on the Lancang (Upper Mekong) and similar plans to develop hydropower on the Nu-Salween River in China, as well as in Myanmar/Burma.

The governments in the Mekong region have often dismissed or constrained critical conversations about water, social change and development. The techniques of constraint can be direct, as with the suppression of dissent in the military regime of Myanmar, or indirect, through the production of knowledge (see Chapters 3 and 12) or *ad hoc* 'participatory' processes rolled out by a variety of agencies (see Chapter 13). But instead of drawing only pessimistic conclusions about democratization in the region, the chapters in this book invite the reader to explore more thoroughly how waterscapes have been, and are being, imagined and transformed.

STRUCTURE OF THE BOOK

This volume is divided into three parts that follow this introductory chapter. The first part focuses on hydropower expansion in the region, the second on issues of livelihoods and local development, while the third part reflects on knowledge, discourses and power.

Chapter 2 by Carl Middleton, Jelson Garcia and Tira Foran focuses on the phenomenon of 'new' hydropower developers and financiers, but also tracks how long-standing actors such as the World Bank and international civil society have responded to the entry of players with apparently lower environmental and social standards. Hydropower dam development in the region is then illustrated by three case studies of dams in various stages of development.

Chapter 3 by Tira Foran and Kanokwan Manorom provides an account of the history and politics of contention over Thailand's Pak Mun Dam. Built between 1990 and 1994 in a context of local support *and* resistance, the Pak Mun case offers a wealth of insights into the challenges of fair compensation, mitigation and participatory management, as well as a window into the complexity of rural livelihoods and democratization. Chapter 4 by Shannon Lawrence reviews the development of the Nam Theun 2 Dam, the largest as well as one of the most publicized and contentious water resource projects in Laos. Containing a trans-basin water diversion, hydropower and rural development scheme of unprecedented size, complexity and aspiration, the project breaks new ground in terms of promises made to better the lives of affected people. The chapter explores the enormous challenges of 'doing dams right', while a different perspective on the same challenge is given by Patchamuthu Illangovan of the World Bank as a chapter appendix. Chapter 5 by Darrin Magee and Shawn Kelly takes us to the Salween River in Myanmar, describing the emerging plans to develop a series of large hydropower dams on both the upper and middle reaches of Asia's longest undammed river. The authors explore, in particular, the 7000MW Tasang Project in Shan State, showing how private enterprise has taken the lead from the Thai state in tapping hydropower from Myanmar. In what has so far been a decidedly non-transparent undertaking, the authors shed light on the project's investors and lenders, the production of feasibility studies, likely impact on local inhabitants, and measures to ensure transparency and accountability.

Part II of the book provides a series of case studies of distinctive Mekong livelihoods, situating them historically and in the context of modern development practices. Chapter 6 by Chu Thai Hoanh and colleagues explores irrigation, an activity that accounts for 80 to 90 per cent of all water abstractions in the Mekong region. Water is considered a key factor for shifting from single-crop, mainly rain-fed rice, to multiple cropping systems and increasing crop yields. Large investments in irrigation systems have been made in all Mekong countries; more effort is also being paid to improving the efficiency of existing schemes. But the rationale that underpinned irrigation development worldwide during the 1960s and 1970s is being increasingly questioned for countries such as Vietnam or Thailand. The potential for poverty alleviation in Laos or Cambodia seems substantial; but recent disappointing experience with projects demands caution.

Chapter 7 by David J. H. Blake, Richard Friend and Buapun Promphakping takes us to the Nam Songkhram, a river basin that drains into the Mekong River south of Vientiane. The Songkhram is Thailand's largest floodplain wetlands in the Mekong Basin. Its fertile flood-dependent waterscape, however, is recurrently the subject of various infrastructure proposals designed to 'develop' a region that authorities classify as infertile and view its population as poor and vulnerable to both flood and drought. On the other hand, environmental organizations have documented and defended the productivity and diversity of the Songkhram's flood and recession hydrology. Countervailing policy narratives, combined with new agricultural practices and markets, make the Songkhram a microcosm of social forces operating in the Mekong more broadly.

Vietnam's Mekong Delta is another microcosm of important social forces. Chapter 8 by David Biggs and colleagues seeks to understand why certain land- and water-use policies prevailed over others and how historical patterns of land development and water use have had an enduring effect in local society and in the physical environment. The authors trace a transition from strategies of adaptation to strategies for regional state-driven technological control of the 'delta machine'. Technology played a very important role in later reclamation efforts and a culture of scientific positivism still largely animates state plans. The chapter considers how this historical trajectory of physical remodelling of the delta has created huge permanent maintenance costs that are likely to increase as sediments are retained by upstream dams and as sea-level rise threatens the stability of coastal areas. The allocation of these costs is central to the current political economy of the delta.

Chapter 9 offers another approach to learning from the past as ecological modellers Juha Sarkkula and colleagues reconstruct the essence of the Mekong flood pulse system using time series data in order to explore the nexus between hydropower development and fisheries impacts. The authors explain how hydropower development changes the natural flood pulse and the hydrograph, directly undermining the productivity of the system by reducing inundated habitats, delaying the onset of flooding and shortening growth periods for aquatic organisms, with negative impacts upon fisheries productivity, nutritional security and economic activity for a significant portion of Cambodians, as well as other populations in the basin. Quantification of fisheries productivity is difficult because of the complex floodplain ecosystem and the diffuse fisheries. With the pace of hydropower development quickening, and with potentially damaging hydropower projects on the Lower Mekong, the authors argue that finding an acceptable balance between dams and productive fisheries is an urgent issue for the region.

Building on the historical, political and ecological case studies presented above, Part III offers a set of analytical perspectives that unpack discursive and ideological dimensions of power and reveal several dimensions of the politics of knowledge.

Chapter 10 by François Molle and colleagues reviews the post-World War II history of Thai water resources development in *Isaan*, the northeast region recurrently cast as overwhelmingly dry, poor, overpopulated, vulnerable to radicalism, and therefore in need of large-scale interventions to secure it and make it prosper. The authors offer insight into what they call 'meta-justifications' – powerful, self-evident, overriding rhetoric that has served as a tool of state-building and elite aggrandizement. Interestingly, they show that both large and small-scale irrigation projects have been proposed by authorities as preferred solutions during the past six decades. Despite repeated setbacks and failed implementation, large projects and basin-scale diversion schemes are perpetuated. They deliver not just loads of wealth, but symbolic advantages irresistible to those who seek power. The authors argue that hegemonic discourses of greening Thailand's *Isaan* have endured even though the evolution of the overall national economic context makes it unlikely that massive injections of capital to grow a second crop of rice (aside from problems of soil salinity and lack of labour force) are the best way to generate growth or alleviate poverty.

In Chapter 11 by Louis Lebel and colleagues, the focus shifts to the region's cities: places such as Bangkok, Hanoi and Ho Chi Minh City, situated in and expanding into flood-prone zones. A shift of focus – but the authors show how authorities have transformed landscapes by repeated appeal to 'promises' no less ideologically charged than the ones reviewed in the previous chapter. The authors demonstrate how difficult it is to keep all people and roads dry in these areas since preferred solutions privilege one area over another and inevitably displace the problem of unwanted water. They argue that better practices are possible, but require both a stronger state, able to restrict land use, and a more secure and reflexive state, able to make more realistic and considered promises.

Is this an impossibly tall order for Mekong societies? Chapter 12 by Richard Friend, Robert Arthur and Marko Keskinen deals with the neglected value of capture fisheries, a case that offers the reader further insight into the challenge of making governance more reflexive. With transformation- and engineering-oriented mindsets dominant, what are the odds that wild fisheries can be sustained at a level meaningful and vibrant enough to offer nutritional security? The authors show that part of the problem is a policy narrative that casts capture fisheries as inevitably in decline as a result of numerous impacts. According to this dominant storyline, capture fisheries can be conserved, but will play no more than a marginal role in livelihoods and waterscapes of the future. The authors review empirical weaknesses in the dominant narrative. The time has come, they argue, for a counter-narrative in which fishers and capture fisheries are recast and reconceptualized as solutions to, rather than inevitable victims of, regional development challenges.

If dominant ways of thinking are to be challenged by new or better ideas, then advocates of alternative water futures might find it useful to understand how certain institutions and ways of knowing bind to and reinforce one another. Chapter 13 by Mira Käkönen and Philip Hirsch offers such an introduction and examines how the production and legitimizing of knowledge is closely linked to interests and power. The example illustrates the role of modelling in the production of knowledge at the Mekong River Commission, how the World Bank and ADB interpret and use that knowledge, and how participatory policies eventually further legitimize rather than challenge it.

Chapter 14 by John Dore and Kate Lazarus offers a governance practitioners' analysis of the Mekong River Commission, an organization subjected to great effort and attention from actors intent both on using it and crippling it. The authors review continuities and contrasts of the MRC from 1999 to 2007, drawing lessons from water-use negotiations and various basin and strategic planning processes. The authors argue that the MRC could play an important role as a space in which action is informed and deliberatively shaped; but in order to do so, its member states need to use it more actively, rather than bypass it, which, of course, entails redistribution of authority and revised decision-making processes.

In the concluding chapter, François Molle, Louis Lebel and Tira Foran offer a synthetic reflection on water governance in the Mekong region. Two worldviews are clearly pitted against each other. One worldview is epitomized by the motto of the Lao official website on hydropower government: 'Powering Progress', which underpins a traditional developmentalist vision that associates capital and infrastructure investments with growth, and growth with poverty alleviation. The Mekong region and its 'exceptional untapped potential' is seen as 'ripe' for massive investments in hydropower, flood control and irrigation infrastructures. On the other hand, civil society groups operate with a more critical worldview, which emphasizes the social and environmental costs of transformations, and how they overwhelmingly benefit political or economic elites. Current project planning and implementation in the region tend to confirm that decision-making processes are often opaque and offer limited support to the claim that 'we have learned from past mistakes'. The authors, however, identify processes that operate between these two divergent worldviews: from examples presented in the book, the conclusion presents five 'pathways' that have the potential to challenge the process of knowledge production, instil a culture of negotiation and social learning, lessen power imbalances, and shift national and regional water governance.

NOTES

1 This 'quite modest' report (Friesen, 1999) stated that 'in the immediate future, power needs could probably best be met by continuing the present programme of addition small thermal and internal combustion plants as needed, and developing attractive small hydroelectric or multipurpose sites that may be found near load centres'. Regarding flood control, the report concluded that 'flood control was of doubtful value except in localized areas'. It noted that 'most of the officials questioned stated that floods were beneficial to agriculture, fish production and high water navigation, and the flood control was of doubtful value except in localized areas'.

2 During the late 1990s, it provided around one quarter of Laos's foreign exchange earnings, as well as most of Laos's domestic electricity (Hirsch, 1998). Yet, if grants and concessional loans had not paid for its construction, and Japanese aid provided for its repairs, it is highly doubtful the project would have been profitable (IRN, 1999).

3 This region includes the catchment of the Sesan, Srepok and Sekong rivers.

REFERENCES

ADB (Asian Development Bank) (2007) *Mid-Term Review of the Greater Mekong Subregion Strategic Framework 2002–2012*, Asian Development Bank, Manila

Bakker, K. (1999) 'The politics of hydropower: Developing the Mekong', *Political Geography*, vol 18, pp209–232

Biggs, D. A. (2006) 'Reclamation nations: The US Bureau of Reclamation's role in water management and nation building in the Mekong Valley, 1945–1975', *Comparative Technology Transfer and Society*, vol 4, no 3, pp225–246

Browder, G. (2000) 'An analysis of the negotiations for the 1995 Mekong Agreement', *International Negotiation*, vol 5, no 2, pp237–261

CNR (Compagnie Nationale du Rhône) and Acres International Limited (1994) *Mekong Mainstream Run-of-River Hydropower*, Mekong Secretariat, Bangkok

Cornford, J. and Matthews, N. (2007) *Hidden Costs: The Underside of Economic Transformation in the Greater Mekong Subregion*, Oxfam-Australia, Australia

de Koninck, R. (2003) 'Les agricultures du Sud-Est asiatique: Interrogations sur l'avenir d'un nouveau modèle', *L'Espace Géographique*, vol 3, pp301–310

ECAFE (Economic Commission for Asia and the Far East) (1952) *The Development of the International River Mekong – Technical Problems Related to Flood Control and Water Resources Development*, Economic Commission for Asia and the Far East, United Nations, Bangkok, Thailand

ECAFE (1957) *Development of Water Resources in the Lower Mekong Basin*, Economic Commission for Asia and the Far East, United Nations, Bangkok, Thailand

Friesen, K. M. (1999) *Damming the Mekong: Plans and Paradigms for Developing the River Basin from 1951 to 1995*, PhD thesis, Faculty of the School of International Service, American University, Washington DC

Greacen, C. (2004) *The Marginalization of 'Small is Beautiful': Micro-Hydroelectricity, Common Property, and the Politics of Rural Electricity Provision in Thailand*, PhD thesis, University of California at Berkeley, Berkeley, CA

Greacen, C. and Palettu, A. (2007) 'Electricity sector planning and hydropower', in Lebel, L., Dore, J., Daniel, R. and Saing Koma, Y. (eds) *Democratizing Water Governance in the Mekong Region*, Mekong Press, Chiang Mai

Grey, D. and Sadoff, C. (2007) 'Sink or swim? Water security for growth and development', *Water Policy*, vol 9, pp545–571

Hirsch, P. (1998) 'Dams, resources and the politics of environment in mainland Southeast Asia', in Hirsch, P. and Warren, C. (eds) *The Politics of Environment in Southeast Asia: Resources and Resistance*, Routledge, London, UK, Chapter 3, pp55–70

Hirsch, P. and Mørck-Jensen, K. (2006) *National Interests and Transboundary Water Governance of the Mekong*, Australian Mekong Resource Centre in association with DANIDA and the University of Sydney, Australia

Hori, H. (2000) *The Mekong: Environment and Development*, United Nations University, United Nations University Press, Tokyo, Japan

Interim Committee (1990) *Mekong Mainstream Development Possibilities: Summary Report*, Mekong Secretariat, Bangkok

IRN (International Rivers Network) (1999) 'Power struggle: The impacts of hydro development in Laos', *Rivers Network*, February 1999

Jenkins, D. (1968) 'The Lower Mekong Scheme', *Asian Survey*, vol 8, no 6, pp456–464

Kummu, M. (2008) *Spatio-Temporal Scales of Hydrological Impact Assessment in Large River Basins: The Case of the Mekong*, PhD thesis, Water Resources Laboratory, Helsinki University of Technology, Helsinki

Le-Huu, T. and Nguyen-Duc, L. (2003) *Mekong Case Study: Technical Documents in Hydrology*, PC-CP series, no 10, UNESCO, Paris

Mekong Secretariat (1970) *Indicative Basin Plan: Committee for Coordination and Investigation of the Lower Mekong Basin*, Bangkok, Thailand

Mekong Secretariat (1988) *Perspectives for Mekong Development: Revised Indicative Plan (1987) for the Development of Land, Water and Related Resources of the Lower Mekong Basin*, Interim Committee for Coordination of Investigations of the Lower Mekong Basin, Bangkok, Thailand

Mitchell, M. (1998) 'The political economy of Mekong basin development' in Hirsch, P. and Warren, C. (eds) *The Politics of Environment in Southeast Asia: Resources and Resistance*, Routledge, London, pp71–89

Muscat, R. J. (1990) *Thailand and the United States: Development, Security and Foreign Aid*, Columbia University Press, New York, NY

Parnwell, M. J. G. and Bryant, R. L. (1996) *Environmental Change in Southeast Asia: People, Politics and Sustainable Development*, Routledge, New York, NY

Ratner, B. D. (2003) 'The politics of regional governance in the Mekong River Basin: Global change', *Peace and Security*, vol 15, no 1, pp59–76

Rigg, J. (1997) *Southeast Asia: The Human Landscape of Modernization and Development*, Routledge, New York, NY

Ryder, G. (2004) *The Greater Mekong Subregion Hydro Grid: A Review of the Asian Development Bank's 2003 Indicative Master Plan for Regional Power Interconnections and Power Trade in the Greater Mekong Subregion*, Probe International, Toronto, Canada

Thi Dieu, N. (1999) *The Mekong River and the Struggle for Indochina: Water, War and Peace*, Westport, Praeger, CT

UNEP and TEI (2007) *The Greater Mekong Environment Outlook*, United Nations Environment Programme and Thailand Environmental Institute

USBR (United States Bureau of Reclamation) (1956) *Reconnaissance Report: Lower Mekong Basin*, Prepared for the International Cooperation Administration, United States Bureau of Reclamation, US

USBR (1970) *Pa Mong Stage One Feasibility Report*, Prepared for the Committee of Investigations of the Lower Mekong Basin and the Agency for International Development, United States Bureau of Reclamation, United States Department of the Interior, US

White, G., deVries, E., Dunkerley, H. B. and Krutilla, J. V. (1962) *Economic and Social Aspects of Lower Mekong Development*, Report to the Mekong Committee Bangkok, Thailand

Part I

Hydropower Expansion in the Mekong Region

2

Old and New Hydropower Players in the Mekong Region: Agendas and Strategies

Carl Middleton, Jelson Garcia and Tira Foran

Introduction

The countries of mainland Southeast Asia and Yunnan Province, China, threaded together by the Mekong River, are currently enjoying a period of stability and rapid economic growth not experienced for centuries. As a result, the region demands increasing quantities of electricity, especially in China, Thailand and Vietnam. Government electricity-demand forecasts and plans to meet this growth are, however, challenged by civil society. Since the early 1950s, frequently controversial and as-of-yet only partly fulfilled plans for extensive large-scale hydropower development have been high on the agenda of the Mekong country governments. Yet, in a region where millions of people depend upon the natural resources that rivers provide, many proposed dams pose risks for the environment and rural communities, as well as, ultimately, for project developers and the host governments.

The World Bank, the Asian Development Bank (ADB), international agencies such as the United Nations (UN), bilateral donors and an entourage of largely Western hydropower companies and consultants have long played a role in pushing forward the hydropower agenda. Their motives have ranged from the ideological to the political to simple financial gain. Yet, as the new century has dawned, new economic realities and political relationships have emerged. Today, private-sector hydropower developers, mainly from Thailand, Vietnam, China,

Malaysia and Russia, have picked up hydropower plans abandoned by Western companies during the Asian financial crisis – often backed by influential political players and their governments' bureaucracies and with the support of financiers from their own countries. These new hydropower proponents appear to hold a new determination to get the job done without becoming entangled in what they consider to be burdensome environmental and social dilemmas that have often dogged dam projects in the past (Middleton, 2008).

This chapter explores how the ADB and the World Bank have influenced the development of dams and electricity infrastructure in the Mekong region, and have attempted to orientate national policies towards private sector-led development. It evaluates to what extent the banks have applied their environmental and social standards in the region, and discusses the implications of the banks' evolving role and declining influence. The chapter identifies the new actors that are now developing, building and financing hydropower projects in each of the Mekong countries. The absence of environmental and social safeguard policies among these new actors, combined with the weak implementation of the host countries' national law, is identified as a threat to the ecological health of the Mekong Basin. The chapter argues that these new actors and the region's governments should adopt international frameworks of best practices that will significantly reduce the risk of developing poorly conceived projects.

Old players and the region's new 'electricity hunger'

Driven by rapid industrialization, export-led economic growth and expanding domestic consumer markets, demand for electricity is growing in the Mekong region, although the magnitude of this growth is contested between government agencies and civil society groups (Greacen and Footner, 2006; VUSTA, 2007). The Thai government estimates that Thailand's electricity demand will approximately double to 58,000 megawatts (MW) by 2021 (EGAT, 2008). In Vietnam, one of the world's fastest growing economies, the government predicts that electricity demand will almost quadruple to 40,700MW by 2015 (EVN, 2006). Myanmar/Burma, Cambodia and Laos have more modest demand growth predictions, although all governments have committed to urgently develop electricity infrastructure to support economic growth and provide electricity services to rural areas.

In the eyes of water engineers and power planners, the limited exploitation of the Mekong River system's hydropower potential – in a region undergoing rapid economic growth – seems a global anomaly (Ratner, 2003). Thailand, which has already developed much of its domestic hydropower potential and faces civil society opposition to further projects at home, plans to import at least 14,000MW of hydroelectricity from Myanmar, Laos and China's Yunnan Province over the coming 15 years (EGAT, 2008). Vietnam plans to develop almost all of its viable domestic hydropower over the next 20 years, and to import hydroelectricity from Cambodia,

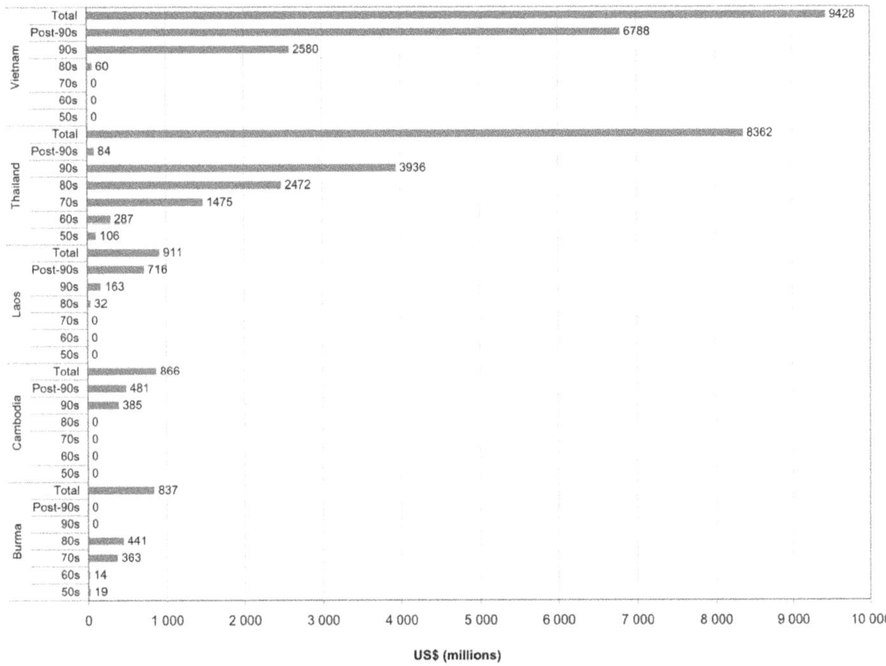

Figure 2.1 *History of World Bank financing in Mekong countries*

Source: World Bank projects portfolio online database

China and Laos (EVN, 2006). Responding to this demand, the governments of Cambodia, Laos and Myanmar are keen to develop their hydropower potential for electricity export and domestic consumption. Other global factors – particularly rising oil prices and the growing carbon offset market – have increased large hydropower's attractiveness in the eyes of the region's governments.

Over the decades, the World Bank and the ADB have played a significant role in shaping the region's electricity sector and in promoting hydropower. They have done this through hosting meetings between key decision-makers; supporting technical studies that promote hydropower development and the regional integration of power systems; offering financial, legal and other forms of expert advice; providing concessional loans,[1] grants, and risk guarantees; and brokering public–private financing deals.

At present, all Mekong countries are members of the World Bank and ADB, although, over the decades, the extent of their interaction has varied as a result of the politics of the region and the banks' policies (see Figure 2.1). The World Bank built a strong presence in Thailand from the 1950s, as did the ADB from the mid 1960s. Their relevance, however, as a major source of development aid has substantially decreased over the last ten years, particularly since the 1997 Asian financial crisis. Cambodia and Vietnam's membership was each suspended in 1975 and 1979, respectively, and was only restored during the early 1990s. Lending to Cambodia and Laos has accelerated since the 1990s, although it is Vietnam that is now the region's largest borrower. The World Bank and ADB ceased their lending to Myanmar in 1987 and 1986, respectively, a consequence of international aid embargos invoked in response to the Myanmar military junta's ongoing human rights violations.

The Western government-backed Mekong Committee (and its most recent guise, the Mekong River Commission) has also actively supported large-scale water resource development (see Chapter 14).

Brief history of dam development in the Mekong region

Early development

Plans for extensive multipurpose dam development in the Mekong region were first conceived during the early 1950s by the US Bureau of Reclamation, which was convinced that the Mekong River's annual flooding was destructive and needed to be tamed to pave the way for modern agricultural techniques (Sluiter, 1992). Inspired by ambitious mega-schemes under way along the major rivers in the US at the time, it envisioned a cascade of mighty dams along the Mekong River that could store water for irrigation and provide cheap hydroelectricity that would power the region's industrialization process (Ryder, 1994; see Chapter 1).

In 1957, the governments of Cambodia, Laos, South Vietnam, and Thailand established the Mekong Committee under the auspices of the UN, which hosted a secretariat to bring the vision to fruition (Bakker, 1999). With the technical support of the US Bureau of Reclamation, as well as funding from the US and other Western countries, the Mekong Committee drafted detailed plans for a cascade of seven massive mainstream multipurpose dams. With a combined reservoir capacity of more than one third the Mekong's annual flow, the dams were conceived to provide 23,300MW of hydroelectricity, and to store water for irrigation, flood control and improved navigation (Ryder 1994). The Mekong Committee also prepared plans for dam cascades on the Mekong's tributaries and large-scale water transfer projects for irrigation, identifying, in total, 180 potential dam sites (Bakker, 1999).

As the Cold War escalated, the work of the Mekong Committee also became a central plank of US and Thai strategy to prevent the Mekong region from slipping into the clutches of communism (Muscat, 1990; Ratner, 2003; see Chapter 1). US and World Bank technical advice and financing supported several large power generation projects in Thailand (including the Bhumipol Dam, in 1964, and the early stages of Thailand's electricity transmission network; Greacen and Greacen,

2004), as well as the establishment of a state-owned electricity utility, the Electricity Generating Authority of Thailand (EGAT), a central player of hydropower development to this day.

The 150MW Nam Ngum 1 Dam, the first large hydropower dam in Laos, was built with technical advice from the Mekong Committee and the World Bank during the late 1960s. Located 90km north of the country's capital Vientiane, the project foreshadowed what would become the predominant development strategy of Laos from the 1980s onwards, with Nam Ngum 1 selling 70 to 80 per cent of its power to Thailand. Inaugurated in 1971, the Nam Ngum 1 Dam became a significant earner for Laos, although the project suffered poor water quality and at least 800 families were resettled to make way for the project, yet received no compensation (Hirsch, 1998). Furthermore, if World Bank and ADB grants and concessional loans had not paid for its construction, and Japanese aid provided for its repairs, it is highly doubtful the project would have been profitable (IRN, 1999).

In Vietnam, throughout the Cold War period, Russia provided support in much the same way that the US and World Bank supported Thailand (Greacen and Palettu, 2007). Russian support for the development of Vietnam's electricity sector was channelled through the state-owned monopoly, Electricity of Vietnam (EVN). Significant technical and financial support was provided for Vietnam's earlier hydropower projects, including the massive 1920MW Hoa Binh Dam (commenced in 1979 but completed in 1994) – still mainland Southeast Asia's largest dam. The project resettled between 50,000 and 60,000 mainly ethnic minority people, the majority of whom continue to suffer impoverishment, as do many more people affected indirectly (Hirsch, 1998).

From the mid 1960s, the Mekong region progressively descended into almost three decades of political instability and conflict. As the war in Vietnam spilled over into Laos and Cambodia, the Mekong Committee's mainstream dam cascade plans were shelved, and the committee itself disintegrated in 1975 (Sluiter, 1992; see Chapter 14).

From 'battlefield to marketplace'

The geopolitical implications of the collapse of the former Soviet Union in 1991 and the weakening of its satellite states since the mid 1980s set in motion substantial shifts in the Mekong region's political and economic landscape. Starting in the late 1980s, as regional stability was largely restored, Western bilateral aid agencies, the World Bank and the ADB once again returned in earnest seeking aid and investment opportunities, and supporting hydropower was high on their agendas.

In 1992, the ADB launched the Greater Mekong Sub-Region (GMS) programme, endorsed by the region's governments, which set a path towards regional economic integration (ADB, 2007a). Orientated around establishing a neoliberal market-based economy, the crux of the GMS programme to date has emphasized the physical interconnectivity of the region, entailing the construction of major infrastructure projects such as transnational highways, railways, hydropower dams and regional transmission lines, as well as programmes that encourage cross-border trade and the integration of markets. The GMS programme has replaced the earlier Mekong Committee as the principal framework for channelling economic development assistance into regional projects (Ratner, 2003).

The integration of electricity markets through a regional transmission grid and the establishment of a regional competitive power market is a priority of the GMS programme. The plan envisages a network of high-voltage transmission lines linking the Mekong countries and opening up mountainous regions mostly in Myanmar, Laos and Yunnan Province of China to hydropower projects, which would be developed mainly by the private sector. A study commissioned by ADB in 1994 (Norconsult, 1994) recommended the development of a series of large hydropower dam and regional transmission interconnection projects, and the formation of an intergovernmental Electric Power Forum (EPF) to coordinate the plan's implementation, first convened in 1995.

In 2002, the ADB consolidated its plan for a regional transmission grid with a second study that recommended a US$43 billion generation and high-voltage transmission system in the Mekong region fuelled exclusively by hydropower, with 12 dams in Cambodia, China, Laos and Myanmar (Norconsult, 2002). The ADB-led plan gained political momentum in 2002 at the first GMS summit when the GMS country leaders signed the Intergovernmental Agreement on Regional Power Trade, committing to establishing a regional power market. The second and third summits, however, signalled a weakening of commitments and a growing reluctance to commit to the plan in full due to the region's utilities' reluctance to commit to privatization, as well as technical concerns.

Numerous criticisms have been raised against the ADB's Mekong Power Grid plan. Chief among them is that the economic benefits appear marginal at best; the ADB's own study estimated that a combined investment in transmission and generation of US$43 billion would reduce investment costs by just over 2 per cent compared to a limited power trading scenario (Norconsult, 2002). Yet, in this plan key costs are not accounted for, such as regional control centre facilities, and costs for hydropower schemes are based largely on assumption, rather than site-specific surveys – a fact pointed out even by the ADB's own consultants – throwing serious doubts on the plan's economic viability (Garrett, 2004; Soluziona, 2004).

The ADB itself has recognized some of the weaknesses of the programme. It has questioned whether achieving competitive regional power trade is realistic given the current governments' reluctance (ADB, 2007a). The ADB has also recognized that more needs to be done to address the social and environmental impacts of hydropower development. Despite these concerns, the ADB, as well as the Japanese, French and Swedish bilateral aid agencies, and the World Bank, all continue to provide financial support to the programme.

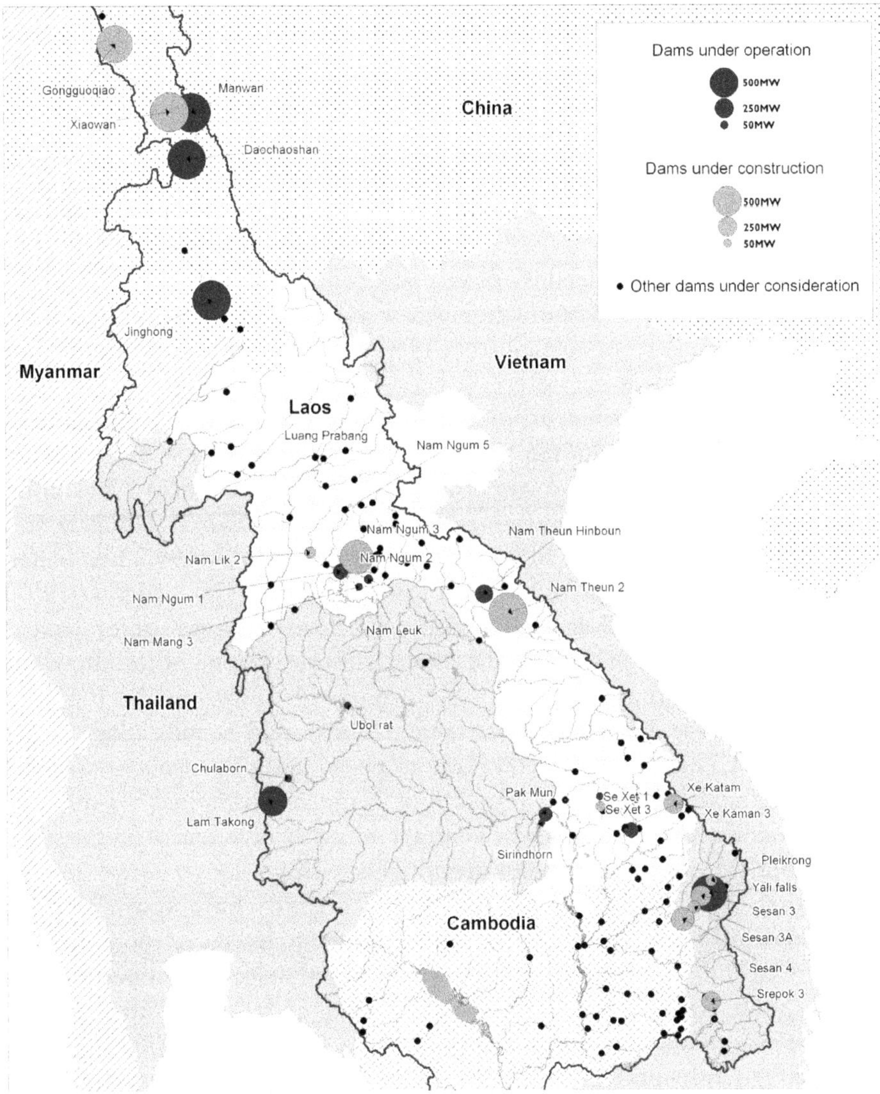

Figure 2.2 *Location of dams in operation and under construction in the Mekong Basin*

Source: Drawn by François Molle

CURRENT TRENDS IN REGIONAL HYDROPOWER DEVELOPMENT

The technical studies, advice and financing of the ADB, World Bank, Mekong Committee and bilateral donors fundamentally shaped the Mekong region's electricity development path during its early stages. This section outlines recent developments in each Mekong country.

Thailand: A voracious power market

During the early 1980s, a credit crisis forced the Thai government to borrow heavily from the International Monetary Fund (IMF) and the World Bank. Conditionalities attached to the loan required the privatization of state-owned enterprises, including EGAT, although this was strongly resisted by Thai labour unions and academics and, ultimately, defeated. As Thailand's economic growth continued and private capital became more readily accessible, the relative importance of the World Bank and ADB as financiers declined (Greacen and Greacen, 2004). The World Bank's final loan to Thailand's domestic hydropower sector was the controversial 136MW Pak Mun Dam project, commissioned in 1994 with co-funding from EGAT (see Chapter 3).

By the early 1990s, a series of pro-market governments increasingly supported power-sector reform, including a role for the private sector. Consequently, Thailand's first independent power producer (IPP), the Electricity Generating Public Company (EGCO), was formed in 1992 from an EGAT subsidiary and commenced trading on Thailand's stock exchange in 1995. By 1997, EGAT had signed contracts with seven IPPs. As the Asian financial crisis struck in 1997, the World Bank and IMF again provided major loans that were accompanied by conditionalities that pushed for the accelerated privatization of the electricity industry, the corporatization of EGAT and a competitive power market (Greacen and Greacen, 2004). However, before these reforms could be fully adopted, the government of Thaksin Shinawatra came to power and once again revised the privatization model, this time to a concept of 'National Champions' – a mode of privatization whereby the state-owned enterprises partly raise capital on the stock markets, but the government retains majority ownership.

Neither the World Bank nor the ADB currently have active lending programmes to Thailand. Both, however, have sought to build a relationship with the government through the transfer of knowledge and skills, rather than financial resources – for example, promoting carbon trading under the Kyoto Protocol's Clean Development Mechanism (World Bank, 2005a; ADB, 2007b). While Thailand's government maintains wariness towards the banks, it has welcomed their efforts in smoothing the way for bilateral power trade, notably their role in pushing through the Nam Theun 2 Hydropower Project in Laos in 2005, which exports 95 per cent of its power to Thailand (see Chapter 4).

EGAT now faces increasing fossil fuel prices, a need to diversify its energy mix (which is currently dominated by natural gas), growing public concern about climate change, and strong opposition to building new large power stations at home. As such, EGAT has increasingly favoured importing hydropower from neighbouring countries. In its 2007 *Power Development Plan*, 4000MW of hydroelectricity imports are planned from Laos between 2008 and 2015, and an additional 8700MW from unspecified neighbouring countries by 2021 (EGAT, 2008). Developing hydropower projects in neighbouring countries – where public opposition is stifled and the rule of law weaker – enables EGAT to export the social and environmental impacts of energy production. Thailand's energy and construction companies, backed by the Thai government, financial institutions and investors, are developing many of these new cross-border hydropower projects that will feed electricity into Thailand's grid.

Laos: The aspiring battery of Southeast Asia

Past projects

Landlocked Laos lies at the heart of the Mekong region, sharing its borders with all of the region's countries. Its mountainous topography offers an estimated 18,000MW of hydropower potential. Generating revenues from power exports has been an ambition of the Government of Laos (GOL) since the 1960s. Although Laos did begin exporting power from the Nam Ngum 1 Dam to Thailand in 1971, it was not until the late 1980s that extensive hydropower exploitation appeared politically realistic (IRN, 1999). Since the late 1980s, representatives from the ADB, World Bank, United Nations Development Programme (UNDP) and bilateral Western donors have consistently advised the GOL that developing the country's hydropower potential was one of its few plausible development options (IRN, 1999). They recommended that smaller projects for domestic power sales should be developed using concessional loans and bilateral aid, and owned and operated by the Laotian state-owned electricity utility, Electricité du Laos (EdL). Larger hydropower projects, mainly for power export, were advised to be developed by the private sector under build–operate–transfer (BOT) contractual arrangements, with the government taking an equity share in the project. In the latter case, the government would benefit from concession royalties, taxes and revenues from power sales, which could be reinvested in funding the development of Laos.

EdL set about developing several smaller projects, with support from the ADB, Japan and Norway, amongst others, including the Nam Song Diversion Dam and the 60MW Nam Leuk Hydropower Project, completed in 1996 and 2000. Both projects sought to address the declining quantities of electricity generated by Nam Ngum 1 since 1982 by diverting additional water into its reservoir. Despite their relatively small size, both dams inflicted serious impacts upon local communities. In 2001, an ADB-commissioned study revealed that Nam Song had affected 13 villages, including severe declines in fisheries for more than 1000 families, the loss of boats and fishing nets, agricultural lands washed away by flooding or erosion, and the deaths of eight people due to sudden releases of water from the project (Watson and Schouten, 2001). The Nam Leuk Dam likewise affected the livelihoods of thousands of villagers (ADB, 2004). Yet, only in January 2007, following prolonged pressure from civil society groups, did the ADB allocate resources for a livelihood restoration package.

The promise of lucrative large hydropower export projects in Laos attracted private-sector hydropower companies from Korea, Australia, Europe and North America. By 1995, memoranda of understanding (MoUs) on 23 feasibility studies had been signed to build dams with a combined capacity of 6676MW (Phonekeo, 1996). Yet, as it turned out, by the end of the 1990s, only two of these BOT projects had been built: the 150MW Houay Ho Dam and the 210MW Theun-Hinboun Dam, both of which imposed heavy costs on local communities (see Box 2.1).

Box 2.1 The injustice of the Theun-Hinboun Hydropower Project

The 210MW Theun-Hinboun Hydropower Project (THHP), commissioned in 1998, is the first build–operate–transfer (BOT) project in Laos. Partially funded by the Asian Development Bank (ADB) and the Nordic Development Fund, the project is owned by Electricité du Laos (EdL) (60 per cent), Norway's Statkraft (20 per cent) and Thailand's Greater Mekong Sub-Region (GMS) Power (20 per cent), and exports 95 per cent of its power to Thailand.

While initially lauded by the ADB as a project with 'little for the environmental lobby to criticize', widespread impacts soon emerged that the ADB later reluctantly acknowledged (ADB, 1999). The project has reduced fishery catches by between 30 and 90 per cent along the three rivers it affected, and has caused extensive river erosion and severe downstream flooding, resulting in repeated loss of wet season rice crops, water contamination, skin diseases and death of livestock from drowning and disease. The net result has been a severe impact upon the livelihoods of 30,000 people living downstream and upstream of the dam (FIVAS, 2007).

After sustained pressure from non-governmental organizations (NGOs), the project's owners released a Mitigation and Compensation Programme in September 2000. While the programme has been able to address some of the material needs of the villagers, such as building wells, its efforts to replace lost livelihoods, such as encouraging villagers to grow dry season rice, cash crops and livestock, have been problematic and are mostly failing (Barney, 2007).

Despite the fact that these problems persist, in April 2008 the project's operators reported strong profits (*Vientiane Times*, 2008). EdL alone had received total dividends of US$145 million since the project was commissioned, which is greater than its initial investment. The Government of Laos has earned about US$27 million as royalty fees and US$9 million in taxes.

As the Asian financial crisis struck in 1997, Thailand's shrinking power market no longer needed hydroelectricity imports from Laos, Vietnam focused on developing its domestic hydropower capacity, and most of the prospective foreign hydropower developers in Laos packed their bags and returned home (IRN, 1999).

A new wave of hydropower developers in Laos

It was not until 2003, when Thailand's economy had recovered sufficiently, that EGAT resumed its commitment to purchase power from Laos by signing a power purchase agreement for the 1070MW Nam Theun 2. The project, now under construction, will export 95 per cent of its power to Thailand. It is owned by Electricité de France (35 per cent), the Electricity Generating Public Company of Thailand (EGCO) (25 per cent), Lao Holding State Enterprises (25 per cent), and Ital-Thai Company (15 per cent), and is financed by shareholder equity and loans from 27 Thai and Western banks, export credit agencies and multilateral development banks. Epitomizing the type of public–private partnership that the World Bank and ADB envision will pull Laos out of poverty, the project constitutes a central pillar of their Lao programme strategies.

The ADB and World Bank claimed that Nam Theun 2 would be a model project that would incorporate lessons learned from past mistakes, that its livelihood programmes would lift those affected by the project out of poverty, and that the revenues the government earned from the project would be reinvested in development programmes. Controversy, however, has continued to rage around the project throughout its implementation as construction deadlines have been prioritized over social and environmental commitments, and it remains uncertain at present whether the project will prove itself successful (see Chapter 4).

The World Bank and ADB also worked with the GOL to establish social and environmental laws and policies to underpin hydropower development. While some laws pre-dated Nam Theun 2, such as the 1999 Environmental Protection Law, others, such as the Decree on Compensation and Resettlement of the Development Project and the National Policy on the Environmental and Social Sustainability of the Hydropower Sector, were adopted in 2005 and were meant to incorporate some of Nam Theun 2's standards to ensure sector-wide implementation.

The economic revival of the Mekong region and Nam Theun 2's approval bought to Laos a new wave of hydropower developers. In contrast to the early 1990s, however, which were dominated by Western hydropower developers, investors from Thailand, China, Vietnam, Malaysia and Russia now lead the hydropower push (see Table 2.1) (International Rivers, 2008). The GOL holds MoUs with Thailand and Vietnam to export 7000MW by 2015 and 3000MW by 2020, respectively.

Thai investors had already joined Western corporations in two major projects in the 1990s – namely, the Theun-Hinboun and Houay Ho hydropower schemes – and two Thai companies are also major shareholders in Nam Theun 2. Yet, it was the construction of the 615MW Nam Ngum 2 Hydropower Project, which broke ground in 2006, that really marked a transition in that it is developed and financed largely by Thai actors. Its shareholders are primarily Thai construction and energy companies, including Ch. Karnchang and Ratchaburi. Thai commercial banks are the main financiers of the US$832 million project, and EdL obtained

Table 2.1 *Large hydropower projects in operation, under construction and planned in Laos**

Status	Number of projects	Total capacity (MW)
In operation	6	660**
Under construction	6	2249
Project Development Agreement, Concession Agreement or Power Purchase Agreement signed or under negotiation	12	4024
MoUs on feasibility studies signed	39	14,155

Notes: * For projects over 10MW.
** Around two-thirds of this operated power generation is exported to Thailand, with the remainder for the domestic market in Laos.
Source: Lao National Committee for Energy (2008)

its equity through a bond issue that was guaranteed by Thailand's Export–Import Bank. Thai power companies – led by Ratchaburi, EGCO and GMS power – and Thai construction companies, in partnership with companies from Malaysia, Japan and Korea, are now conducting studies on at least 15 new hydropower schemes in Laos, including 2 that are part of a list of controversial projects on the Mekong mainstream (see Table 2.2). As a result of Thailand's partial privatization process, EGAT remains a major shareholder in Ratchaburi and EGCO, two of Thailand's largest 'independent' power producers (Greacen and Greacen, 2004). As such, EGAT's key role in determining future power-sector investments has been flagged as a conflict of interest by Thai civil society groups.

Chinese companies are currently involved in two hydropower projects that are under construction in Laos, the Xeset 2 Dam and Nam Lik 1-2 Dam, and have secured MoUs to conduct feasibility studies on at least ten more projects. Sinohydro Corporation, a Chinese state-owned enterprise (SOE) and China's largest hydropower construction company, has spearheaded this push, signing five MoUs, including one for a 1100MW cascade on the Nam Ou River and the Pak Lay Dam proposed on the Mekong River mainstream. Sinohydro Corporation has an extremely weak environmental and safety record; in 2004, 2005 and 2006, it was reprimanded by the Chinese government due to construction and environmental accidents (Haggart, 2006).

Meanwhile, the Vietnam–Laos Joint Stock Company (VLPC)[2] began construction of the 250MW Xekaman 3 Project in southern Laos in 2006. Financing for the project was largely provided by Vietnamese financial institutions, including the Vietcom Bank and the Bank for Investment and Development of Vietnam. The consortium is currently studying four more hydropower projects in the Sekong and Xekaman basins in southern Laos for electricity exports to Vietnam.

Table 2.2 *Proposed dams on the mainstream Mekong River*

Project	Capacity (MW)	Project sponsor	Status
Pak Beng Dam (Oudomsay Province, Laos)	1300	Datang International Power Generation Company *(China)*	MoU for feasibility study signed on 29 August 2007
Luang Prabang Dam (Luang Prabang Province, Laos)	1410	PetroVietnam Power Corporation *(Vietnam)*	MoU for feasibility study signed on 14 October 2007
Xayabouri Dam (Xayabouri Province, Laos)	1260	Ch. Karnchang Public Company Ltd *(Thailand)* and the Government of Laos	MoU for feasibility study signed on 4 May 2007
Pak Lay Dam (Xayabouri Province, Laos)	1320	Sinohydro Corporation, China National Electronics Import *(China)*	MoU for feasibility study signed on 11 June 2007
Pak Chom Dam (Loei Province, Pak Chom district, Thailand)	1500	Has not been officially announced, but consultants have been observed undertaking surveys in Pak Chom district	
Ban Koum Dam (Ubon Ratchatani Province, Thailand)	1800	Ital-Thai Development Plc *(Thailand)*	Pre-feasibility study completed April 2008. Feasibility study underway
Don Sahong Dam (Khong district, Champasak Province, Laos)	240 or 360	Mega First Corporation Berhad *(Malaysia)* and Government of Laos	MoU for feasibility study signed on 23 March 2006 Project Development Agreement signed 13 February 2008
Sambor Dam (Sambor district, Kratie Province, Cambodia)	465 or 3300	China Southern Power Grid Company *(China)*	MoU for feasibility study signed in November 2007

Source: News reports from *Bangkok Post*, *Cambodia Daily* and *Vientiane Times*

Environmental and social safeguard standards in Laos

In Laos, a one-party socialist state, government criticism is rarely tolerated, press freedoms are curtailed, independent civil society organizations are restricted, and corruption is high (Stuart-Fox, 2006). These circumstances significantly enhance the risks associated with hydropower development – particularly for the hundreds of thousands of villagers poised to lose land, fisheries and other resources as a result. Hydropower development, while generating revenue for the government and generally proving profitable for private-sector investors, has incurred major environmental and social costs, a legacy of damage that remains largely unaddressed.

Many of the laws, regulations and policies developed in preparation for Nam Theun 2, with support from the ADB and World Bank, contain important provisions to ensure participation, consultation, information disclosure, compensation and resettlement with livelihood restoration for affected communities. However, in practice, these provisions are often not being followed by the new developers and are not being enforced by the GOL (International Rivers, 2008). These implementation failures are most evident during the development and review of the environmental impact assessments (EIAs) and resettlement action plans for hydropower projects, which have generally not been disclosed to the general public and are often of questionable quality. Although the Nam Theun 2 Project has surpassed the standards of early Lao hydropower projects and can be credited with piloting several innovative aspects in Laos, such as the presence of independent monitors, a revenue management framework and a commitment to public reporting, implementation within both the Lao context and the tight timeframe of investors is problematic (see Chapter 4).

For many of the new hydropower developers, the ADB and World Bank's environmental and social safeguards are viewed as burdensome, time consuming and costly.[3] Now that private sources of finance are more readily available, these hydropower developers are less inclined to seek the banks' financial support (World Bank, 2007a). Despite this, both the ADB and World Bank plan to remain involved in the hydropower sector in Laos, principally through their ongoing support for the Nam Theun 2 Project and for the development of several regional transmission lines (World Bank, 2005b; ADB, 2007c). Both banks also plan to build the capacity of the GOL to manage hydropower development and the public finances that will be generated, and to support policy reform that will further facilitate private sector-led hydropower development. The ADB is considering supporting two new hydropower projects in Laos – Nam Ngum 3 and Nam Ngiep 1 – and the World Bank Group's Multilateral Investment Guarantee Agency may provide a guarantee to Sinohydro's Nam Ngum 5 Project.

Vietnam: Racing to meet power demand

The rapid economic growth of Vietnam has seen a massive forecasted increase in demand for power – currently growing at 16 per cent per year – which the state-owned electricity utility, Electricity of Vietnam (EVN), is struggling to meet (World Bank, 2006). The World Bank estimates that investments of US$3 billion annually are required for new generation and transmission infrastructure until 2010 alone, significantly exceeding EVN's own resources.

Until the late 1990s, hydropower constituted the backbone of Vietnam's power supply. More recently, however, fossil fuel-fired power stations have become predominant. At present, around one quarter of Vietnam's economically viable hydropower is in operation and efforts are well under way to exploit the remaining 17,000MW potential by 2025 (EVN, 2006).

Since the World Bank and ADB resumed operations in Vietnam during the early 1990s, they have loaned heavily to EVN (ADB 2007d; World Bank, 2007b).

At the same time, they have also pushed for sweeping power-sector reforms, calling for a greater role for private-sector power generators, the restructuring of EVN into shareholding companies, and the establishment of a competitive power market. The reforms cumulated in the promulgation of a new Electricity Law in 2004 and the establishment of the Electricity Regulatory Authority of Vietnam.

As such, since 2004, EVN has undergone a process of corporatization that will ultimately involve selling shares in up to 50 of its power plants and other subsidiary enterprises, while maintaining a state monopoly over the transmission network and the largest dams such as Son La, Hoa Binh and Yali Falls (Reuters, 2007). The utility hopes to earn more than US$700 million by selling shares in many of its power plants. EVN is also borrowing heavily from commercial sources, export credit facilities, bond issues and overseas development aid to fund its expansion plans. A growing number of private-sector independent power producers are also operating in Vietnam, mainly developing profitable thermal power stations, leaving the development of less-profitable hydropower stations to EVN (World Bank, 2006).

To secure its electricity supply, Vietnam has also looked to its neighbours, partly facilitated by the ADB's GMS programme. Since September 2004, Vietnam has imported growing quantities of electricity from China and has exported increasing volumes of coal (Bo, 2008). From Laos, by 2010, Vietnam will import power from the 250MW Xekaman 3 Dam, the first of at least nine hydropower projects it is considering in Laos (Lao National Committee for Energy, 2008). In June 2007, EVN agreed with Cambodia's government to undertake feasibility studies on the Lower Sesan 2, which would export some of its electricity to Vietnam.

Large hydropower dams in Vietnam have often caused serious social upheaval and high environmental costs (Hirsch, 1998; CRES, 2001; VUSTA, 2006). The planned Son La Dam, for example, requires the resettlement of up to 100,000 mainly ethnic minority people (VUSTA, 2006). Although Vietnam passed a Law on Environmental Protection in 2005 and despite a growing recognition amongst hydropower proponents in Vietnam of the need to fairly address the environmental and social impacts of large dams, experience has been limited and implementation weak (World Bank, 2006).

The ADB and World Bank have been involved in various hydropower planning studies in Vietnam since the early 1990s. The World Bank, for example, conceived the preparation of Vietnam's *Hydropower Master Plan*, completed in 2001 with funding by the Norwegian and Swedish development agencies (Lang, 2000). The ADB funded the *Sekong-Sesan and Nam Theun River Basins Hydropower Development Study*, completed in 1998, that prioritized six dams for further development within the three river basins shared by Cambodia, Laos and Vietnam (Lang, 1998). The ADB would have gone on to fund the Sesan 3 Dam in Vietnam, but its own project preparation technical study prepared in 2000 revealed the 'severe to catastrophic' impacts of the Yali Falls Dam (commissioned in 2000, located upstream of the Sesan 3) on tens of thousands of Cambodian villagers downstream. EVN ultimately acquired funding from Russian sources rather than conduct a transboundary study required by ADB that ultimately could have required them to pay compensation to the affected villagers (Hirsch and Wyatt, 2004).

In June 2008, the ADB financed its first hydropower project in Vietnam, the US$270 million 156MW Song Bung 4 Hydropower Project in Quang Nam Province, central Vietnam. To give grounds for its involvement, the ADB established a river basin management organization and undertook a cumulative impact assessment; yet it approved the project despite the fact that its reservoir will submerge 143ha of Song Thanh National Park and the resettlement consultation process was found to fall short of ADB safeguard standards (RDSC and BIC, 2008). The World Bank is also seeking to provide a loan for the US$310 million 250MW Trung Son Hydropower Project in Thanh Hoa Province, north-western Vietnam.

Both the ADB and World Bank have also supported transmission infrastructure in Vietnam, including regional interconnections. The ADB's US$360 million support for transmission lines in northern Vietnam connects the controversial Son La Hydropower Project to its domestic load centres. The ADB has justified its association with the Son La Project – which is not subject to and certainly does not attain the ADB safeguard standards – by stating that the dam would have gone ahead anyway, with or without the bank's support for the associated transmission infrastructure. Instead, the bank has claimed to try to improve the project by providing technical assistance for its resettlement and environmental programmes, although these have found limited success. Whether projects such as Son La should be considered as 'associated projects' and, therefore, subject to the ADB's safeguard standards before supporting loans are supplied is questioned by civil society groups and is a grey area within the bank's policy. Under similar dubious circumstances, the World Bank funded the transmission line that connected the Yali Falls Dam to Ho Chi Minh City (Trandem, 2008).

EVN has also welcomed other foreign assistance for its dam projects, especially from those whose financing does not come with rigorous social and environmental conditionalities. In 2006, for example, partnering with EVN, China Southern Power Grid Company invested US$28 million in the 21.4MW Lao Cai Hydropower Station (Bo, 2008). In January 2008, the Indian Export–Import Bank provided a US$45 million concessional loan for the 200MW Nam Chien Hydropower Plant, complementing the US$156 million provided for the project mostly by Vietnamese banks. Russian financial and technical support has continued to be important, especially on controversial projects such as Son La and Sesan 3.

Cambodia: Big plans for hydropower

Cambodia is on the threshold of an extensive domestic hydropower development programme, backed mainly by Chinese developers and financiers. In Cambodia, the cost of electricity is amongst the highest in the world and electricity infrastructure remains rudimentary, a result of decades of fighting and political turmoil (ADB, 2005a). Cambodia's political elites have expressed strong support for large-scale hydropower projects, citing the need to secure access to cheap electricity to supply Cambodia's expanding economy (International Rivers and RCC, 2008).

Until recently, Cambodia has struggled to attract investment for major hydropower development. Western bilateral donors and the multilateral development banks have proven reluctant to provide support, in part over concerns about environmental and social impacts. The World Bank's and ADB's initial efforts focused on rehabilitating power supplies in Phnom Penh and the provincial centres, and have since focused mainly on expanding electricity transmission and distribution, rural electrification, and developing Cambodia's electricity institutional and legal framework (ADB, 2005a; World Bank, 2005c). As a result, the Electricity Authority of Cambodia was established in 2001. In line with ADB and World Bank policy, this framework also places the private sector centrally to developing Cambodia's power system.

Over the past several years, China's political and economic ties with Cambodia have strengthened and the Chinese government has indicated high-level support for Cambodia's hydropower plans. To date, deals have been finalized on four major hydroelectric projects, all of which will be built by Chinese companies. Construction of the 193MW Kamchay Dam, located in Bokor National Park, Kampot Province, by Sinohydro Corporation commenced in late 2007 (see Box 2.2). The 120MW Stung Atai project, to be developed by the China Yunnan Corporation for International Techno-Economic Cooperation, was approved in February 2007, followed in June 2008 by the 246MW Stung Tatay Dam and the 338MW Stung Russey Chrum Krom Dam (Associated Press, 2008). Reportedly, a further nine hydropower projects are planned to be built in Cambodia by 2019 (*XFN–ASIA*, 2008).

In contrast to the intense competition between hydropower developers in Laos, Chinese hydropower companies appear to have largely cornered Cambodia's hydropower market. Aside from Chinese companies, the only other hydropower developer known to be conducting a feasibility study in Cambodia is a subsidiary of EVN for the (controversial and risk-prone) Lower Sesan 2 Dam.

The ADB and World Bank, although not directly funding hydropower projects, have subsidized their development through supporting the construction of several domestic and regional high-voltage transmission lines (ADB, 2005a).

Myanmar: Thai and Chinese companies move in

Myanmar has plans for extensive hydropower development to generate electricity for domestic use, and for export to Thailand and China for revenue generation (Win Kyaw, 2006). Myanmar's total hydropower potential is a substantial 39,720MW, of

Box 2.2 Kamchay Dam, Cambodia

In April 2005, the Cambodian government awarded Sinohydro Corporation a contract to develop the Kamchay Hydropower Scheme – Cambodia's first large dam. High-level Cambodian and Chinese government officials pushed forward the Kamchay Dam in closed-door negotiations that largely left other stakeholders, including the local authorities and the public, out of the process. Financing for the Kamchay Dam was secured in April 2006 from China Exim Bank. The project had previously been considered by the Canadian International Development Agency (CIDA) a decade earlier, but was eventually dropped because of social and environmental concerns.

The 110m high dam, now under construction, is located in Bokor National Park, southwest Cambodia, and will flood 2000ha of protected forest. According to a 2002 survey, this forest is the habitat of 31 mammals and 10 endangered species, including Asian elephants, leopard cats and tigers. This area is also an important source of non-timber forest products for local residents, many of whom depend upon the income earned through selling the forest products. It is not known if Sinohydro Corporation will provide compensation or support the development of alternative livelihoods. There are also concerns that poor river water quality could devastate the local tourism industry, pollute irrigation water that feeds the abundant durian orchards and rice fields nearby, and contaminate Kampot town's water supply extracted just downstream of the planned dam site. Shortly after construction commenced, the *Cambodia Daily* reported that water contamination from construction activities and untreated sewage discharges from the workers camp into the Kamchay River had caused tourism to plummet from 60,000 people in February to 7700 in March at the popular Touk Chuu rapids immediately downstream (*Cambodia Daily*, 2008).

Sinohydro Corporation will build, own and operate the Kamchay dam for 44 years, despite the unusual contract length having been questioned in political debates.

which approximately 25,000MW is large-scale hydropower dams (approximately 35 projects). Myanmar plans to bank-roll these projects mainly through loans and suppliers' credits from China, through government funds, and by encouraging private-sector investment. As of 2006, there were only ten hydropower stations larger than 10MW in operation in Myanmar with a total capacity of 745MW supplying domestic demand.

In Myanmar, many major development projects, including large dams, take place in ethnic minority areas. The country's laws allow for no public participation in decision-making, require no environmental, social or human rights impact assessments, and effectively offer no access to justice (BRN, 2008). Increased militarization around project areas often results in the use of forced labour and forced portering, forced relocation and other human rights abuses. Increased troop presence leaves women particularly vulnerable to abuse, including to sexual violence. Large dams in Myanmar benefit foreign investors while continuing to support Myanmar's military junta financially and politically.

Between 1997 and September 2007, at least 14 Chinese companies became involved in at least 40 hydropower projects in Myanmar (EarthRights International, 2007a). Major hydropower dams being planned with strong Chinese backing include the Hutgyi and Tasang dams on the Salween River, the Shweli River cascade, and seven dams on the Ayeyarwady (Irrawaddy), N'Mai Hka and Mali Hka rivers, the first of which will be the 3600MW Myitsone Dam (see Chapter 5).

The Government of Thailand signed an MoU with Myanmar in July 1997 to purchase up to 1500MW of power by the year 2010. In May 2005, Thailand's Ministry of Energy signed an MoU with Myanmar's Ministry of Electric Power to develop hydropower projects on the Salween and Tanintharyi rivers (EGAT, 2008). Thai companies have actively sought joint ventures with Chinese partners, such as Sinohydro and Gezhouba, to develop the hydropower dams on the Salween River.

Although the ADB and World Bank currently do not provide loans or grants to Myanmar, representatives of Myanmar's regime continue to join regional meetings hosted by the banks, particularly those of the ADB's GMS. This, in the eyes of some, constitutes a dubious interpretation of the bank's current embargo on Myanmar (EarthRights International, 2007b). By attending the GMS's regional electricity meetings and participating in its studies, the Myanmar military junta has gained the opportunity to further Myanmar's integration into GMS, undermining Western-led pressure for political reform.

A CHANGING WORLD: THE BANKS SEEK TO REINVENT THEMSELVES

In the aftermath of the 1997 Asian financial crisis, Mekong governments made deliberate efforts to expand their options to finance development plans. Except in Myanmar, the economies of the Mekong countries, in general, are exhibiting strong growth, and, especially in Thailand and Vietnam, there is increasing domestic liquidity. The growing influence of international private capital and bilateral funds from new economic powers, such as China and India, are radically shifting the architecture of international finance and are increasingly predominant over development bank financing.

Shifting roles and expanded lending instruments

These new sources of project financing have forced the ADB and World Bank to reconfigure their operations. An examination of the banks' key operational and sector strategy papers reveals that the banks refuse to become totally redundant in the Mekong region's project financing. Although bypassed by private and institutional banks, it seems too early to dismiss them as irrelevant because they appear to be making inroads into reinventing their roles to maintain regional influence.

In middle-income countries, such as Thailand, the banks have sought to redefine their role from project-led financiers to 'knowledge solution' providers (ADB, 2008). In order to achieve this, the banks' country offices now host more in-house experts that help, if not initiate, the identification of priorities and reform targets for countries to access more funding. The banks claim to root their policy advice and analytical studies in alignment with the borrowers' own development plans.

The ADB and World Bank agenda also remains unchanged. The banks justify their engagement in the Mekong hydropower industry on the basis of fulfilling their self-assigned mandate to reduce poverty. As energy is seen as a prerequisite to economic growth, the banks view the Mekong region's water resources as a vast opportunity for investment in hydropower, and their support for the sector as critical to alleviating poverty.

The banks have refined their overarching strategy towards promoting private sector-led investment, using their expertise and finance to build public–private partnerships or otherwise lever private capital (ADB, 2001, 2005b). The banks do not necessarily see private equity funds, commercial banks and independent power producers as competitors. Instead, the banks are confident that they can adapt to the entry of new players by striking co-financing agreements, direct lending, investing in equity or providing risk guarantees.

To build these new partnerships with the private sector, as well as appear more attractive to low- and middle-income countries, the ADB has sought to repackage itself, especially through expanding its lending instruments (ADB, 2007e). They now include loans in local currency rather than in US dollars, and a multi-tranche financing facility under which the bank commits to financing an entire sector or multiple tranches of a large project. The ADB is also now offering loans and guarantees to sub-sovereign government agencies, such as provincial governments, and state-owned enterprises, without requiring guarantees from central government. The ADB is moving ahead to mainstream its lending instruments despite concerns about possible loopholes that would avoid rigorous application of environmental and social safeguards, lessen compliance with information disclosure and anti-corruption policies, and reduce the bank's board and management's project oversight (Fried et al, 2008).

Pressure on the ADB to keep its business afloat has led the bank to push for the dilution of its environmental and social safeguard policies, and in 2005 the ADB launched a process to redraft the policies. A consultation draft released in October 2007, in the eyes of civil society organizations, essentially eviscerated the bank's current environmental, indigenous peoples and involuntary resettlement safeguards policies, and replaced them with far shorter and more vague 'policy principles', together with a weakened commitment to information disclosure for lending operations to the private sector (Fried et al, 2008).

Since 2005, the World Bank has piloted a Country Safeguard Systems (CSS) approach that relies more heavily on national procedures and laws instead of the bank's own policies. Principles rather than policies and mandatory procedures are used in the CSS approach. This approach is considered applicable if the host country's laws are judged by the bank to be equivalent in content, intent and spirit to its own safeguard policies. Responding to demands from borrowing countries, the CSS approach is intended to shorten the bank's loan transaction period and to reduce project costs. The ADB now plans to adopt an equivalent Country Safeguard Systems approach, and intends to pilot the method in Vietnam.

Independent evaluation of the World Bank's application of the CSS approach, however, has revealed that environmental and social standards can slip because:

- the principles used to implement the CSS approach can be less stringent than the bank's original environmental and social policies;
- the borrower's national laws, policies or measures can be inconsistent with or weaker than the bank's previous project management standards;
- national capacities to implement safeguards may not be sufficient and are not realistically addressed by the bank; and
- an affected community's ability to invoke the World Bank's 'accountability mechanism', which can halt a project if the bank's safeguard policies are violated, is more difficult because it can be unclear whether a policy has been violated (CIEL, 2007).

The World Bank has yet to prove that its CSS approach – still at pilot stage – ensures that environmental and social standards are of equal quality to the bank's existing safeguard policies.

Regional initiatives: A key role for the development banks

The ADB and World Bank, in the face of declining demand for their conditionality-tied project financing loans, have attempted to recast themselves as purveyors of international best practice for the region and as 'honest brokers' of regional cooperation initiatives.

In a working paper released in June 2006, the World Bank, the ADB and the Mekong River Commission (MRC) outlined their major new collaborative initiative: the Mekong Water Resources Assistance Strategy (MWRAS) (World Bank and ADB, 2006). It promotes the construction of controversial water infrastructure projects in three sub-regions of the Mekong Basin where transboundary impacts would occur that include dams, irrigation schemes and water transfer projects – namely, north-eastern Thailand and north-western Laos, where large-scale trans-basin water transfers are proposed; the Sesan, Srepok and Sekong ('3S') river basins where the interests of Cambodia, Vietnam and Laos coincide under extensive hydropower development plans; and the Mekong Delta shared by Vietnam and Cambodia to address flood/navigation/agriculture and wetlands-associated projects. The strategy claims that livelihood restoration programmes for affected communities could mitigate any negative impacts from the projects, suggesting that affected communities might even benefit from the new river flows, leading to potential win–win situations.

The MWRAS project drew wide criticism from civil society groups (IUCN et al, 2007). Key concerns included:

- The MWRAS claimed that economic and other pressures on each of the Mekong countries mean that it is inevitable that large-scale water infrastructure projects will go ahead. The MWRAS did not question whether the infrastructure projects themselves are the most effective way to reduce poverty in the region or if they are sustainable.
- The MWRAS misleadingly extrapolated the results of a hydrological model to suggest that the Mekong River could accommodate further extensive infrastructure development. The model's results were narrowly hydrological and failed to account for ecological or socio-economic impacts, particularly the subtleties of the flood pulse ecosystem (Lamberts, 2008; see Chapter 13).
- The MWRAS calls for closer collaboration between the banks, the MRC and the four member states to develop new infrastructure projects. It encourages reorienting the MRC's role from that of a basin *management* organization to that of a basin *development* organization. Given that numerous actors, including the banks, are already heavily promoting infrastructure-oriented development in the basin, civil society groups and some of the MRC's donors argue that the MRC should work to emphasize the joint management and conservation of the river basin, embracing local participation and diverse perspectives (Hirsch and Mørck-Jensen, 2006).

Independently from the ADB, the World Bank has also led recent regional initiatives. In September 2007, the World Bank hosted the Thai–Lao Sustainable Hydropower Forum in Bangkok, which invited senior representatives from the governments of Laos and Thailand, existing project operators, project developers, financiers and civil society to discuss working towards a triple bottom line approach (economic, environmental and social) for the Laos hydropower sector (World Bank, 2007a). The forum indicated a move by the World Bank to address the fact that hydropower projects subsequent to Nam Theun 2 were failing to replicate its standards. The forum issued a joint communiqué, co-signed by Thailand's minister for energy and the Laos minister of energy and mines, which indicated a commitment to work towards 'enhancing the quality of investments to make the hydro power sector both environmentally and socially responsible and sustainable in Laos'. The governments agreed to form a bilateral task force to develop an action plan, and a second forum is planned for late 2008. Whether these commitments will translate into action on the ground remains to be seen.

DISCUSSION AND CONCLUSIONS

We are witnessing a revived rush to develop hydropower in the Mekong region, a rush fraught with pitfalls for project developers, financiers, host governments and, most of all, for communities whose livelihoods would be affected. The legacy of many earlier projects already stands testament to the environmental and social costs of large dams.

Despite this, the new hydropower proponents have so far demonstrated little commitment to social and environment standards or to public participation in decision-making.

Have the ADB and World Bank raised hydropower project standards?

While the ADB and World Bank claim to have strong environmental and social policies, as well as commitments to public participation, in reality these measures have often proven inadequate to mitigate the risks of large dams. Existing hydropower projects backed by the ADB and World Bank have failed to ensure that project impacts were mitigated and livelihoods restored, let alone that the project's benefits were shared. Recent examples include the Theun-Hinboun and Nam Song dams in Laos and the Pak Mun Dam in Thailand (where impacts were lessened only after multiple rounds of protest by affected people and their advocates; see Chapter 3). While the banks' most recent project, the Nam Theun 2 in Laos, has substantially raised the bar compared to earlier project studies of environmental and social impacts, it remains to be seen whether promises made to affected communities can be kept or will be adequate. Early indications are that they will not be (see Chapter 4).

Both the ADB and World Bank have invested considerable time and resources into hydropower development in the Mekong region, including cross-border power trade through the GMS programme. Hydropower helps to provide electricity necessary for economic growth, but also undermines the livelihoods of the rural affected communities. The failure of the GMS programme to adequately address the environmental and social impacts of its projects, as well as the issues of equity (between GMS countries and between rural and urban), ultimately undermines sustainable development and remains its biggest shortcoming (Cornford and Matthems, 2007; UNEP and TEI, 2007).

Indeed, in some instances, the ADB's GMS programme is directly incongruous with the ADB's own country-level programmes. In Cambodia, for example, the ADB's country programme has identified the Tonle Sap Basin as its geographical focus, recognizing the area as one of Cambodia's poorest and most environmentally sensitive regions (ADB, 2005a). Yet, extensive regional hydropower development promoted under the GMS programme constitutes a serious threat to the lake's ecosystems and its fishery productivity – and, ultimately, therefore, to efforts to alleviate poverty (CNMC and WorldFish Centre, 2007).

The evolving role of the old players

The entry of new hydropower developers and financiers into the Mekong diminishes the influence of 'old actors' such as the development banks and donor agencies, as well as some NGOs. But it does not make them wholly irrelevant.

As reviewed above, both the ADB and World Bank have new hydropower projects in their lending pipelines. Yet, given the large number of projects planned in the Mekong region by the new hydropower developers and the wide array of private-sector and government-backed financing options available to them, it is clear that the role of the ADB, the World Bank and Western bilateral donors as project developers and financiers has significantly lessened.

The efficacy of the development banks in their self-assigned new niche as knowledge providers and purveyors of best practice remains to be seen. As relatively minor and, now, replaceable financiers, it will be challenging for the banks to make 'best practice standards' attractive enough to appeal to the short-term interests of the new hydropower proponents and, perhaps even, the region's governments.

A recommendation, for example, of an ADB-supported cumulative impact assessment (CIA) completed in early 2008 for the proposed Nam Ngum 3 Dam was that several smaller dams proposed for the Nam Ngum Basin not be built in order to protect fisheries vital to livelihoods. This advice, however, has been ignored by other hydro-developers in the basin (and apparently also by the GOL), who are pushing forward with construction of contentious projects on the Nam Lik River (Norplan, 2008).

In addition to hydropower dams, the ADB and World Bank are financing or planning to finance several regional transmission lines that will facilitate bilateral power trade between Laos, Thailand, Vietnam and Cambodia. These transmission lines will export power from numerous private-sector hydropower projects that do not meet each country's regulatory standards, let alone the banks' safeguard policies. The banks are able to indirectly support (and therefore subsidize) these hydropower projects because of the lack of clarity surrounding how their safeguards apply to 'associated facilities'. Exploiting this loophole, the development banks appear increasingly keen to support associated transmission lines rather than hydropower projects directly.

The ADB and World Bank, and its donors, also continue to support regional-level meetings and to undertake studies that promote regional integration of the electricity sector, to which hydropower development is central. An ADB study, now being finalized, recommends that regional integration should be expanded to the wider energy sector, including natural gas, coal and oil (Nangia, 2008). These studies largely build on existing government plans, providing justification to move mostly large infrastructure-intensive plans forward, yet incorporate the same lack of vision of alternative scenarios to meet energy needs.

The Mekong River Commission has struggled to define its role as a basin management organization, especially in the face of the current wave of extensive hydropower development plans. The MRC has only recently funded its hydropower programme in 2008, and, alongside other programmes such as the Environment Programme and Fisheries Programme, it is uncertain to what extent the MRC can project its scientific knowledge to influence the politicized decision-making process.

The need for international standards for new hydropower proponents

The new Mekong hydropower developers – predominantly energy and construction companies from Thailand, Vietnam, China, Russia and Malaysia – have yet to commit to international best practice standards, such as those outlined in the recommendations of the World Commission on Dams (WCD) report (WCD, 2000). These developers are not even striving to meet the ADB and World Bank's social and environmental standards and commitments to public participation, which are notably weaker than the WCD's recommendations. Very few project developers have developed and published corporate social responsibility (CSR) policies. The companies that have, such as Thailand's EGCO and Ratchaburi, have adopted a very narrow interpretation of CSR that provides only limited support for affected communities (and, apparently, only for those in Thailand).

Amongst the commercial banks from Vietnam and Thailand that are known to be financing hydropower projects, none have adopted environmental and social standards, such as the Equator Principles, or hold an equivalent set of standards.[4] In China, the Equator Principles are only just beginning to gain momentum: the Ministry of Environmental Protection has embraced them as part of its domestic green credit policy. Amongst the new export credit agencies actively supporting hydropower projects in the Mekong region, only China's Exim Bank is known to have an environmental policy and guidelines, released in April 2007 and August 2007, respectively. There is little evidence, however, of its rigorous implementation on the ground.

These frameworks could reduce the risk of developing poorly conceived projects. Given the massive interest in developing hydropower throughout the region, the region's governments are in a strong position to only select those developers of sound reputation. In general, best practices address issues of concern to wider society through eliminating or minimizing externalities and sharing project benefits. While, in principal, such practices also reduce project developer risk – for example, from protests or legal measures that could delay project construction or add unforeseen additional cost – where the rule of law is weak, corruption high or local protest stifled, such risks appear smaller to project developers; therefore, they have less impetus to implement best practices. Commercial or strategic short-term interests favour poor practices that constantly override consideration or application of precautionary measures or standards. Past dam projects, unfortunately, confirm that all the compensation schemes and other concessions from dam builders and governments have been secured only after substantial mobilization or protest.

The need for better electricity and water planning practices

Environmentally sustainable and socially desirable solutions to meeting the Mekong region's energy needs do exist, although, at present, they are not a part of any regional energy plan (Greacen and Footner, 2006; VUSTA, 2007). The planning processes currently in place both at the national and regional levels fall well short of international standards in electricity planning.

Hydropower projects begin as abstractions, as a series of numbers, drawings and equations shaped by experts (Foran, 2006; Greacen and Palettu, 2007). How these coalesce over time to establish a particular hydropower plant as an attractive option to expanding energy supply is one of the most difficult and important questions. This difficulty stems from a number of factors.

First, the financial incentive structure of power utilities is a 'rate-of-return' structure. All utility costs are periodically submitted for review to a regulator; if approved, the organization will be permitted to recover its costs, plus a profit margin, by passing them on to captive customers (Foran, 2008a). This incentive structure rewards utilities for investing in power plants, not for saving energy through energy efficiency programmes.

Second, electricity planning is done on behalf of society, not by society. Planners report to state-owned power utilities or state energy agencies, not to other branches of the state (such as legislative committees or an independent regulatory body). These two forces – institutionally shaped interests and practices – combine with the aura of technical complexity promoted by insiders. It leads to relatively closed 'state-knows-best' planning processes.

Civil society groups have questioned Thailand and Vietnam's power development plans, which heavily promote the development of new large-scale electricity generation plants (Greacen and Footner, 2006; VUSTA, 2007). They claim that future electricity demands are overestimated, and the role that energy efficiency measures, renewable energy and decentralized energy options could play is downplayed. In Thailand, they argue that existing plans mostly serve the interests of the state-owned electricity utilities, energy companies and the construction industry, rather than the needs of Thailand's electricity consumers. Civil society groups in Thailand are calling for reform of the power planning process towards integrated resources planning (IRP), a process that considers a full range of feasible supply- and demand-side options, as well as the full cost to society – including social and environmental costs, as well as risk – rather than the lowest commercial cost to investors.

Electricity savings programmes implemented by EGAT between 1995 and 2006 have reduced actual peak demand by 6 per cent. Studies indicate that by 2018, Thailand should be able to avoid adding 7900MW of generation capacity through further savings, offering slightly higher tariffs to renewable energy projects and accepting bids to sell electricity from both renewable and high-efficiency natural gas cogeneration (Foran, 2008b, Tables 31–32; cf. Greacen and Footner, 2006). Vietnam and China likewise have significant energy efficiency and renewable energy potential (USAID, 2007).

Furthering the IRP concept, the WCD put forward a wider framework in the form of a *Comprehensive Options Assessment* that combines sustainable water and energy planning practices with public participation to prepare congruous, sustainable and publicly acceptable electricity- and water-sector plans.

Coupling electricity and water planning is critical to determining the true cost of hydropower development. It is therefore surprising and of serious concern that despite more than 15 years of ADB support for extensive hydropower development and regional power trade, the ADB has failed to evaluate the cumulative impacts of widespread hydropower development on the Mekong River's ecosystems and its people, which are anticipated to be severe (Ratner, 2003; UNEP and TEI, 2007; Lamberts, 2008).

Recognizing the river's existing values

The Mekong region's rivers continue to provide abundant natural resources for the region's riparian peoples, as well as the wider basin population, as they have done for millennia. By changing the river's hydrology, blocking fish migration and affecting the river's ecology, the extensive construction of hydropower dams throughout the Lower Mekong Basin – especially on the mainstream – is likely to have serious repercussions throughout the entire basin, both on the region's economy and its food security (see Chapter 9). According to the MRC, the economic value of the Mekong River's fisheries alone is in excess of US$2 billion per year (MRC, 2005). The value of this natural resource is largely unrecognized in regional infrastructure development plans.

Where a comprehensive and participatory assessment of all options has concluded that a hydropower project is the best option to meet water and energy needs, all parties involved should commit to implementing international best practice standards. An atmosphere that encourages a race to the top, not to the bottom, needs to be fostered. Currently, the region is far from that vision. As new hydropower proponents (such as those from China, Thailand and Vietnam) become increasingly influential in the Mekong region and step onto the global stage, they should accept their international responsibilities and adhere to international standards when developing and financing large infrastructure projects.

Notes

1 These are loans that are extended on terms substantially more generous than market loans. The concessionality is achieved either through interest rates below those available on the market or by grace periods, or a combination of these (see http://stats.oecd.org/glossary/detail.asp?ID=5901).
2 VLPC is a Vietnamese consortium, formed of Song Da Corporation (49 per cent), PetroVietnam, the Bank for Investment and Development of Vietnam, the PetroVietnam Finance Company and the BIDV Securities Company.
3 This observation is based on comments by Thai bankers during a 2007 forum on Thai–Lao hydropower development sponsored by the World Bank (see World Bank, 2007).
4 The Equator Principles (EP) are a voluntary set of environmental and social standards in development project finance globally that have been adopted by 60 private banks around the world.

References

ADB (Asian Development Bank) (1999) *Loan No 1329-LAO (SF): Theun Hinboun Hydropower Project*, Asian Development Bank, Aide Memoir: Special Review Mission, 9–18 November 1999

ADB (2001) *Private Sector Operations: Strategic Directions and Review*, Asian Development Bank, Manila

ADB (2004) *Project Performance Audit Report on the Nam Leuk Hydropower Project in the Lao PDR*, Operations Evaluation Department, Asian Development Bank, Manila

ADB (2005a) *Country Strategy and Program 2005–2009: Kingdom of Cambodia*, Asian Development Bank, Manila

ADB (2005b) *Private Sector Development*, Asian Development Bank, Mandaluyong

ADB (2007a) *Mid-Term Review of the Greater Mekong Subregion Strategic Framework 2002–2012*, Asian Development Bank, Manila

ADB (2007b) *Thailand Country Partnership Strategy 2007–2011*, Asian Development Bank, Manila, April 2007

ADB (2007c) *Country Strategy and Program 2005–2009: Lao People's Democratic Republic*, Asian Development Bank, Manila

ADB (2007d) *Country Strategy and Program 2007–2010: Viet Nam*, Asian Development Bank, Manila

ADB (2007e) *Financing Instruments and Modalities: Flexibility in Development Finance*, ADB, Mandaluyong

ADB (2008) *Strategy 2020: The Long Term Strategic Framework of the Asian Development Bank 2008–2020*, Asian Development Bank, Manila

Associated Press (2008) 'Cambodian Cabinet approves 2 hydroelectric dam projects to be built by Chinese companies', Associated Press, 14 June 2008

Bakker, K. (1999) 'The politics of hydropower: Developing the Mekong', *Political Geography*, vol 18, pp209–232

Barney, K. (2007) *Power, Progress and Impoverishment: Plantations, Hydropower, Ecological Change and Community Transformation in Hinboun District, Lao PDR*, York Centre for Asian Research Paper No 1 (June 2007), Centre for International Forestry Research, Probe International, Rights and Resources Initiative and the York Centre for Asian Research, Toronto and Washington DC

Bo, K. (2008) *Sino-Vietnamese Energy Cooperation: Promises and Pitfalls*, Chinastakes.com, 30 January 2008

BRN (Burma Rivers Network) (2008) 'Standards for hydropower development: A Burmese perspective' in Bosshard, P. and Brewer, N. (eds) *New Financiers and the Environment: Ten Perspectives on How Financial Institutions Can Protect the Environment*, International Rivers, Berkeley, CA

Cambodia Daily (2008) 'Kampot tourism takes hit amid construction', *Cambodia Daily*, 24 March 2008

CIEL (Centre for International Environmental Law) (2007) *An Evaluation of the Asian Development Bank's Approach to Country Safeguard Systems*, CIEL, Washington DC

CNMC (Cambodia National Mekong Committee) and WorldFish Centre, (2007) *Infrastructure and Tonle Sap Fisheries: How to Balance Infrastructure Development and Fisheries Livelihoods? The Challenge Facing Decision-Makers in Cambodia: Policy Brief*, WorldFish Centre, Phnom Penh

Cornford, J. and Matthems, N. (2007) *Hidden Costs: The Underside of Economic Transformation in the Greater Mekong Subregion*, Oxfam, Australia

CRES (Centre for Natural Resources and Environment Studies) (2001) *Study into Impact of Yali Falls Dam on Resettled and Downstream Communities*, Centre for Natural Resources and Environmental Studies, Vietnam National University, Hanoi, Vietnam

EarthRights International (2007a) *China in Burma: The Increasing Investment of Chinese Multinational Corporations in Burma's Hydropower, Oil and Gas, and Mining Sectors*, EarthRights International, September 2007

EarthRights International (2007b) 'Correspondence with ADB Re: Information disclosure request to the Asian Development Bank's Southeast Asia Department regarding the ADB's involvement in Burma via the Greater Mekong Subregion Program Regional Technical Assistance Grants', 5 April–10 May 2007

EGAT (Electricity Generating Authority of Thailand) (2008) *Thailand Power Development Plan 2007–2021: Revision 1*, Systems Planning Division, EGAT, January 2008

EVN (Electricité du Vietnam) (2006) *Master Plan IV on Power Development of Vietnam in Period 2006–2015 with the Vision to 2025*, Institute of Energy, Vietnam

FIVAS (The Association for International Water Studies) (2007) *Ruined Rivers, Damaged Lives: The Impacts of the Theun-Hinboun Hydropower Project on Downstream Communities in Lao PDR*, Association for International Water Studies, Oslo, November 2007

Foran, T. (2006) *Rivers of Contention: Pak Mun Dam, Electricity Planning, and State–Society Relations in Thailand, 1932–2004*, PhD thesis, Division of Geography, School of Geosciences, University of Sydney, Australia

Foran, T. (2008a) 'Better financial regulation could make energy organizations more sustainable', *Generating Dialogue, A Forum on Sustainable Energy, Good Governance, and Electricity Regulation*, World Resources Institute, Centre on Asia and Globalization, National University of Singapore, 16–18 March 2008

Foran, T. (2008b) *Analysis of Thailand's Electricity Planning Process and Demonstration of Integrated Electricity Planning*, USER Working Paper, Unit for Social and Environmental Research, Chiang Mai University, Chiang Mai (in Thai)

Fried, S., Herz, S., Leake, H., Levitt, J. and Perault, A. (2008) *Summary of Concerns Regarding the ADB's Draft Safeguard Policy Statement*, Written on behalf of Environmental Defence Fund, Bank Information Centre, Forest Peoples Programme, International Accountability Project and Centre for International Environmental Law, Morton-in-Marsh, New York, San Francisco and Washington DC, January 2008

Garrett, B. (2004) *Comments on Indicative Master Plan on Power Interconnection in GMS Countries: Discussion Document*, Commissioned by Palang Thai, June 2004

Greacen, C. S. and Footner, J. (2006) *Decentralizing Thai Power: Towards a Sustainable Energy System*, Greenpeace SE Asia, Bangkok

Greacen, C. S. and Greacen, C. (2004) 'Thailand's electricity reforms: Privatization of benefits and socialization of costs and risks', *Pacific Affairs*, vol 77, no 3, pp517–541

Greacen, C. and Palettu, A. (2007) 'Electricity sector planning and hydropower', in Lebel, L., Dore, J., Daniel, R. and Koma, Y. S. (eds) *Democratizing Water Governance in the Mekong Region*, Mekong Press, Chiang Mai

Haggart, K. (2006) 'Sinohydro "downgraded" over accident record', *Financial Times*, 1 September 2006

Hirsch, P. (1998) 'Dams, resources and the politics of environment in mainland Southeast Asia', in Hirsch, P. and Warren, C. (eds) *The Politics of Environment in Southeast Asia: Resources and Resistance*, Routledge, London, Chapter 3, pp55–70

Hirsch, P. and Mørck-Jensen, K. (2006) *National Interests and Transboundary Water Governance of the Mekong*, Australian Mekong Resource Centre in association with DANIDA and the University of Sydney, Australia

Hirsch, P. and Wyatt, A. (2004) 'Negotiating local livelihoods: Scales of conflict in the Sesan River Basin', *Asia Pacific Viewpoint*, vol 45, no 1, pp51–68

International Rivers (2008) *Power Surge: The Impacts of Rapid Dam Development in Laos,* International Rivers, Berkeley, CA, September

International Rivers and RCC (Rivers Coalition of Cambodia) (2008) *Cambodia's Hydropower Development and China's Involvement*, International Rivers and the NGO Forum on Cambodia, January 2008

IRN (International Rivers Network) (1999) *Power Struggle: The Impacts of Hydro Development in Laos*, International Rivers Network, Berkeley, CA, February

IUCN, TEI, IWMI and M-POWER (2007) *Exploring Water Futures Together: Mekong Region Waters Dialogue*, Report from Regional Dialogue, Vientiane, Laos

Lamberts, D. (2008) 'Little impact, much damage: The consequences of Mekong River flow alterations for the Tonle Sap ecosystem', in Kummu, M., Keskinen, M. and Varis, O. (eds) *Modern Myths of the Mekong*, pp3–18

Lang, C. (1998) 'Se kong-Se San and Nam Theun: A fistful of studies', *Watershed*, vol 3, no 2, pp40–44

Lang, C. (2000) 'The National Hydropower Plan Study: Planning and damming in Vietnam', *Watershed*, vol 5, no 3, pp48–51

Lao National Committee for Energy (2008) *Power Projects in Lao PDR (Update on 7 July 2008)*, Lao National Committee for Energy, Vientiane, www.poweringprogress.org

Middleton, C. (2008) 'Perspective from the Mekong Region: New financiers and familiar problems', in Bosshard, P. and Brewer, N. (eds) *New Financiers and the Environment: Ten Perspectives on How Financial Institutions Can Protect the Environment*, International Rivers, Berkeley, CA

MRC (Mekong River Commission) (2005) *Annual Report of the MRC Programme for Fisheries Management and Development Cooperation*, April 2004–March 2005, Mekong River Commission, Vientiane, Laos

Muscat, R. J. (1990) *Thailand and the United States: Development, Security and Foreign Aid*, Columbia University Press, Columbia, US

Nangia, R. (2008) *Building a Sustainable Energy Future: The Greater Mekong Sub-Region*, Final Workshop Report, 5–6 June 2008, Asian Development Bank, Manila

Norconsult (1994) *Promoting Subregional Cooperation among Cambodia, Lao PDR, Myanmar, Thailand, Viet Nam and Yunnan Province of the People's Republic of China: Subregional Energy Sector Study for Asian Development Bank*, Asian Development Bank, Manila

Norconsult (2002) *Indicative Master Plan on Power Interconnection in GMS Countries (TA No 5920 REG)*, Asian Development Bank, Manila

Norplan (2008) *Nam Ngum 3 Hydropower Project Public Consultation Workshop: Handout Presentations*, Vientiane, 8 February 2008

Phonekeo, K. (1996) *Country Paper: National Hydropower Sector, Lao PDR*, Hydropower Forum, Hanoi, Vietnam, 16–18 July 1996

Ratner, B. (2003) 'The politics of regional governance in the Mekong River Basin', *Global Change*, vol 15, no 1, pp59–76

RDSC (Rural Development Services Centre) and BIC (Bank Information Centre) (2008) *Trip Notes on Song Bung 4 Hydropower Project in Vietnam*, Bank Information Centre, 24 June 2008

Reuters (2007) 'Vietnam utility EVN plans 1st overseas bonds issue', Reuters, 21 March 2007

Ryder, G. (1994) 'An overview of regional plans', *World Rivers Review Mekong Special*, vol 9, no 4, pp7–13, www.internationalrivers.org/files/WRR.V09.N4.pdf

Sluiter, L. (1992) *The Mekong Currency*, Project for Ecological Recovery/TERRA, Bangkok, Thailand

Soluziona (2004) *Study for a Regional Power Trade Operating Agreement in the Greater Mekong Subregion (TA 6100-REG)*, Asian Development Bank, Manila

Stuart-Fox, M. (2006) 'The political culture of corruption in Lao PDR', *Asian Studies Review*, vol 30, pp59–75

Trandem, A. (2008) 'A Vietnamese/Cambodian transboundary dialogue: Impacts of dams on the Se San River', *Development*, vol 51, pp108–113

UNEP (United Nations Environment Programme) and TEI (Thailand Environment Institute) (2007) *Greater Mekong Environment Outlook*, UNEP and TEI, Bangkok

USAID (US Agency for International Development) (2007) *From Ideas to Action: Clean Energy Solutions for Asia to Address Climate Change*, US Agency for International Development: Regional Development Mission for Asia, Bangkok

Vientiane Times (2008) 'Laos profits from Theun Hinboun hydropower', *Vientiane Times*, 2 April 2008

VUSTA (Vietnam Union of Science and Technology Associations) (2006) *A Work in Progress: Study on the Impacts of Vietnam's Son La Hydropower Project*, VUSTA, Vietnam

VUSTA (2007) *Assessment of Vietnam Power Development Plan*, VUSTA, Vietnam

Watson, S. P. and Schouten, R. (2001) *Nam Song Diversion Project: Draft Impact Analysis Report and Action Plan*, Asian Development Bank, Manila, October

WCD (World Commission on Dams) (2000) *Dams and Development: A New Framework for Decision Making*, Earthscan Publications, London and Sterling, VA

Win Kyaw (2006) *Overview of Water Resources Development in the Union of Myanmar*, Department of Hydropower Planning, Myanmar

World Bank (2005a) *Thailand–World Bank Group Partnership: Country Dialogue Monitor January 2004–June 2005*, World Bank, Washington DC

World Bank (2005b) *International Development Association Country Assistance Strategy for the Lao People's Democratic Republic (Lao PDR)*, World Bank, Washington DC

World Bank (2005c) *International Development Association and the International Finance Corporation Country Assistance Strategy for the Kingdom of Cambodia*, World Bank, Washington DC

World Bank (2006) '*Vietnam Infrastructure Strategy: Power Strategy – Managing Growth and Reform*', World Bank, Washington DC

World Bank (2007a) *High Level Forum on Lao-Thai Partnership in Sustainable Hydropower Development*, Summary of Break-out Discussion Groups, Group 1, World Bank, Washington DC, 7 September 2007, http://go.worldbank.org/RQR6YSCJI0, accessed 11 September 2008

World Bank (2007b) *International Development Association and International Finance Corporation Multilateral Investment Guarantee Agency Country and Assistance Strategy with the Socialist Republic of Vietnam for the Period 2007–2011*, World Bank, Washington DC

World Bank and ADB (2006) *WB/ADB Joint Working Paper on Future Directions for Water Resources Management in the Mekong Basin: Mekong Water Resources Assistance Strategy*, www.mpowernet.org/mweb.php?pg=133

XFN-ASIA (2008) 'Cambodia govt approves 1bln USD Chinese dams', *XFN-ASIA*, 13 June 2008

3

Pak Mun Dam: Perpetually Contested?

Tira Foran and Kanokwan Manorom

Introduction

Ever since its approval in 1989, Pak Mun Dam has sparked controversy, linked to resistance networks that have sought to defend, mitigate and restore fisheries-dependent livelihoods. This chapter provides a history of the Pak Mun Project and summarizes its main features – ecological, engineering and governance. It explores the project's trajectory in the context of wider changes in Thai state–society relations. We review important debates between proponents and opponents, and unresolved controversies and risks. Finally, we reflect on the consequences of the dispute over Pak Mun Dam and discuss critical lessons from the case.

The first publicized dispute in Thailand about a large dam was over the Nam Choan Project (on the Upper Kwae Noi River), proposed in 1982 by the state-owned Electricity Generating Authority of Thailand (EGAT). Nam Choan Dam would have flooded 223km^2 of a wildlife sanctuary in Kanchanaburi Province northwest of Bangkok. The same year, a coalition of Bangkok environmentalists, students, local middle-class people, non-governmental organizations (NGOs) and some of the up to 2000 Karen minority people who would be displaced joined in protest against the project. Opponents argued that the project imposed unacceptable social costs, would deplete forests and harm wildlife. Twice during the 1980s, the government of Prime Minister Prem Tinsulanonda reviewed and shelved the Nam Choan Project. This signalled new complications for the dam-building programme of EGAT and the Royal Irrigation Department (RID), the main builders of large dams in Thailand. Partly as a product of the Nam Choan Dam campaign, Project for Ecological Recovery, a new NGO, emerged in 1986 to defend forests and resource-dependent livelihoods. Some observers have interpreted the state's determination to proceed with Pak Mun Dam in 1989 as a reaction to EGAT's earlier setback at Nam Choan.

Pak Mun Dam is located approximately 80km downstream from the provincial centre of Ubon Ratchathani and 5.5km upstream of the confluence of the Mun and the Mekong rivers (see Figure 3.1). Constructed during 1990 to 1994, the dam is 17m high, 300m wide, with eight radial gates that can be fully opened to release water.

The Mun's living aquatic resources are noted for their high biodiversity and contribution to subsistence and trade (Roberts, 1993; Sretthachau, 2002). When the gates of Pak Mun were opened between 2001 and 2002 for a year-long experiment, two studies counted more than 150 species of fish (Sretthachau, 2002; UBU, 2002). Fishermen use a variety of gear, including hook and line, traps, nets and beach-haul seines. Total catch has not been estimated for a number of reasons, including the large number of landing sites, subsistence consumption and – most importantly for sustaining important fish populations – lack of a long-term fisheries assessment programme. A concrete fish ladder was installed in 1996, but its design does not allow significant upstream migration (Roberts, 2001). Instead, in a 2003 decision that we explore below, EGAT was requested to fully open the dam's gates during the annual wet season, nominally for four months beginning in June.

Because it is a 'run-of-river' dam operated for power generation, Pak Mun cannot also be used to store significant amounts of water. Yet, since the early 2000s – partly as a result of populist development policy – the state has expanded small pumped irrigation systems near and upstream from the dam. Critics view this initiative as a justification for not opening the dam gates beyond four months (Foran, 2006, Chapter 8), reserving the dam instead for power generation at least eight months per year, including during Thailand's peak power demand hot season.

Our Pak Mun Dam 'case' (see Table 3.1 for chronology) consists of a complex series of interactions between local people (including anti-dam campaigners), civil society organizations and state agents during the period of 1989 to 2008. Resistance against Pak Mun began in 1989 to 1990 with informal networking among villagers who opposed the dam and the state's process. The state worked through local authorities such as district and sub-district officers and village headmen.[1] Prior to construction, they tried to elicit public support at meetings which they summoned. The state's paternalistic process and threats of repression failed to intimidate a few articulate and confident middle-aged women. They helped to form a larger network and sought advice from a small civil liberties NGO in Ubon Ratchathani (Missingham, 2003). Opposition spread to town people: first to vendors opposed to the flooding of Kaeng Saphue, a large rapids and tourist attraction. Later it spread to a segment of the middle class in Ubon Ratchathani.

In the earliest stages, people seemed to be responding to a lack of information and fear of widespread impacts. When it became understood that the run-of-river

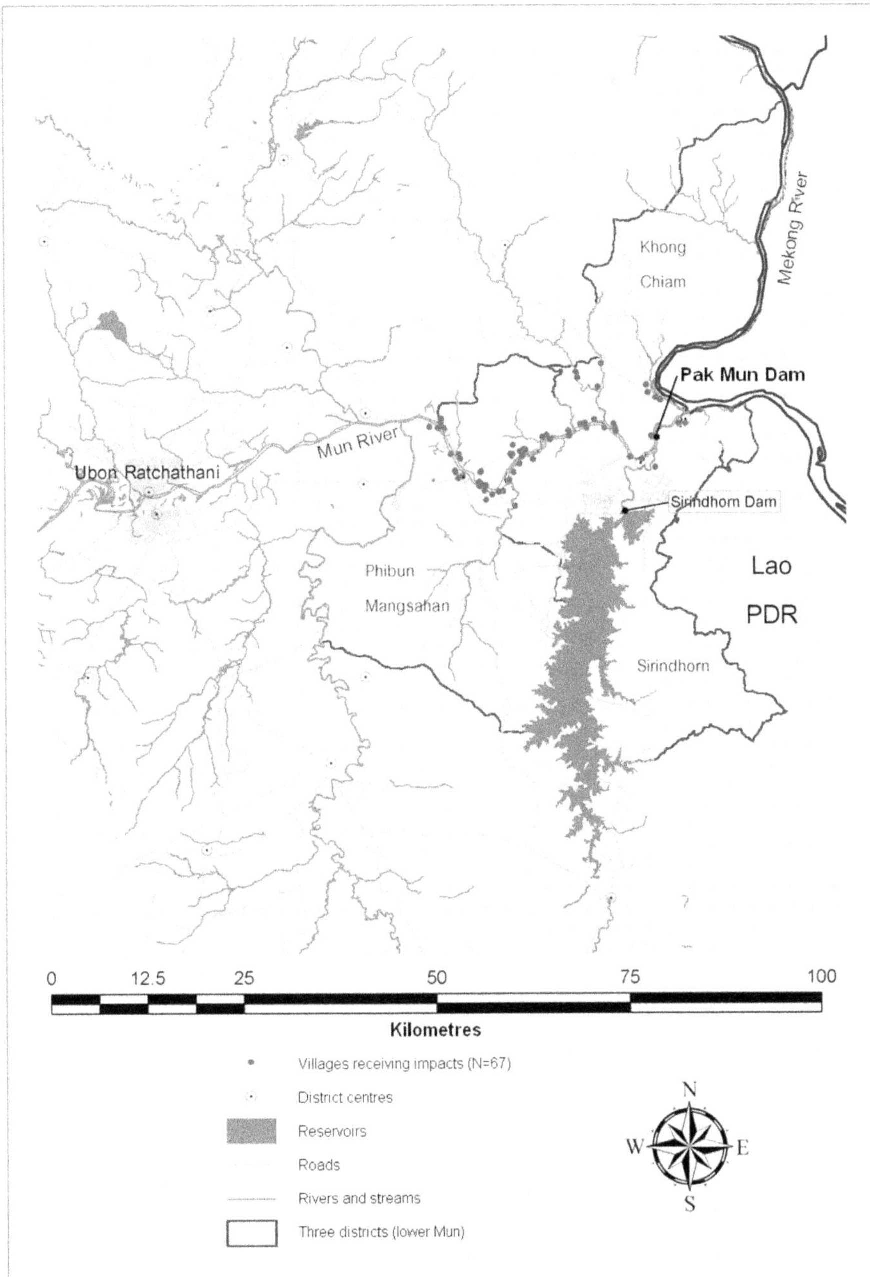

Figure 3.1 *Lower Mun Basin and Pak Mun Dam*

Source: village data from EGAT (undated)

design chosen by EGAT would lead to a much smaller area inundated, and when EGAT undertook not to flood Kaeng Saphue, most of the opposition dissipated. The protest campaign narrowed to a core of villagers and alternative development NGOs. Local people were apparently the first to raise the concern that the Pak Mun Dam would destroy wild fisheries harvests after they witnessed blasting of the river bed during construction in 1991. By 1994, a villagers' organization representing some 2500 families from more than 50 villages had formed to press for compensation in terms of land and fisheries. A pro-dam network led by village headmen and other local authorities also emerged, promising equal compensation to people without the need to protest. In 2008, Pak Mun's 15th year of operations, pro- and anti-dam coalitions still existed.

Regardless of their level of engagement in dam disputes, villagers in the Lower Mun River Basin pursue similar livelihood strategies. They are smallholder farmers who grow one main crop of rice during the May to October monsoon season. Holdings are typically 5ha to 7ha; but soils near the dam are often poor. By Thai standards, many households are income poor. Almost all households supplement their income by off-farm labour, with significant rates of seasonal and long-term out-migration, especially among younger people (UBU, 2002).

The degree to which local people derive benefit from living aquatic resources, especially wild fisheries, has been a topic of multiple rounds of dispute. Although marginalized in state-sponsored livelihood surveys, and difficult to quantify because diffuse, variable and politicized, living aquatic resources make meaningful contributions to livelihoods (Sretthachau, 2002; UBU, 2002; Foran, 2006, Chapter 8). This is especially true for land-poor farmers.

Pre-operational patterns of contention, 1989 to 1994

By the 1980s in Thailand, the public sphere vital to engage state decision-making had distinctly increased as a result of contentious democratization during the 1970s (Foran, 2006, Chapter 4). Yet, advocates calling for more deliberation over Pak Mun Dam in the early 1990s met with predominantly aloof responses from state agents (Foran, 2006, p193). As in many other conflicts over rural development at the time, Thai NGOs stepped in to help villagers publicize their grievances.

NGOs sensed that EGAT's application for World Bank financing generated opportunities to amplify their concerns internationally and to mobilize transnational opposition (Hubbel, 1992).[2] They cast the problem in terms developed since the late 1960s. A 'community culture' school of thought portrayed the village economy as inherently superior to dependent capitalist development (Phongpaichit and Baker, 1995, p387). It defended local peoples' common natural resources, such as wild-capture fisheries and tourist attractions against resource-grabbing by state agencies and private interests.

Table 3.1 *Pak Mun Dam case: Key events (1982 to 2007)*

Year	Actions
Events under Prime Minister Prem Tinsulanonda (March 1980–April 1988)	
1982	Environmental impact assessment (EIA) for a dam located above confluence of Mun and Mekong rivers at 112m above mean sea level (amsl) crest indicates 4000 households would need resettlement.
1983–1987	1983: Kaeng Tana National Park declared; the Electricity Generating Authority of Thailand (EGAT) conducts new feasibility study for dam located upstream. 1985: EGAT lowers water retention level to 108m amsl, relocates site to lower inundation impact. 1987: Pak Mun Dam appears in EGAT's *Power Development Plan*.
Events under Prime Minister Chatichai Choonhavan (April 1988–February 1991)	
1989	April: Pak Mun Project first approved by cabinet; first protest at Ubon Ratchathani provincial hall.
1990	EGAT applies for World Bank loan to develop power system, including Pak Mun. February: campaigns begin to increase transparency about the number of affected households; three-day anti-dam rally/confrontation with supporters' rally. May: cabinet approves Pak Mun budget of 3.88 billion baht (US$155.2 million); sets up committees for compensation and resettlement; 262 households understood affected. May: northeast NGOs support anti-dam villagers. June: site preparation work commences. August: government releases environmental mitigation plan. October: World Bank completes pre-investment staff appraisal report.
1991	January: completion of preliminary site works.
Events under Prime Minister Anand Panyarachun (national peacekeeping council junta) (February 1991–April 1992)	
1991	February: military coup led by General Suchinda Kraprayoon ousts Chatichai government, appoints Anand as prime minister; EGAT sticks to its estimate of 262 affected households; proceeds with construction. March: 12,000 petition World Bank against making loan. May: two-week rally ends with agreement to establish participatory impact assessment committee. June: Anand government appoints multi-stakeholder problem-solving committee. October: World Bank directors meet with Pak Mun opponents. December: World Bank approves loan, with two objections and one abstention.
1992	March: EGAT, Royal Forest Department, National Parks Department and Fine Arts Department defend rapids blasting in Kaeng Tana National Park as legal and not harmful. April: 200 villagers protest against rapids blasting, claiming damage to fisheries migration. April–May: Bangkok demonstrations against Suchinda assuming prime ministership; Suchinda resigns after large-scale demonstrations turn violent; Anand reappointed as prime minister. September: Chuan Leekpai (Democrats) win elections.

Table 3.1 *(continued)*

Year	Actions
Events under Prime Minister Chuan Leekpai's first administration (September 1992–July 1995)	
1993	February: oppositional villagers begin new campaign for just compensation.
1994	June: dam commissioned, impoundment begins, fish pass completed. April–June: rallies against dam in Bangkok. October: 2000 villagers rally for fair fishing livelihood compensation at Ubon Provincial Hall; after two weeks, march to dam site. December: 300 villagers affected by Sirindhorn Dam march to join Pak Mun rally, demanding compensation for the earlier project.
1995	January: government approves new consensus-based participatory fisheries compensation committee, chaired by Plodprasop Suraswadi. March: government agrees to pay for three years' lost fishing income; 157-day rally ends. May–November: Plodprasop committee approves 2932 out of 4530 applications for fisheries compensation. June: first round of compensation to 571 households; agricultural co-operative formed to hold two-thirds of payment in trust. May: Chuan dissolves parliament amidst corruption scandal; Chart Thai Party wins elections.
Events under Prime Minister Banharn Silpa-Archa (July 1995–November 1996)	
1995	December: Assembly of the Poor (AOP) announces *Mun River Declaration*; 600 demonstrators present this sustainable development manifesto to ASEAN leader in Bangkok.
1996	March: AOP stages multi-issue farmers' rally with up to 12,000 people. April: participatory problem-solving committee set up with Prime Minister Banharn as chair. Sirindhorn Dam villagers' movement for retroactive compensation grows to 2500 households. September: Banharn's coalition government collapses; Chavalit's New Aspiration Party wins subsequent elections.
Events under Prime Minister Chavalit Yongchaiyudh (November 1996–November 1997)	
1997	January: AOP begins 1999-day multi-issue rally. April: government agrees to pay perpetual fisheries compensation for 3080 Pak Mun fishermen. July–November: financial crisis erupts; Chavalit devalues baht and resigns in November; Chuan assumes prime ministership.
Events under Prime Minister Chuan Leekpai (second administration November 1997–February 2001)	
1998	April: Chuan cabinet refuses to honour Chavalit government resolution of 29 April 1997. July: Thaksin Shinawatra launches Thai Rak Thai Party.
1999	March: campaign against Pak Mun demands dam decommissioning, builds protest village adjacent to dam. Late 1999: World Commission on Dams (WCD) multi-stakeholder process begins.

Table 3.1 *(continued)*

Year	Actions
2000	March: WCD draft report criticizes performance of dam and fish pass. May: AOP villagers blockade Pak Mun and Rasi Salai dams. June: government establishes 'neutral' problem-solving committee, chaired by Bantorn Orndam. July: Bantorn committee proposes four-month trial opening; local university to study social, economic and ecological impacts; Chuan government declines; AOP villagers scale walls of Government House, resulting in mass arrests.
Events under Prime Minister Thaksin Shinawatra (February 2001–September 2006)	
2001	April: government accepts trial opening; commissions Ubon Ratchathani University (UBU) study.
2002	June: after Pak Mun open for 12 months, EGAT Governor Chalermchai Ratanarak offers four months per year seasonal opening policy. September: problem-solving committee chaired by Deputy Prime Minister Pongpol Adireksan votes for four-month seasonal opening. October: cabinet resolution ratifies Pongpol committee decision. Late 2002: UBU Pak Mun study publishes interim and final conclusions; Tai Baan research completed. December: Thaksin chairs roundtable with academics and anti-dam villagers; National Statistics Office (NSO) surveys the opinions of 3750 household heads in 150 villages.
2003	January: government reconfirms final decision of four-month seasonal opening; Bangkok governor evicts 500 demonstrators. March: Ministry of Agriculture announces Pak Mun Irrigation Project and fish-stocking investments.
2003–2006	Four-month seasonal opening policy implemented.
Events under Prime Minister Surayud Chulanont (National Security Council junta) (September 2006–February 2008)	
2007	February–May: local authorities organize villager survey (n = 20,592); claim overwhelming support for a year-round dam closure policy; results unpublished. March: 3000 villagers hold a pro-dam rally. June–July: government abandons, then readopts, the four-month opening policy; devolves operational management to multi-stakeholder provincial committee.

Source: adapted from Foran (2006)

NGOs also presented an injustice narrative: the state trampled over basic rights such as the provision of transparent information and wider participation in project planning. Opponents complained about the state's closed practices, its lack of reliable information and its occasionally arrogant handling of public meetings. They decried plans to resettle inundated farmers on land often of marginal value and with less secure titles.

How did villagers' opposition emerge and sustain itself? Missingham (2003) credits the strength and efficacy of an NGO-led internally democratic social change network. However, the movement's internal structure was not the only cause of sustained mobilization. Events during 1989 to 1994 suggest that one important reason people kept on challenging the state's handling of Pak Mun was that state agents persistently responded with a *mixture* of concession and repression. Process concessions made during Anand Panyarachun's government needed to be fought for again during the subsequent Chuan Leekpai administration. Chuan's government first granted these concessions in the form of a multi-stakeholder committee to review the scope and eligibility of compensation. Later it revoked the multi-stakeholder process when senior decision-makers perceived it as too threatening to established practices. This tantalizing and frustrating dance of concession and denial spurred opponents to keep struggling. Defining and identifying affected households was a point of contention.

State response to dam opponents

The collective action that emerged was perceived as very threatening to established notions of political order. Many immediate responses by state agents were repressive.[3] The police described people who distributed leaflets, wrote letters and attended demonstrations as a 'minority', as 'paid' agents or even as 'communists' (*Bangkok Post*, 1991). A second dimension of repression was state-owned radio and television coverage that failed to report on anti-dam activism, and consistently reported favourably on public meetings in Ubon Ratchathani. Provincial media also gave the project favourable coverage (Arthit, interview 2 July 2002). A third dimension of repression involved mobilizing local support. EGAT and the state instructed local leaders to speak out in favour of the dam and to strongly discourage villagers from voicing dissent (Missingham, 2003, pp73–76).

Impacts and consequences of protest

Until 1993, the civil society opposition campaign had limited impact. The government of Prime Minister Chatichai Choonhavan moved forward with approval and construction during 1989 to 1991. The military-appointed government of Prime Minister Anand Panyarachun did not halt construction during 1991 to 1992 when the dam was less than 10 per cent complete. Some World Bank directors voted against the project in late 1991, but not a majority.

Opponents, nonetheless, opened up spaces for debate, and notable concessions were made. Early discursive concessions underpinned subsequent campaigns. Under public pressure, the governments of Prime Minister Anand and, later, the first administration of Chuan Leekpai set up committees to review flooding and livelihoods compensation. A 1991 committee under the Anand government validated protesters' claims that EGAT was *not* handling Pak Mun as transparently as they and the government desired. It also helped to legitimate the idea of problem-solving committees with ordinary villager participation. Campaigns in 1993 to expand recognition of affected households built directly on promises made in Anand's committee in 1991. By December 1993, the state had accepted demands for a process devolved down to a set of village-level committees, with villager participation, to help process residents' grievances (Foran, 2006, p191).

A devolved process helped to deliver compensation but was not sufficient to hold the state accountable for all types of impact. The state and protesters were particularly in dispute over land higher than 108m above mean sea level (the nominal maximum water level). EGAT initially claimed that it would pay only for earthworks to raise affected structures. Opponents claimed their homes would be surrounded by water. After more demonstrations at the construction site, they prevailed in getting the state to pay the costs of moving and re-erecting houses (see Table 3.2; Foran, 2006, p192).

Table 3.2 *Categories of households recognized for compensation of structures and fixed assets*

Set (1): Recognized in 1982–1983 studies by Team Consulting Engineers Ltd	Number
1.1 Affected by construction, Ban Hua Haew village	11
1.2 Living below 108m amsl	136
1.3 Living 108m–108.5m amsl	96
Sub-total	243
Set (2): Recognized in 1994 by civil society campaigns, 1990–1994	
2.1 Affected by river bed blasting	227
2.2 Agriculture land inundated	706
2.3 Living above 108.5m amsl, chose to relocate	473
Sub-total	1406
Total	1649

Notes: Non-fisheries impacts only. 'Recognized' refers to recognition of categories; numerical estimates vary. Original set (1) estimates ranged from 241 to 379, but at least 1821 households eventually received compensation (Missingham, 2003, p72).
Source: Amornsakchai et al (2000, p58)

Dam opponents did not succeed in defeating some powerful pragmatic arguments, such as the argument that the dam had already been approved, construction had started and, therefore, the dam must proceed. In light of this discursive and institutional context, the fact that activists successfully forced the state to recognize broader categories of people and impacts was a very important outcome. It delivered collective benefits to those otherwise invisible to the state.

Protest also produced important unanticipated consequences. The mobilization of pro-dam villagers frequently led to violent encounters. Displacing protesters from a given site could trigger sympathetic media coverage. Protesters' non-violent forms of contention, when met with violence, tended to generate social movement and media frames of innocent and displaced victims, which governments subsequently had trouble dismissing.

A second outcome unanticipated by activists was that local communities divided. As Chanchai, a pro-dam local leader, remembers it, generous compensation divided local people between people whose land or structures would be inundated and those not:

> *Land around here was cheap before the compensation process because it wasn't great paddy land. It was worth only about 500 baht per rai [US$125 per hectare]; it would flood almost every year. The state announced it would compensate at US$8750 per hectare, up to 1.6ha, and would provide another 1.6ha of land. It would compensate trees on flooded land and structures. This led to envy. People were divided into two factions: those getting impacts, and those not. Everyone wanted to receive flooding impacts. The protesters joined [anti-dam rallies] out of envy. 'Nam ko tong thuam khoi bang' ['The water has to flood me as well'].* (*Kamnan* [Sub-District Officer] Chanchai, interviews 2 June 2002 and 12 November 2005)

During subsequent years, these divisions and pressures for a wider distribution of compensation benefits proved to be a great challenge both to authorities and activists.

To sum up the pre-operational period, campaigns against Pak Mun produced potent discourses of transparency and accountability. Occasionally, sympathetic and open-minded policy-makers exercised agency, but were constrained by institutions and associated discursive practices. Opponents did not change these powerful practices and institutions, but won some important concessions, such as recognition of a broader range of affected people (see Table 3.2). Doing so required strong and sustained collective action, as well as the ability to broaden problem definitions and solutions. The state's responses to the strategic actions of its opponents ranged from repression, to opening negotiation, leading to concessions. Outcomes hinged on decision-makers' reactions to the *emergent* process of sustained collective action. Two plausible processes are involved in those outcomes:

1 deliberative processes that changed problem definitions and solutions; and
2 conciliatory openings offered by elites, after peak protest events or unexpected episodes of violence.

Post-operational patterns of contention, 1994 to 2003

After completion of the dam in 1994, a new round of collective action emerged around a discourse to hold the state explicitly accountable for impacts upon fisheries. Fisheries impacts had been argued about since 1991, but were only beginning to be processed by a provincial-level sub-committee two years later. In late 1993, Maliwan and Pho Siang, two protest leaders, emerged from jail to demand that the state pay fisheries compensation of 35,000 baht ($US1400) per household for each year of the three-year constructions (Buchita, 1997; Foran, 2006, p197).[4]

After Pak Mun's commissioning in June 1994, the state's response to dam opponents continued to range from repression to negotiated concessions. As during the prior period, violence against the protesters (e.g. police crackdowns on direct actions) provoked a temporarily empowering media backlash. Despite these similarities, the post-operation period is qualitatively different. The main movement organizations opposed to Pak Mun – the Mun River Villagers' Committee and, later, the Assembly of the Poor (AOP) – began to stage larger demonstrations. This was, in part, because the new campaign for fisheries compensation appealed to a larger set of villager beneficiaries, and also because of proven successes in earlier campaigns and the entry of new aggrieved groups into the AOP.[5]

Faced with the challenge of sustained mobilization, the state, beginning during the first Chuan Leekpai administration, deliberately stalled and otherwise refused to budge on protesters' demands. The Prime Minister's Office devolved conflict management responsibilities to provincial government, but did not, despite claims to the contrary, devolve authority adequate to resolve those conflicts (Foran, 2006, Chapter 7).

Opposition to Pak Mun triggered defensive action. In 1994, the year Pak Mun was commissioned, *Kamnan* Chanchai (the leader quoted above) formed a new group to oppose the protesters. He did so, he said, out of loyalty to the '80 per cent' of villagers who still respected their leaders, and found the protesters' behaviour outrageous. Chanchai remembers the protesters as aggressive, wilful and immoral: 'If they felt like blocking a road to demonstrate, or a district office, they just did it.' He found that his peers in three local districts felt the same way: villagers increasingly viewed them as having lost their power. District officers and the provincial governor supported his effort, as did senior EGAT management. 'They gave me a green light', said Chanchai (interview, 12 November 2005).

Thus emerged the Kamnan and Village Headmen's Group (KVHG). Lacking a broad change agenda, its primary objective was to dissuade villagers from joining anti-dam action. KVHG did this by making claims on behalf of villagers who stayed out of protests. It administered interim fisheries compensation claims for more than 2000 villagers, without their having to join any of the anti-dam campaigns. It reasoned that whatever claims anti-dam protesters established would eventually be granted, on equity principles, to other fishing households. But some members of the KVHG also organized counter-demonstrations against the anti-dam villagers.

Meanwhile, faced with protests, the central government learned to withhold force. It let protesters languish and, especially post-1997, espoused their right to protest. Such a response helped to produce marathon demonstrations for livelihoods restoration: in 1994 to 1995 (157 days); 1997 (99 days); and a 'protest village' adjacent to the dam during 1999 to 2002. A counter-response by the protesters was to stage actions in ways calculated to maximize the odds of favourable media coverage. Outcomes were mixed: the 157-day sit-in (during the first Chuan government) and the 99-day rally in Bangkok (during the government of Prime Minister Chavalit Yongchaiyudh) led to negotiations resulting in unprecedented agreements to compensate for damage to fishery-dependent livelihoods (see Table 3.2). By contrast, the protest village sit-in campaign during the second Chuan government yielded 15 months of impasse.

Reinvigorated protest campaign and World Commission on Dams Assessment, 1999 to 2000

During 1999 to 2000, as the protest village campaign wore on, the Pak Mun conflict was transformed. This was partly as a result of an expert assessment made on behalf of the World Commission on Dams, and partly as a result of the second Chuan Leekpai government's hardened stance towards protesters. After the Thai financial crisis and regime change in late 1997, Chuan's incoming government decided to withdraw the concessions made by the previous Chavalit government for lost fisheries income. In April 1998, after several weeks of another large dry season rally, Chuan's cabinet resolved to not pay any compensation for past development projects, arguing that this would open a never-ending series of claims and that the government was broke. Essentially, it refused to honour any of the commitments to the AOP made by the preceding Chavalit government. These reversals meant wider setbacks – for Thai highlanders who had won some rights to live in protected areas, for opponents of two other dams in northeast Thailand, and for villagers claiming compensation for Sirindhorn and Pak Mun dams (Missingham, 2003).

In March 1999, the AOP launched a new campaign. It established a protest village occupying several hectares of a public park and riverbank immediately adjacent to the Pak Mun Dam site. The assembly announced that it was abandoning its previous claim for permanent loss of fisheries income. In 1997, Chavalit's administration had agreed to provide 2.4ha of land (or the monetary equivalent at US$8750 per hectare) for 3080 Pak Mun fisher households. After the Chuan II government refused to compensate, the assembly demanded that the government decommission the dam in order to restore fisheries to the river.

In 1999, another process began that was to prove influential. The World Commission on Dams (WCD) – a multi-stakeholder process funded by a range of development and private-sector donors – was a sophisticated attempt to conduct a series of participatory studies about the performance of large dams worldwide. For its eight in-depth case studies, the WCD asked governments, including Thailand, for permission to study the economic, environmental and social impacts; the benefits, costs and distribution of these impacts; and the decision-making processes for these dams. WCD chose to study Pak Mun in part because its sponsors considered it an exemplary project. In June 1998, the World Bank's Operations Evaluation Department released a report stating that Pak Mun's resettlement programme was 'overly generous' and denied that the dam caused any decline in the fish population in the Mun (World Bank, 2000). On the other hand, members of the WCD, such as the International Rivers Network, had helped to campaign against the dam.

Thailand has limited experience with formalized knowledge-building multi-stakeholder processes. The only such process during the 1990s was the Constitutional Drafting Assembly of 1996 to 1997 (Phongpaichit and Baker, 2000).[6] Considering the immediate political context at Pak Mun, the WCD assessment was ambitious. Thailand appears to have been the only WCD case study that proceeded while dam opponents staged ongoing protests; tensions required the WCD to hold separate meetings with EGAT and civil society.

In November 2000, the WCD released its Pak Mun case study. The evaluation was critical – of its intended hydropower benefits, the dam delivered only 21MW of actual dependable capacity versus 75MW planned. Thus, its economic cost-benefit ratio, calculated from the higher number, had been overestimated. Furthermore, despite installation of a fish pass, Pak Mun had reduced the diversity and overall supply of fish to income-poor, labour-exporting rural households (Amornsakchai, et al, 2000, Chapter 4). The report included dissenting reviews from the World Bank and EGAT, and responses to those reviews. It was a dense multi-vocal compilation of knowledge.

Unfortunately, despite its well-designed process, the WCD Pak Mun study ended in acrimony. EGAT steadfastly argued that:

- The dam produced peaking power benefits of 126MW to 136MW, consistent with the original feasibility studies justifying the project (EGAT, 2000a, pp102–103).
- The study over-exaggerated the decline in the number of fish species found in the Mun after construction of the dam; the decline resulted from multiple causes and should not be attributed to the Pak Mun Dam alone (EGAT, 2000a, pp105–111).
- By 2000, it had compensated more than 6200 families for fisheries impacts, paying out more than 989 million baht (EGAT, 2000b).
- Grievances were driven by villagers' material incentives for compensation and were manufactured by Thai and foreign environmental NGOs (EGAT, 2000a, p111).

In short, EGAT congratulated itself for producing power benefits and compensating generously, while limiting its responsibility for fisheries decline. Its response to WCD repeatedly took the form of categorical assertions that certain methods and studies were credible, while other studies and methods were invalid (EGAT, 2000a).

The World Bank defended the project, but conceded deficiencies in resettlement planning, the failure of the EIA to account for local fisheries dependence, and lack of consultation with affected people (World Bank, 2000).

A debate between WCD and project sponsors EGAT and the World Bank over Pak Mun's 'dependable capacity' goes to the heart of the dam's benefits. Dependable capacity refers to a threshold level of peak power output, which is exceeded by a given proportion of all peak power output values. It is a measure of reliability. During the early 1980s, EGAT defined dependable capacity of hydropower plants as the value of power production that will be ensured (or exceeded) half of the time.

Based on this definition, EGAT presented Pak Mun's dependable capacity as 75MW to the Chatichai government in 1988 for approval. It took the 75MW figure from SOGREAH consulting engineers (SOGREAH, 1985). To get this result, SOGREAH estimated the dam's average monthly energy production. Averaging, however, leads to overestimating dependable capacity because high values in the time series bias it upwards (Kansuntisukmongkol, 1994, pp51–52).

During the late 1980s, EGAT toughened its definition of dependable capacity, specifying it as power production that will be exceeded 90 per cent of the time, based on long-term hydrological records (Amornsakchai et al, 2000, p26). Based on this later definition, the WCD found that dependable capacity was less than 45MW. Furthermore, analysis of operating performance during 1995 to 1998 revealed:

> *The actual dependable capacity of Pak Mun project calculated from daily power output between 1995–1998, assuming that all available power gets assigned to a 4-hour peak demand period, is only 20.81MW. This 21MW is what the Pak Mun project [reliably] offsets in gas turbine capacity.* (Amornsakchai et al, 2000, pv).

The 21MW dependable capacity was only 15 per cent of Pak Mun's total generation capacity. Lower dependable capacity means lower benefit-cost ratios. The WCD argued that if Pak Mun was treated as a 21MW peaking power plant, its economic rate of return would be less than 8 per cent, below the opportunity cost of capital in Thailand, and, hence, uneconomic (Amornsakchai et al, 2000, pv).

The dispute over Pak Mun's 'dependable capacity' was, thus, partly about which method of computing reliability was most appropriate for estimating the power benefits of a hydropower plant. Interestingly, SOGREAH estimated Pak Mun's total energy production accurately, at least during the period of 1995 to 1999, when the dam was operated for maximum hydropower benefit (WCD, 2000, p22). But EGAT planning practice did not adequately take into account the effect of low flows during dry months. These depress peak power output. They also lower the dependable capacity (as defined by a 90 per cent probability of occurrence standard).[7]

Adverse media publicity surrounded successive drafts of the WCD report (e.g. Chang Noi, 2000). Drafts were leaked on several occasions to the press by dam opponents participating in the WCD process. Finally, conflict over Pak Mun intensified in May 2000, when protesters blockaded the powerhouse. Both the WCD report and the media coverage caused project proponents EGAT and, to a lesser degree, the World Bank to lose face. All of these factors provided EGAT motive and ammunition to attack the conduct and integrity of the process.

For encamped protesters, however, the emerging WCD findings provided a significant morale boost (AOP, 2000). In May 2000, after more than a year of government inattention to their protest village campaign, the Assembly of the Poor dramatically escalated its campaign. It launched a sit-in demonstration, disrupting access to the Pak Mun Dam powerhouse, as well as a simultaneous blockade of Rasi Salai, an upstream irrigation dam in Srisaket Province. At Pak Mun, the protesters denied EGAT staff access to the powerhouse for a number of days. They later agreed to move aside a few metres so that access could continue.

This dam blockade achieved what months of sit-in demonstrations outside Government House since 1994 could not: it conveyed to EGAT senior management that they needed to take much more active measures to resolve the conflict (Surapong, interview, 20 August 2004). In June 2000, Chuan's cabinet established a bilateral Neutral Committee to Solve Problems of the Assembly of the Poor. It was chaired by Bantorn Ondam, a former academic and respected social activist. Bantorn had previously served on the 1995 fisheries compensation negotiating committee chaired by Plodprasop Suraswadi.[8]

The committee's findings were 'overwhelmingly in support' of the assembly's positions on all disputed issues, which included land tenure, just compensation and the need for further impact assessments at several large dams (Missingham, 2003, p207). For Pak Mun, Bantorn's committee recommended a four-month experimental opening to restore fisheries migration (NC–AOP, 2000).

Chuan initially downplayed Bantorn's committee findings, treating them as non-binding advice. However, he and his advisers revised their positions one month later, when a contingent of assembly demonstrators again rallied outside Government House. They staged a night scaling of the perimeter walls on 16 July 2000, an event that ended in bloodied heads, several hundred arrests and condemnation in the print media about police violence (*Nation*, 2000; Chalermsripinyorat, 2004). A week later, Prime Minister Chuan ordered EGAT to open the gates of Pak Mun. But the government justified the action as a special operation to manage unusual flooding that year, and EGAT closed Pak Mun in late October once the flooding subsided.

During the remainder of 2000, a small contingent of protesters remained outside Government House to pressure Prime Minister Chuan to reopen negotiations, but without success. By this time the economic crisis had truly set in; Chuan faced regular calls from critics to dissolve parliament and call fresh elections.

Contention under the Thaksin government, 2001 to 2003

In January 2001, Thaksin Shinawatra toppled Chuan in the national elections. In March, acting on campaign promises, he visited AOP protesters encamped outside Government House. His government quickly established a Committee to Resolve Problems of the Assembly of the Poor led by Deputy Prime Minister Pongpol Adireksan; but the committee included no representatives or observers from the AOP.

In April 2001, three days after the final contingent of protesters returned home, Thaksin's cabinet accepted the recommendations originally made by the Bantorn committee: it ordered EGAT to open all eight sluice gates of Pak Mun for four months, during May to August, and for Ubon Ratchathani University (UBU) to conduct a multidisciplinary study.

Pongpol's committee set up several sub-committees. The university study was to be submitted to a task force chaired by the university's president. This group included representatives from the university, EGAT and the AOP. It was supposed to report directly back to Pongpol's committee. In addition, EGAT commissioned its own study, led by the Thailand Institute of Scientific and Technological Research (TISTR et al, 2003). A notable component of this study consisted of questionnaires administered to 94 per cent of the 6176 households that had received fisheries compensation. Villagers themselves, coordinated by Southeast Asia Rivers Network (a Thai NGO that campaigns against large dams), initiated the participatory *Tai Baan* research project to document all fish species caught by villagers, along with other evidence of ecological change in river condition (Sretthachau, 2002; see also Chapter 7 in this volume).

The new studies were attempts to generate different knowledge discourses from which to argue competing options: should Pak Mun Dam open indefinitely, as opponents demanded? Should it stay closed to generate hydropower, as EGAT would prefer? Should it, as a compromise, open seasonally and, if so, during what months and based on what evidence?

By the end of the first four-month trial opening period, the AOP felt that it had strong evidence that the opening had allowed fish migrations to occur. Some activists embarked on a long march to publicize the good news. In December 2001, the four-month experiment was extended to *one year* after the trial dam opening task force accepted an argument from its AOP member that the study needed a full year to observe all seasonal effects. In June 2002, a few days before the one-year opening of the dam was to expire, EGAT offered to open Pak Mun Dam seasonally, from July to October, ceding the option to generate hydropower from approximately 52 per cent of the river's average annual flow.[9]

UBU began presenting findings in September 2002. It reported that although households interviewed wanted irrigation water in the dry season, new river-pumped irrigation systems would have a minimal positive impact. Soils were poor, pumping costs were high and farmers lacked capital inputs needed to grow high-value dry season crops. For at least another five years, the dam's chief benefit – improving electric power reliability in the lower northeast – could be substituted by increasing electricity imports. Technical substitutes existed for goods provided by the dam; but none existed for improving the security of community-based livelihoods (UBU, 2002).

Nevertheless, in October 2002, Thaksin's cabinet, acting on the recommendation from Pongpol's committee, resolved that Pak Mun would henceforth be operated with a four-month seasonal opening. The AOP quickly denounced this decision, taken without benefit of public deliberation, and prior to final submission of the university's government-commissioned report.

The following month, on petition by the AOP and its allies, the Senate Committee on Public Participation held a hearing. EGAT Governor Sitthiporn Rathanopas conceded that EGAT could reliably supply the lower northeast's growing power needs by expanding transmission lines – hence, hydropower from Pak Mun was not indispensable. Based on this admission, university President Mongkhon Visetsuk reversed his position and backed a year-round opening for Pak Mun (Foran, 2006, Chapter 7). In December 2002, following unexpected harassment of demonstrators outside Government House, Prime Minister Thaksin intervened in the case, ordering the National Statistics Office (NSO) to survey occupations and attitudes towards dam management of residents in the Lower Mun Basin. NSO reported that among 3750 householders sampled from 150 villages, the least disruptive and most favoured option was a four-month dam opening. Only 4 per cent stated that fishing was their primary 'occupation' (*achip*); less than 7 per cent stated it was their secondary occupation (NSO, 2003).

Several weeks after the poll, NSO held a public meeting about its survey. Dam opponents argued that in the context of rural livelihoods, it would have been more accurate to ask villagers about their fishing activities, not if they regarded fishing as their 'occupation'. One villager asked: 'Why didn't you gather information using wording such as "*Pho Yai*" [grandfather], do you have children or grandchildren that fish?' This implied that the response to this question would have been different than to questions based on *achip*.[10]

In any case, in January 2003 the cabinet reiterated its resolution to operate Pak Mun Dam with a four-month opening, while offering a package of limited support for fisheries-dependent villagers. On 29 January 2003, some 500 villagers outside Government House were evicted by the Bangkok governor. As of 2008, the 14 January 2003 cabinet resolution still represents Thailand's basic policy statement regarding the value of Pak Mun and its future mode of operation.

Pak Mun politics since the decision on the four-month opening

As part of the 2003 cabinet resolution, the Royal Irrigation Department announced an 807 million baht (US$20 million) five-year pumped-water project to expand existing stations and to build new works. Investment focused on villages in the upstream vicinity of Pak Mun Dam, but would eventually extend almost 80km upstream towards the provincial centre. The Pak Mun Irrigation Project was framed as a special development project and did not require a cost-benefit test. The project would build a constituency of local people interested in dam-induced high water levels.

Pumped-water irrigation supplies river water to fields by large electric pumps mounted on floating platforms.[11] First provided during the 1980s, about 70 such systems exist on the Mun River in Ubon Ratchathani. But as of 2003, the three Lower Mun districts of Phibun, Khong Chiam and Sirinthorn had only 16. Unlike gravity-fed irrigation, which is currently supplied without user charges in Thailand, farmers using pumped water had to pay up to US$2 per hour in 2002.

As part of the UBU study discussed above, a team from the Faculty of Agriculture studied farming practices in four districts in the Lower Mun. They found that pumped water was used primarily at the end of the dry season to raise rice seedlings for the main rain-fed rice crop, and, second, to grow higher-value crops, such as watermelon and chillies, and to stock fish ponds in the dry season. However, in 2000 to 2002, the average usage rate in the existing scheme was only 14 per cent of the total projected irrigable area (UBU, 2002, p*khor*-6). Dam supporters argued that during the trial dam opening, water levels in the Mun were too low to operate the pumps, but UBU (2002) concluded that such problems could be solved with relatively minor retrofits.

To better accommodate wet season upstream fish migrations, the AOP requested the Thaksin government to allow Pak Mun's annual four-month opening to begin slightly earlier, in May rather than June. The change was agreed and announced in June 2004. But since then implementation of the dam's four-month opening policy has been far from smooth.

In April 2007, six months into the military-appointed government of Prime Minister Surayud Chulanont, 3000 pro-dam villagers (mobilized by the KVHG) rallied at the provincial hall to keep the gates closed and, thus, to overturn the 2004 Thaksin cabinet resolution. As well, local leaders, allegedly with the backing of the National Security Council junta and EGAT, organized a survey of 8091 Lower Mun households (AOP, 2007; Sangsok, 2007). Administered by village headmen and household heads, the survey asked for the name, identification number and signature of each household member, and for a simple yes/no response to the question of whether EGAT should store water at 106m to 108m amsl (i.e. normal operating levels for power generation).

In late May 2007, on the recommendation of the Ministry of Energy, Surayud's cabinet resolved to open Pak Mun in June. But shortly after this announcement, results of the new survey were presented privately to cabinet, claiming overwhelming support for dam closure from 20,592 people (8091 households). On the basis of this unpublished survey (see discussion below), Surayud's cabinet then reversed its earlier decision and decided on 23 June 2007 to keep Pak Mun *closed*. This triggered another protest rally in Bangkok by the AOP. After pressure from NGOs, academics and criticism in the broadsheet print media, the Surayud government finally resolved in July 2007 to delegate decision-making about Pak Mun's opening and closing to the provincial governor.

Understanding operating decisions during the 2000s

For dam opponents, the 1997 economic crisis ushered in the unfriendly second Chuan Leekpai administration, but also hastened the passing of the 1997 Constitution, which protected a much higher level of civil liberties. The crisis increased calls from farmers and business people alike for governance reform. These events were conducive to the rise of Thai Rak Thai, the first political party offering coherent policies to benefit both constituencies.

The manner in which activists' claims were processed depended upon framing contests between dam opponents and the state. These unfolded over time and were contingent on micro- and macro-political contexts. Some contexts, such as norms of conflict management and participation, were durable. Some were novel, such as the widespread groundswell for reform after the 1997 crisis and Thai Rak Thai's populist policy initiatives (initially well received by anti-dam villagers).

After Thaksin's unprecedented decision to have a trial opening in 2001 to 2002, EGAT proposed a four-month opening policy. It made the offer in 2002, prior to a formal decision from government, to pre-empt more drastic concessions. Although activists rallied against it, and academics urged Prime Minister Thaksin to declare a year-round opening during a televised hearing in late 2002, his administration regarded this as too regressive and institution-bending for the state. By allowing only a face-saving four-month opening, Thaksin and his men defended their party's vision of development and their authority to rule. In doing so, they also upheld dominant institutions.

The January 2003 cabinet resolution was followed by a five-year period in which Pak Mun's seasonal opening and closing were periodically disputed at the local level. Operating rules were not significantly elaborated upon. This pattern shattered in early 2007 under the military-appointed Surayud government. EGAT and its allies in the military intervened; the government reversed its operating policy. After renewed criticism, it devolved detailed management responsibility to committees reporting to the provincial governor. Such *ad hoc* problem-solving characterizes Thai policy-making when elites face popular pressure (Foran, 2006).

Both the four-month seasonal opening decision under Thaksin (2003) and the short-lived year-round dam closure decision under Surayud (2007) were justified based on surveys of citizens' preferences. The validity of rapid, high-n, non-confidential sampling on a politically sensitive issue is debatable (Foran, 2006). However, assuming that the responses accurately captured local people's opinions, why did local households who won fisheries compensation during the 1990s and early 2000s turn away from the assembly's position in subsequent years?

We suggest an explanation that involves:

- the power of broad development discourse (hegemonic storylines promising water security);
- the power of specific counter-framing rhetoric (e.g. 'the dam has already been built, so why not use it'); and
- the unpopularity of the AOP (see Foran, 2006, for detailed political analysis).

This unpopularity was a contingent and emergent outcome of multiple rounds of struggle, during which time state agents ultimately prevailed in framing dam critics as a disruptive social force, even as they set new agendas and delivered unprecedented benefits to protesters and free-riders alike.

During the 1990s, authority delegated to provincial-level committees failed to resolve conflicts over compensation, particularly fisheries compensation. In early 2008, however, the Provincial Pak Mun Dam Commission chaired by Governor Chuan Sirinuntaporn stressed reconciliation and participation of affected people (supporters and opponents), local NGOs and academics. The second author (Kanokwan) is a member of a steering committee on quality of life, development and resilience of affected people. In May 2008, this sub-committee recommended rigorous monitoring of dam opening and closing, and quantification of fisheries and agriculture benefits. It also raised the larger question of how to improve developmental outcomes for the Lower Mun River Basin people during the eight months of the year the dam is closed.

Will devolution lead to structured deliberation? Detailed and participatory monitoring of livelihood outcomes might weaken the authority of EGAT and RID. On the other hand, it might institutionalize more effective use of knowledge in decision-making, and contribute to conflict resolution via structured deliberation.

PAK MUN DAM: PERPETUALLY CONTESTED

This chapter presented Pak Mun as an important case in dam decision-making, one that has mobilized large numbers of supporters and opponents, and contributed to the reshaping of state–society relations in Thailand. A dam planned and implemented with low transparency and accountability helped to trigger an unfolding, emergent series of disputes. Disputes over Pak Mun attempted to democratize an authoritarian state. The movement against Pak Mun has helped to socialize Thai society in order to tolerate, and participate in, vigorous street demonstrations. It helped to open up new spaces for deliberative politics: on campuses, in the offices of independent organizations set up under the 1997 Constitution (e.g. the National Human Rights Commission), and – when routine politics fails ordinary people, as it often does in Thailand – on the street.

Pak Mun offers lessons about rhetoric contests in a democratizing setting. The manner in which activists' claims were processed depended upon framing contests between dam opponents and the state. By 'framing' we refer to rhetorical work deployed to champion a particular interpretation. Such discourse ranges from terse speech acts (e.g. 'Pak Mun opponents [or supporters] are paid to protest') to more elaborate rhetoric in policy statements and scientific reports. Framing by power-holders is particularly worth tracking. It justifies government inaction on activists' demands. In closed venues of final decision-making, such 'counter-framing' disarms radical policy narratives. One important example is the shifting justification of Pak Mun heard repeatedly over the years of the project. It took the form: 'The project has already been approved', or 'Construction has already started', or 'The dam has already been built' so 'therefore the project must proceed'. Framing contests are driven by competing interests; but as discourse they also constrain what can be imagined and what is reasonable.

In addition to authorities and activists, mass media and technical experts also engaged in framing disputes. Dam opponents presented themselves as worthy citizens and courted media coverage; but coverage, following norms of news reporting, required campaign escalation. Media framing was divided: more serious broadsheets provided detailed and sympathetic coverage. High-circulation papers have been sites of hostile counter-framing (Chalermsripinyorat, 2004). Most technical studies were commissioned by EGAT or the state in an attempt to inform or legitimize decision-making. When experts were called in to assess debates over Pak Mun, the knowledge they produced was not neutral and immune from attack by contending parties, including the original research sponsors.

Pak Mun offers sobering lessons about politics of knowledge. Sustained production of knowledge for dispute resolution (e.g. the WCD study) occurred relatively late in time. This meant that knowledge production did not always contribute in a 'rational' way to informed negotiation. New knowledge could, instead, trigger reactive framing, as, for example, when Thaksin's advisers disputed the findings of Ubon Ratchathani University (2002) and ordered an opinion poll. Concepts such as dependable capacity, occupation, fisher and farmer were contested by laypeople and by experts using different methodologies. Contending research sponsors required simple conclusions on key issues such as the importance of wild-capture fisheries to local livelihoods. They ignored the nuances of smallholder livelihood strategies. With authority highly concentrated in the state's executive branch, knowledge production was manipulated. The 'politics of knowledge' thus should not be abstracted from the politics of blame, threat and other forms of contention present in a particular dispute (McAdam et al, 2001; Foran, 2006, p6).

What practical lessons does Pak Mun offer? Proponents of large water infrastructure in Thailand such as EGAT have been compelled to move to neighbouring countries such as Laos or Myanmar/Burma to build projects that can supply power and water to Thailand. Analysts and advocates for affected people

– not just in Thailand, but as a result of Pak Mun's international prominence – have learned to question all project studies from their inception (the fundamental need for the projects), extending to ramifying impacts (WCD, 2000). Civil society actors have learned to mobilize, often in cross-scale coalitions, and to re-politicize knowledge and capture public arenas of deliberation by undertaking, compiling and publishing their own data and research. After 20 years of debate over Pak Mun, some infrastructure sponsors have learned to approach complex questions of livelihoods restoration with more humility.

Far from disappearing under agricultural modernization, Pak Mun shows that dependence upon wild-capture aquatic resources persists. This important finding from relatively 'modern' Thailand implies that hydropower development will lead to even stronger negative impacts for small farmers elsewhere in the Mekong region. With national economic development as the overriding priority, rural people face a spate of large new water proposals, wrapped in powerful discourses of modernization and poverty alleviation. In practical terms, how might reformists encourage better decision-making? Reforming water and energy governance challenges power interests and institutions; therefore, advocates encounter resistance. They can expect counter-framing, oppositional elite intervention and rejection of dialogue. Advocates of improved governance and sustainability could, nonetheless, promote particular combinations of processes. One idea is to promote processes that are scientifically credible and legitimate to different stakeholders, while savvy about the many faces of power.

Will the provincial-level management initiated in 2007 lead to reasoned problem-solving? The answer hinges on stakeholders' ability to agree upon and formalize decision-making processes. Otherwise, as we saw, concessions such as the seasonal opening can be withdrawn. For any operational policy at Pak Mun to work, it also needs to be presented and run as an experiment aimed at delivering meaningful livelihood outcomes to supporters and opponents alike. If not, both factions – which claim to represent poor farmers – will abandon it in favour of prior understandings.

In 2008, Thailand's instantaneous peak demand was less than 21,395MW. Pak Mun running at 136MW would have lowered it by 0.6 per cent, equivalent to the peak demand of two large commercial buildings in Bangkok.[12] The dam has made a slim contribution to energy security, but generated two decades' worth of hardship for those who dared question its value, spoke up about its impacts upon their way of life, and pushed for a better deal from the state.

If Pak Mun's fate is to be perpetually in dispute, it is for several good reasons. There are competing interests and contested interpretations. A nuanced understanding requires going beyond an engineer's worldview, optimizing trade-offs between power generation, wild-capture fisheries and water for pumped irrigation. It also demands that we understand rural development as an ongoing intensely political conversation. During two decades of such conversation, some of the most articulate voices on how to sustain resource-dependent livelihoods in the face of change and regional integration have come from the activists at Pak Mun.

NOTES

1 Elected village headmen report to appointed sub-district officers (*kamnan*). Both serve as line officers of the Ministry of Interior.
2 In addition to its long-standing role articulating Thai post-World War II development policy, the World Bank has been a significant lender to Thai energy projects, though not always the majority financier. In the eyes of commercial banks, the bank's various project review processes reduce political risk. Associated with World Bank project review are channels for foreign donor and transnational advocacy (Fox and Brown, 1998).
3 By repression we mean any deliberate action by authorities or bystanders that increases the difficulties of collective action (della Porta et al, 1996).
4 The two leaders were charged with offences related to a construction site protest occupation in early 1993 and released on bail.
5 The Assembly of the Poor, a national social movement organization, emerged in late 1995. Resistance against Pak Mun constituted one of its core local networks (Missingham, 2003).
6 In late 1987 the Prem government commissioned a multi-stakeholder process chaired by General Tienchai Sirisamphan to review Nam Choan Dam (Foran, 2006, Chapter 4).
7 The World Bank (2000, p127) claimed that it had anticipated the intermittent nature of Pak Mun's hydropower production. It claimed that the energy Pak Mun generated during the wet season allowed Thailand's large storage dams to save water, which they could release for power production during the dry season. In response, the WCD noted that EGAT had presented no evidence of coordinated inter-seasonal power production.
8 Plodprasop served as director general of the Department of Fisheries during 1989 to 1997.
9 Critics argued that less than 100 per cent of total annual flow was available for power generation to begin with. Some rainy season peak flow events force EGAT to spill flood water (Amornsakchai et al, 2000), so EGAT's 52 per cent figure is an overestimate. EGAT has, however, ceded the option to generate electricity during those months.
10 Foran (2006, Chapter 8) provides an extended discussion.
11 Canals are concrete lined, approximately 2m wide, and run inland with occasional branches for a total length of 3km to 4km. Water reaches fields through simple openings that can be raised by hand. Requesting water usually requires agreement among three or four farmers, and the pump is operated by a resident employee of the RID.
12 Chuenchom S. Greacen, pers comm, 10 September 2008.

REFERENCES

Amornsakchai, S., Annez, P., Vongvisessomjai, S., Choowaew, S., Thailand Development Research Institute, Kunurat, P., Nippaporn, J., Schouten, R., Sripatrprasite, P., Vaddhanaputhi, C., Vidthayanon, C., Wirojanakud, W. and Watana, E. (2000) *Pak Mun Dam, Mekong River Basin, Thailand: A WCD Case Study Prepared as an input to the World Commission on Dams*, Final draft, November, World Commission on Dams, Cape Town

AOP (Assembly of the Poor) (2000) *Rebel With a Real Cause*, Video in Thai with English subtitles, Friends of the People, Bangkok

AOP (2007) *Thalaeng kan sammacha kon chon korani khuean pak mun. Pid khuean pak mun kho mo cho saek saeng ratthaban pheua rang kae kon chon* [*Pak Mun Dam Case: Closing the Gates of Pak Mun – National Security Council Intervenes with Government to Harass the Poor*], AOP Press Release, www.newspnn.com/detail.php?dataid=4779&code=m3_18062007_01&mode=th, accessed 11 June 2008

Bangkok Post (1991) 'Bank chiefs hear both sides of dam issue', *Bangkok Post*, 19 October

Buchita, S. (1997) *Fish Culture and the End of the Mun River Fisher's Story*, Committee to Publicize and Promote Development Work, Bangkok

Chalermsripinyorat, R. (2004) 'Politics of representation: A case study of Thailand's Assembly of the Poor', *Critical Asian Studies*, vol 36, no 4, pp541–566

Chang Noi (2000) 'Fish and fiddles at Pak Mool', *Nation*, 29 January 2000, pA5

della Porta, D. (1996) 'Social movements and the state: Thoughts on the policing of protest', in McAdam, D., McCarthy, J. D. and Zald, M. N. (eds) *Comparative Perspectives on Social Movements: Political Opportunities, Mobilizing Structures, and Cultural Framings*, Cambridge University Press, Cambridge

EGAT (Electricity Generating Authority of Thailand) (2000a) 'EGAT Comments on June 2000 Version of the Pak Mun Report', in Amornsakchai, S., Annez, P., Vongvisessomjai S., Choowaew, S., Thailand Development Research Institute, Kunurat, P., Nippaporn, J., Schouten, R., Sripatrprasite, P., Vaddhanaputhi, C., Vidthayanon, C., Wirojanakud, W. and Watana, E. (eds) *Pak Mun Dam, Mekong River Basin, Thailand: A WCD Case Study Prepared as an Input to the World Commission on Dams,* Final Draft, November, World Commission on Dams, Cape Town, Chapter 9

EGAT (2000b) *Khwam ching thi khuean pak mun* [*The Truth at Pak Mun Dam*], EGAT, Public Relations Division, Nonthaburi, Thailand

EGAT (undated) *Rai chue chumchon thi dai rap phon kratop chak kan ko sang kheuan pak mun* [*List of Communities Receiving Impacts from Pak Mun Dam*], Unpublished report provided to authors, EGAT, Nonthaburi, Thailand

Foran, T. (2006) *Rivers of Contention: Pak Mun Dam, Electricity Planning, and State–Society Relations in Thailand, 1932–2004,* PhD thesis, Division of Geography, Department of Geosciences, University of Sydney, Australia, http://hdl.handle.net/2123/1984

Fox, J. A. and Brown, L. D. (1998) *The Struggle for Accountability: The World Bank, NGOs, and Grassroots Movements*, MIT Press, Cambridge, MA

Hubbel, D. M. (1992) *Antithetical Perceptions of Development and Environment: Village People in the State and Rural Thailand*, York University, Toronto, Canada

Kansuntisukmongkol, C. (1994) *Reappraisal of Cost-Benefit Analysis: Pak Mun Dam Project*, MA thesis, Faculty of Economics, Thammasat University, Bangkok

McAdam, D., Tarrow, S. and Tilly, C. (2001) *Dynamics of Contention*, Cambridge University Press, New York

Missingham, B. D. (2003) *The Assembly of the Poor in Thailand: From Local Struggles to National Social Movement*, Silkworm Books, Chiang Mai

Nation (2000a) 'Violent clashes spark rallying cry', *Nation*, 18 July

NC–AOP (Neutral Committee to Solve Problems of the Assembly of the Poor) (2000) *Khor saneu korani khuean pak mun lae sip hok korani panha sammacha khon chon* [*Recommendations for the PMD Case and 16 Problem Cases of the Assembly of the Poor*], Presented to His Excellency Deputy Prime Minister and Minister of the Interior, Mr Banyat Bantatthan, NC–AOP, Bangkok

NSO (National Statistics Office) (2003) *Rai ngan phon kan samruat khwam kit hen khong prachachon kiew kap khuean pak mun nai amphoe phibhunmangsahan khongjiam sirindhorn changwat ubon ratchathani* [*Report of Public Opinion Survey about Pak Mun Dam in Phibun Mangsahan, Khong Jiam, and Sirindhorn Districts, Ubon Ratchathani Province*], National Statistics Office, Ministry of Information and Communications Technology, Bangkok

Phongpaichit, P. and Baker, C. (1995) *Thailand: Economy and Politics*, Oxford University Press, Kuala Lumpur

Phongpaichit, P. and Baker, C. (2000) *Thailand's Crisis*, Silkworm Books, Chiang Mai, Thailand

Roberts, T. R. (1993) 'Just another dammed river? Negative impacts of Pak Mun Dam on fishes of the Mekong Basin', *Natural History Bulletin of the Siam Society*, vol 41, pp105–133

Roberts, T. R. (2001) 'On the river of no returns: Thailand's Pak Mun Dam and its fish ladder', *Natural History Bulletin of the Siam Society*, vol 49, pp189–230

Sangsok, S. (2007) 'Tin chang yiab pak nok: "Cut shoe" rue "combat" luan nak pho kan' ['Elephant feet step on birds' mouths: "Cut shoes" are just as heavy as combat boots'], *PNN News*, www.newspnn.com/detail.php?dataid=4838&code=n6_13072007_01&mode=th, accessed 23 June 2008

SOGREAH (1985) *Electricity Generating Authority of Thailand: Pak Mun Multipurpose Development Project*, Feasibility studies, Final report, Executive summary, SOGREAH Consulting Engineers, Grenoble, France

Sretthachau, C. (2002) *Mae Mun Kan Klup Keun Ma Khong Kon Ha Pla. Bot Sarup Lae Khwam Ru Rueang Pla Khong Kon Pak Mun* [*Moon River – Return of the Fishers: Summary and Fisheries Knowledge of Pak Mun People*],Tai Ban Research Project, Southeast Asia Rivers Network –Thailand, Chiang Mai, Thailand

TISTR (Thailand Institute of Scientific and Technological Research), Rajabhat Institute Ubon Ratchathani, Kasetsart University Department of Aquatic Resources Engineering and Chulalongkorn University Faculty of Commerce and Accountancy (2003) *Electricity Generating Authority of Thailand: Social-Environmental Monitoring and Establishment of Quality-of-Life Development Plan for People and Community in the Lower Mun River Basin*, Final report, TISTR, Bangkok

UBU (Ubon Ratchatani University) (2002) *Khrong kan sueksa naew tang fuen fu rabop niwet withi chiwit lae chumchon thi dai rap phon krathop chak kan sang khuean pak mun*

[*Project to Study Approaches to Restoration of the Ecology, Livelihood and Communities Receiving Impacts from Construction of Pak Mun Dam*], Ubon Ratchatani University, Ubon Ratchatani

WCD (World Commission on Dams) (2000) *Dams and Development: A New Framework for Decision-Making*, Earthscan Publications, London

World Bank (2000) 'World Bank comments on June 2000 version of the Pak Mun Report', in Amornsakchai, S., Annez, P., Vongvisessomjai, S., Choowaew, S., Thailand Development Research Institute, Kunurat, P., Nippaporn, J., Schouten, R., Sripatrprasite, P., Vaddhanaputhi, C., Vidthayanon, C., Wirojanakud, W. and Watana, E. (eds) *Pak Mun Dam, Mekong River Basin, Thailand: A WCD Case Study Prepared as an Input to the World Commission on Dams,* Final draft, November, World Commission on Dams, Cape Town, Chapter 10

4

The Nam Theun 2 Controversy and Its Lessons for Laos

Shannon Lawrence

Introduction

The US$1.45 billion Nam Theun 2 (NT2) project is the biggest hydropower project under development in Laos, with costs approaching one third of the country's gross domestic product (GDP). When it reached financial closure in May 2005, NT2 became the largest ever foreign investment in Laos and the largest hydroelectric project with private-sector financing (NTPC, 2005d). The project will export 95 per cent of its 1070MW power output to neighbouring Thailand starting in late 2009.

NT2's electricity sales will generate revenue for Laos, considered in conventional development indicators as one of the economically poorest countries in Asia. But these revenues come with a big trade-off: NT2 is flooding 450km^2 of land, river, forest and wildlife habitat, and displacing thousands of predominantly ethnic minority peoples. As a trans-basin diversion project, NT2 will have significant impacts upon two river basins, reducing fish catches, affecting water levels and water quality for tens of thousands people downstream (ADB, 2004; International Rivers, 2008b).

In 2005, the NT2 project received support from the World Bank and other investors, becoming the first major dam approved by the World Bank in almost a decade. At that time, NT2 was lauded by the World Bank, the Asian Development Bank (ADB) and other project proponents as a 'model' that could pave the way for best practice dam development in the region. But NT2 has attracted controversy ever since it was first proposed in the 1980s. As the project's construction nears completion, the debate about its costs and benefits – and about large dam development, more generally – continues.

In order to overcome the challenges of developing such a high-risk hydro project in a country characterized by low capacity, poor public financial management and weak governance, NT2's primarily Western backers insisted on a number of commitments from the Government of Laos (GOL) and the Nam Theun 2 Power Company (NTPC). Through these commitments, they attempted to turn a private-sector energy project into a development initiative – an experiment that NTPC and the GOL were prepared to go along with in order to access concessional funding and guarantees from the multilateral development banks (MDBs).

In a country with a one-party authoritarian government, no independent judiciary or independent civil society organizations, no free press, and a ranking as one of the world's 25 most corrupt countries by Transparency International (Transparency International, 2008; US State Department, 2008), dams have left a legacy of broken promises and uncompensated losses. As a result of these projects, tens of thousands of Laotians lack sufficient food to eat, clean water to drink and income to meet their basic needs.

But river-rich Laos, which contributes about one third of the Mekong River's flow, is experiencing a dam-building boom. In its bid to become 'the battery of Southeast Asia', the GOL has signed memoranda of understanding (MoUs) with foreign companies to build more than 45 dams on Mekong tributaries, and is even considering at least six projects on the Mekong River. Power from these projects will be sold primarily to neighbouring Thailand as well as Vietnam and, possibly, China.

As the Lao hydropower sector takes off, observers are wondering just what sort of lessons may be learned from NT2, the country's most controversial large dam. Is NT2 a model for how to plan and build sustainable hydropower projects? Are NT2's social and environmental risks being managed adequately during the construction stage, and planned for during the operations stage? If not, why not? How is the NT2 experience affecting other hydro projects in Laos? This chapter will seek to answer these questions by examining NT2's planning and implementation and drawing lessons for the future.

The chapter starts with an overview of NT2's history and the debate focused on World Bank support for NT2. It then examines the planning process for NT2 and describes the project and its notable features, including social and environmental commitments and revenue management plans. Problems with implementing NT2's social and environmental mitigation and compensation measures three years into its construction phase are also discussed. Finally, the chapter examines why problems have occurred with NT2 and what lessons can be learned, and concludes with general recommendations to promote better outcomes in Laos. An alternative perspective from the World Bank is included as an appendix to the chapter.

THE HISTORY OF NAM THEUN 2 (NT2)

By the time NT2 lined up financing in 2005, the project had already been in the GOL's and the World Bank's sights for more than two decades. World Bank support for NT2 began in the mid 1980s when it financed a feasibility study for the project (GOL, 2002). During those early years, Thailand was identified as the likely market for NT2's power.

In 1993, the GOL awarded Transfield Holdings Ltd, one of Australia's largest construction companies, the rights to develop NT2 as a build–own–operate–transfer (BOOT) project for a 25-year concession period. Transfield established a consortium called the Nam Theun 2 Electricity Consortium (NTEC) with Electricité de France (EDF), Ital-Thai Development (ITD), Jasmine International, Merrill Lynch Phatra Thanakit Securities and the GOL. Electricité de France later took over the lead shareholder spot from Transfield, acquiring a 35 per cent stake in NT2.

By the mid 1990s, concerns about the economic viability of NT2 and its social and environmental impacts were being raised by international non-governmental organizations (NGOs) and some academics (International Rivers Network et al, 1997). Conscious of the growing public opposition to the project, in 1995 the World Bank called for additional economic, environmental and social studies (Scudder, 2005) and urged the GOL to conduct public workshops to discuss NT2. The importance of World Bank support for NT2 was already apparent; David Iverach, NTEC's director in Laos, emphasized in January 1997: 'If the World Bank does not proceed with the project, nor would we' (Nette, 1997).

But the Asian financial crisis put the brakes on NT2's development in late 1997 (Manolom, 2002) and the lull lasted for a few years. As Thailand's economic situation improved, their interest in NT2 returned. An electricity tariff MoU between the Electricity Generating Authority of Thailand (EGAT) and NTEC was signed in 2000 (GOL, 2002). Meanwhile, in 1999 and 2000, Jasmine International and Merrill Lynch Phatra Thanakit sold their equity stakes, and the Electricity Generating Company of Thailand (EGCO) joined NTEC.

In 2002, a Concession Agreement was finally signed between the GOL and the renamed Nam Theun 2 Power Company (NTPC). The World Bank, waiting for the Power Purchase Agreement (PPA) between Thailand and Laos to be finalized, was on the verge of initiating formal appraisal of the NT2 project. But in July 2003, a day before the PPA was to be signed, lead shareholder EDF announced that it was withdrawing from NT2. EDF attributed its pull-out to 'advice from French government officials' (Reuters, 2003) and a 'strategy to consolidate assets and re-focus priorities in Europe' (Praiwan, 2003).

According to news reports, the GOL gave the consortium three months to identify a new shareholder to take over EDF's 35 per cent stake (Lang, 2003). For a short while, NT2 once again seemed to be in jeopardy. But EDF decided to rejoin the NT2 consortium – with little explanation – and in November 2003,

Figure 4.1 *Nam Theun 2 Transmission tower*

the PPA between EGAT and NTPC was finally signed and the search for project financing began.

The US$1.45 million NT2 was expected to be financed with approximately 30 per cent equity from the shareholders and 70 per cent international loans and guarantees from the World Bank, the ADB, the European Investment Bank, the Agence Française de Développement (AFD), export credit agencies and commercial banks. Given the size of the investment and the risks of operating in Laos, commercial banks were unlikely to fund the project without guarantees from the World Bank. The potential financiers also primarily relied on the World Bank to lead the economic, social and environmental due diligence for NT2.

WAITING FOR THE WORLD BANK: THE NT2 DEBATE IN BRIEF

In June 2002, the World Bank outlined a 'Decision Framework' for NT2 that identified three criteria that would have to be met for the GOL to receive the World Bank's support. The Decision Framework clarified that:

> *... the project must be embedded in a development framework aimed at poverty reduction and environmental conservation; must be technically, financially, managerially and economically sound and adhere to the Bank's environmental and social safeguard policies; and must have greater understanding and wider support within the international donor community and civil society.* (World Bank, 2002).

More economic, social and environmental studies were required of NT2's developers, as well as local and international consultations.

Civil society organizations, including International Rivers Network, Towards Ecological Recovery and Regional Alliance (TERRA), Environmental Defense, Mekong Watch, Friends of the Earth–France and Probe International, had by this time launched a concerted campaign against World Bank support for NT2, asserting that the dam's risks would outweigh its benefits and that the GOL did not have the political will or capacity to implement such a large and complex project. NGOs and academics in Thailand questioned whether or not NT2's power was needed, citing Thailand's faulty energy planning process (Permpongsacharoen, 2004; Greacen and Sukkamnoed, 2005). These and other NGOs had been raising concerns about NT2 since the project was first proposed, but the intensity of their campaign increased once the PPA was signed and the World Bank's support for NT2 became more likely.

The civil society campaign against NT2 was founded largely on concerns about the GOL's track record and the lack of legal and political protections for affected communities in Laos. Smaller hydropower projects in the country, such as Houay Ho, Nam Song, Nam Leuk and Theun-Hinboun (most of which had been financed by the ADB), had been poorly managed from an environmental and social standpoint. Tens of thousands of Laotians were still awaiting compensation, livelihood restoration and mitigation measures to replace the fisheries, land, drinking water and other resources that they had lost to these dams (International

Rivers Network, 2004). Even the ADB noted in its technical assistance paper for NT2 that 'the Government's capacity to implement large-scale complex hydropower projects still remains a major concern' (ADB, 2003, p3).

NGOs, donors, the World Bank and others also cited concerns about the GOL's lack of transparency and poor performance in terms of public expenditure management and other governance indicators. A June 2004 World Bank evaluation of Bank-supported financial management reforms in Laos found that 'budget discipline [had] not improved significantly' over the previous decade and – in addition to a lack of implementation capacity – there was a 'weak commitment to reform at the level of the true policy-makers' (World Bank, 2004a, p8). The World Bank's December 2004 *Country Economic Memorandum* for Laos warned that without significant governance improvements upfront, hydropower revenues would not result in good development outcomes (World Bank, 2004b, p72).

Flaws in the NT2 planning and preparation process

In response to the concerns raised by civil society organizations, NT2 promoters argued that the project's preparation was a model for future hydropower development and could be used to strengthen the GOL's capacity to manage new dam projects. The World Bank and others touted the transparency and participation of the process – especially compared to that of other dam projects in Laos – and pointed to the volumes of studies that had been completed to assess NT2's environmental and social impacts and plan for mitigation measures.

But NT2's critics were not convinced. An analysis by International Rivers Network and Environmental Defense in 2005 found that the NT2's planning process violated six of the seven strategic priorities of the World Commission on Dams (WCD), including priorities on gaining public acceptance, comprehensive options assessment, and sustaining rivers and livelihoods (Imhof and Lawrence, 2005). Another analysis documented numerous violations of World Bank safeguard policies, including on environmental assessment, involuntary resettlement, indigenous peoples and natural habitats, as well as the failure to comply with World Bank guidelines on procurement and economic evaluation. Finally, NGOs asserted that the requirements of the NT2 Decision Framework – namely, concerning adherence to the World Bank's policies and support from civil society – had not been met. These concerns and others were outlined in a letter sent to then World Bank President Wolfensohn from 153 civil society organizations in 42 countries in March 2005 (Friends of the Earth–Japan et al, 2005). Some of these issues are elaborated upon further in Table 4.1.

As Table 4.1 demonstrates, sponsors argued that NT2 was needed, was exemplary in the quality of its studies, and was based on extensive consultations with local people. Critics vigorously debated all three claims; but their arguments did not prevent the project from moving forward. Pressure from NGOs and others

Table 4.1 *Problems with the Nam Theun 2 planning process*

NT2 claim	NGO response
NT2 achieved public acceptance through a highly participatory consultation process. World Bank and other NT2 supporters claimed the project had achieved public acceptability in Laos through consultation processes conducted throughout the project development period.	*True participation is limited in Laos; consultation was late and focused on mitigation options.* Although NT2 involved more consultation than other Lao dam projects, the political climate in Laos does not allow for genuine participatory processes. Access to independent sources of information is limited. There are no independent local NGOs and there is no independent media. The GOL continues to arrest and imprison critics (Amnesty International, 2003; United Nations Committee on Elimination of Racial Discrimination, 2003; US State Department, 2005). The legal system is at a rudimentary stage of development. There is no independent judiciary: it is impossible for affected communities to bring legal actions to protect their rights. *Nakai Plateau:* villagers to be resettled for NT2 were consulted on numerous occasions, but logging commenced and the decision to build the dam was taken before any participation processes were conducted. Most discussions with villagers focused on improving resettlement outcomes, not debating whether or not the project was appropriate or desirable. Affected communities had no access to independent legal or other professional support (Imhof and Lawrence, 2005, pp4–5). *Downstream areas:* discussions along the Xe Bank Fai were only initiated in mid 2004, so many people were not consulted prior to project approval (NTPC, 2005b). Villagers were given misleading information about NT2's risks and benefits, being told repeatedly and erroneously that 'more water means more fish' (Les Amis de la Terre et al, 2003, p6; NTPC, 2005f; Scudder, 2005). Most of the 1500 families living along the Theun River were also not consulted before project approval (NTPC, 2005b, Vol 3, Chapter 6, pp7–8). Consultations with villagers whose lands and assets would be affected by NT2 construction were only initiated at the end of 2004, and were not completed prior to NT2 approval (NTPC, 2005c, Vol 4, Chapter 6, pp2–3).
NT2 is the best option for Laos and for Thailand. The World Bank had promoted NT2 as an important revenue earner for Laos since 1986, claiming that Laos had few other means, aside from timber exports, to generate revenue and reduce the country's dependence upon foreign aid.	*No comprehensive options assessment was conducted.* These claims were not based on comprehensive and participatory analyses of options and alternatives. NT2 revenues were estimated to be no more than 5% of total annual government revenues over the life of concession (World Bank, 2005c). There were probably other development options for Laos since the direct contribution of natural resources (such as hydropower) to GDP growth is relatively small (World Bank, 2004b): for example, broadening the tax base, improving revenue administration nationally (AusAid, 2005) and investing in agriculture (World Bank, 2004b, p17) would increase GOL revenue and support poverty reduction without NT2's significant social/environmental impacts.

Table 4.1 *(continued)*

NT2 claim	NGO response
NT2 is the least expensive means of meeting Thailand's energy needs. The World Bank states that Thai energy demand is growing and NT2 will be easily absorbed. NT2 will not compete with renewable energy and demand-side management options (World Bank 2005a).	*There is no evidence that NT2 is, undeniably, the least-cost option for meeting Thailand's energy needs.* A World Bank-commissioned study – not disclosed until *after* NT2 was approved – showed that feasible demand-side management, energy conservation measures and renewable energy generation in Thailand would 'exceed the output of NT2 and would provide energy to the customer at a cost approximately 25% less than NT2' (duPont, 2005). Thai energy projections have historically overstated future energy demand estimates (Kuankachorn, 2005).
NT2 is the most studied hydropower project ever. NT2 proponents highlighted the large number of environmental, social and economic studies produced over the decade-long project development period. The World Bank asserted: 'sound methodologies were utilized to address data gaps commonly found in lower-income developing countries', and that there would be additional time during the construction phase for technical support, improving social programmes and identifying markets (World Bank, 2005b).	*NT2 studies have critical gaps in data and analysis; some livelihood programmes are infeasible.* Hydrological data and analysis were deficient: it is difficult to predict how much water will be available for power generation (Willing and Knoop, 2005). Inadequate data was used to characterize baseline water quality: it is impossible to accurately predict water quantity and quality changes in the reservoir and in downstream rivers (Lanza, 2005). The proposed agriculture programme for resettled villagers relied on heavy inputs of fertilizer and experimental cropping systems, and was of questionable long-term viability. The market for the sale of cash crops to buy rice was unclear (International Rivers Network, 2005).

did help to encourage NT2 developers to make important changes, however, before the project proceeded, such as increasing the budget for downstream mitigation and compensation to US$16 million and incorporating a purpose-built downstream channel to reduce erosion and sedimentation downstream.

NT2 MOVES FORWARD

In March and April 2005, the boards of directors of the World Bank and the ADB approved loans and guarantees for NT2 totalling US$270 million and US$107 million, respectively. The US was the only MDB shareholder that did not support the project, citing concerns about environmental and social risks, the macro-economic environment in Laos, and the lack of potential recourse measures (US Treasury Department, 2005).

With the World Bank and the ADB's endorsement, other lenders – such as the European Investment Bank, the Nordic Investment Bank, the Swedish, Norwegian, French and Thai export credit agencies, AFD and a number of private banks – committed to finance NT2. Construction had already commenced a year earlier, in 2004, and power production was scheduled for December 2009.

About NT2

NT2 is a trans-basin diversion project that is dramatically altering not one, but two, key Mekong River tributaries. A 39m high dam has blocked the Theun River to form a 450km^2 reservoir on the Nakai Plateau where 6200 predominantly ethnic minority people have been resettled. Habitat for the endangered Asian elephant and other wildlife is being inundated by the NT2 reservoir. Downstream from the dam, only 2 cubic metres per second (m^3/s) of water will be released.

Once the reservoir has been filled, water will be directed down a 350m drop to the power station before being transferred to the Xe Bang Fai River via a 27km downstream channel. According to independent research (Shoemaker et al, 2001), approximately 120,000 people living in the Xe Bang Fai area will be affected by increased water flows, resulting in fisheries and aquatic resources losses, erosion, flooding and sedimentation. Water quality problems, caused by anoxic water released from the reservoir area, are also anticipated. NTPC asserts that approximately 75,000 villagers in the Xe Bang Fai region will experience negative impacts as a result of NT2.

NT2 will operate under an intermediate peaking power regime, and the resulting weekly fluctuations in water levels will exacerbate erosion and pose safety hazards to downstream villagers. NTPC has proposed both engineering and operational measures to try to reduce the negative effects downstream. However, while the NT2 regulating pond, aeration weir and concrete-lined downstream channel will help to mitigate some of these impacts, their design is insufficient to prevent these problems (Mekong River Commission, 2007). NTPC has also committed to stop power production (and therefore water releases) when the Xe Bang Fai is close to over-bank flooding at Mahaxai town.

In order to secure the World Bank's approval and to meet the standards of other lenders, NTPC and the GOL made a number of social, environmental, disclosure, accountability and revenue management commitments. While these commitments generally surpassed those made by other hydro developers in the region, many NGOs argued that they would still be insufficient to address the significant risks that NT2 would pose to tens of thousands of Lao people. Project critics were sceptical of NTPC's and the GOL's ability and will to implement

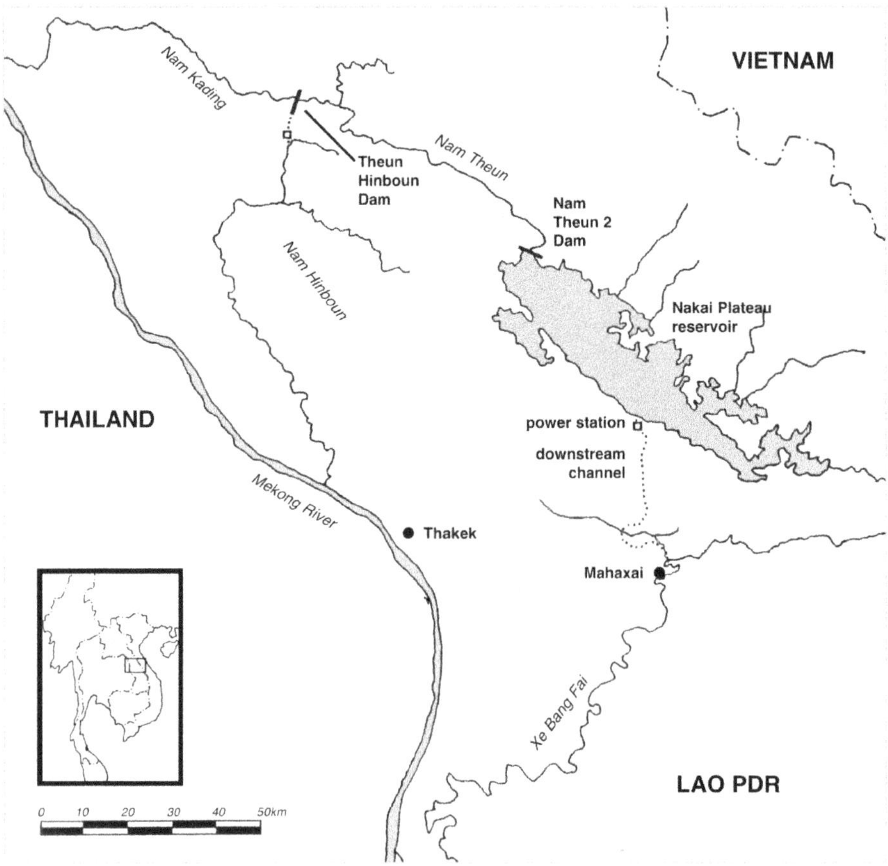

Figure 4.2 *The Nam Theun 2 project area*

Source: International Rivers

these commitments based on experience with other dam projects in Laos. The commitments made by NTPC and the GOL, and the response of project critics, are highlighted in Table 4.2.

The commitments outlined in Table 4.2 were trumpeted as part of an aggressive and extraordinary communication strategy led by the World Bank to rebrand a large, destructive dam project as a 'poverty-reduction' and 'conservation' initiative. These communication efforts were necessary to convince the World Bank's member governments, as well as the general public, that the Bank had learned from its dam mistakes of the past.

Table 4.2 *Summary of key Nam Theun 2 commitments and NGO response*

NT2 commitment	NGO response
Social and environmental	
US$90.5 million provided for social and environmental mitigation and compensation, including US$31.5 million over 25 years to support protection of the Nakai-Nam Theun watershed area.	Inadequate social and environmental budget, particularly to address downstream impacts.
NTPC and GOL required to comply with World Bank and ADB safeguard policies, as well as environmental and social commitments in the Concession Agreement.	Lack of any legal enforcement measures to ensure that agreements are complied with.
Grievance procedure would allow affected people to submit complaints.	Since the grievance mechanism relies on official district/provincial government channels in a closed political climate, villagers may be unwilling or unable to submit grievances – fearing retribution instead of redress.
NTPC committed to raise resettled villagers' income to national poverty level within five years after resettlement.	Livelihood plans for resettled villagers are inappropriate and unrealistic given poor-quality soils, reduced availability of land for farming and grazing, and high risk that village forestry and reservoir fisheries components will not meet targets.
NTPC acknowledged late in the planning process that villagers dependent upon the Xe Bang Fai River would suffer from destruction of riverbank gardens, loss of buildings close to the riverbanks due to erosion, 'a collapse in the aquatic food chain', and impacts upon domestic water supply and transportation difficulties (NTPC, 2005b, Chapter 5, pp12–15). NTPC committed to 'at least restore' livelihoods on a sustainable basis by year nine of project implementation. US$16 million would be provided for the downstream programme to address these impacts upon 75,000 people (NTPC's estimate).	Key baseline information on the pre-project livelihoods of downstream villagers was lacking, and no clear compensation plan was provided. The proposal to replace freshwater fisheries with aquaculture will be difficult in Laos: uptake is slow and the poorest people often lack necessary land and capital to develop and sustain fish ponds. More money will be required to address downstream impacts.
External monitoring	
An International Panel of Environmental and Social Experts (PoE) would continue to visit the project area at least once per year and advise the GOL on social and environmental issues. An International Advisory Group (IAG) would also visit the project regularly and advise the World Bank on implementing environmental, social and revenue management measures.	External project monitoring is critical; but there are no enforcement mechanisms to ensure that the recommendations of the PoE and IAG, in particular, are implemented, which limits monitors' effectiveness. A more streamlined but transparent and enforceable system of external monitoring would better ensure that NTPC and the GOL meet their NT2 commitments.

Table 4.2 *(continued)*

NT2 commitment	NGO response
Independent monitoring agencies (IMAs) would monitor resettlement, watershed management and downstream compensation activities, reporting to the GOL. A Lenders' Engineer would visit NT2 quarterly and provide confidential reports to NT2 lenders on construction, environmental and social issues until the commercial debt was repaid. A Dam Safety Review Panel would advise project developers and the World Bank on safety issues.	
Information disclosure	
Key documents have been disclosed during project preparation and implementation, such as project updates and social and environmental plans required by World Bank and ADB disclosure policies. The World Bank and the ADB would also produce semi-annual implementation updates.	While NT2 released more information than many other projects, complete analyses of potential economic and financial risks and benefits of NT2 have never been disclosed. The PPA and the complete Concession Agreement are not public. Key underlying studies on hydrology and water quality were not made available. Reports of the IAG, PoE and the IMAs are made public only several months after their visits once they have been reviewed by NTPC, the GOL and the World Bank.
Revenue management	
Given the weaknesses in Laos's public expenditure management system, the World Bank insisted on a revenue management framework for GOL's NT2 revenues (estimated to be US$250 million net present value over the 25-year concession). The framework includes the identification of eligible programmes based on GOL poverty reduction strategies. NT2 revenues are channelled through a dedicated Lao Treasury account so that the transfer can be verified and funds withheld from programmes if they fail to meet standards. Budget execution reports, financial statements, summaries of internal audits and audits of NT2-eligible programmes will be published (World Bank, 2005d, pp17–18).	Although a revenue management plan is welcome and necessary, critical gaps in the framework minimize the likelihood of success. According to World Bank indicators, Laos rates below most countries in terms of control of corruption, citizen participation and freedom of expression. The revenue management system would not provide adequate transparency and anti-corruption controls to ensure that these constraints would be overcome. NT2 revenues will be 'co-mingled' with other revenues and managed following standard procedures (World Bank, 2005c). Revenue allocation, monitoring and reporting will be left to the Ministry of Finance and the fledgling State Audit Organization. The revenue management arrangements specifically reject the use of an independent oversight body or the external independent auditing of NT2 revenues.

NT2's implementation

The NT2 legal framework includes the Concession Agreement between the GOL and NTPC, and loan agreements with project financiers such as the World Bank and the ADB. The World Bank and ADB loan and guarantee agreements with the GOL and NTPC, in turn, require compliance with these institutions' own policies. Essentially, these legal agreements constitute the promises made to Laotian villagers regarding compensation and mitigation measures, and the allocation of responsibility amongst NTPC and the GOL.

In the first three years of project construction, NT2's developers have already failed to meet some of their social and environmental obligations. Dated covenants have been missed and NTPC and the GOL have not complied with key provisions of the World Bank's policies and the Concession Agreement, though neither party has been penalized (International Rivers Network, 2007a). The provisions that have not been adhered to include paying compensation before the taking of land and assets, timely disclosure of project information, and provision of irrigation systems for resettled villagers, among others (International Rivers Network, 2007a; International Rivers, 2008a; McDowell et al, 2008).

NT2's construction has proceeded largely on schedule, and the project is on track to start power production by December 2009. However, soon after construction was initiated, it became clear that social and environmental programmes – often more challenging and time consuming than engineering works – were falling behind schedule.

In 2006, the PoE raised the possibility of risks posed to affected villagers and to NT2's reputation of this two-speed process continuing (McDowell et al, 2006a): villagers would lose land and natural resources upon which they depend before compensation and alternative livelihood programmes are in place, causing income declines and other hardship.

The NT2 project did attempt to minimize the risk of construction overtaking social and environmental programmes by including a provision in the Concession Agreement requiring that all Nakai Plateau resettlement infrastructure be completed and resettled villagers moved to their new sites before reservoir flooding began. This provided an important incentive to NTPC and the GOL to achieve these targets, as delays in reservoir filling could impact upon the NT2's ability to deliver power to Thailand on time. The PoE was given the important responsibility of evaluating whether physical resettlement and other Concession Agreement requirements had been met before recommending that reservoir flooding proceed.

All 17 villages on the Nakai Plateau were initially to be resettled by the 2006 to 2007 dry season. But due to delays in permanent housing construction (International Rivers Network, 2007a, p23; International Rivers, 2008a, p16), this deadline was not met. Instead, the first villages were relocated in mid 2006. They, and many of those that followed, moved into temporary houses in their new village sites. It was expected that villagers would remain in these temporary houses for just a few months, and they were instructed to build their houses with that in mind.

In the end, many villagers had to spend two rainy seasons in temporary substandard housing. Two field visits in 2007 by staff of International Rivers documented villagers' frustration with this arrangement (International Rivers Network, 2007a; International Rivers, 2008a). As late as March 2007, NTPC said that only 20 to 30 permanent houses had been completed (Salignat, 2007), and by December 2007, approximately 30 per cent of new houses were built (World Bank and ADB, 2007, p7).

In February 2008, NT2's resettlement and other social and environmental programme delays reached a critical stage. The PoE submitted a report (McDowell et al, 2008) to the GOL and NTPC that raised doubts about the project's ability to proceed with reservoir filling as scheduled, given the serious ongoing delays. This threat helped to galvanize NTPC and the GOL to action. By the time the PoE returned to Laos in April 2008, the company and the GOL completed (or were likely to complete) the resettlement infrastructure, move the remaining villagers, and meet the other minimal requirements of the Concession Agreement. The PoE issued a follow-up report that commended NTPC and the GOL for the progress made since February and noted that the sturdy houses constitute 'one of the most impressive aspects of the entire project' (McDowell et al, 2008, p39). As a result of these achievements, the PoE recommended that the first step towards reservoir impoundment could proceed as scheduled. On 10 April 2008, NT2's diversion tunnel was sealed and preparations began for closure of the dam's spillways in July 2008. By June 2008, nearly all the 1272 houses on the Nakai Plateau had been completed (Gasparini and Rex, 2008).

Although physical resettlement and reservoir filling represents an important milestone for NT2, problems with livelihood restoration programmes for resettled villagers on the Nakai Plateau, downstream villagers along the Nam Theun and Xe Bang Fai, and villages affected by downstream channel construction remain. In recent reports (McDowell et al, 2007a, 2007b, 2008), the PoE has been critical of NTPC's and the GOL's performance in these areas, while praising their achievements in terms of the physical resettlement of Nakai Plateau villagers (McDowell et al, 2008). Even as the water began rising on the Nakai Plateau, the PoE warned that the main challenges lay ahead:

> *In a very real sense, the more easily accomplished phase of the project's environmental and, particularly, the social programmes is coming to an end and the hard part is upon us. By comparison with the largely physical tasks of infrastructure building, the challenge now is [to] move on to help convert these aggregations of houses, community facilities, cleared patches of still smoking vegetation and disrupted families into self-managed and self-sustaining communities, viable in all dimensions.* (McDowell et al, 2008, p48)

Livelihood programmes for Nakai Plateau resettled villagers

Before reservoir flooding began, villagers in the resettlement sites were primarily surviving on rice and protein supports from NTPC, income earned from the project for land clearance and other work, fishing, and forest product collection. During visits by International Rivers in 2007, a number of villagers reported that the amount of rice provided each month is often not enough for large families. There was also concern about premature cut-offs of rice and protein support for resettled villagers (International Rivers Network, 2007a; International Rivers, 2008a). Some people said that they missed the fruit trees in their old villages, and still returned there to get fish from the river and bamboo and vegetables from the forest. A number of villagers worried about how they would find food in their new sites, especially 'after the flood'.

Despite these apprehensions, many villagers said that they are pleased with the better houses, improved water supply and sanitation, electricity and roads to the new villages. Health improvements, particularly as a result of access to safe water and better sanitation, were quickly noted by the PoE and other monitors. The income from project-related jobs also contributed to initial increases in villagers' living standards.

However, the greatest challenge for NT2 continues to be developing and implementing sustainable livelihood programmes for Nakai Plateau villagers. Resettled villagers have been moved to what will become the reservoir shores so that they can remain, by their request, on their traditional lands. However, soil quality is generally poor on the Nakai Plateau, and two-thirds of the land that villagers once used for farming, grazing livestock and collecting non-timber forest products (NTFPs) will be flooded by the large reservoir. The water buffalo and cattle populations on the Nakai Plateau can no longer be maintained, and the total herd will need to be reduced by approximately half (NTPC, 2008b, p29).

NTPC has committed to raise resettled villagers' income to the national poverty level within five years. To this end, resettled villagers will be provided with:

- house gardens;
- 0.66ha plots (to be irrigated by the end of 2009) for growing some rice, fodder and vegetables;
- use of the reservoir drawdown zone for rice cultivation and grazing land;
- a community forest area for collecting NTFPs and sustainable timber extraction (some of which will also be used for grazing and fodder cultivation); and
- boats for fishing in the reservoir, an area to which they've been granted exclusive access rights for ten years (NTPC, 2005a).

NGOs and academics began to point out shortcomings in these livelihood restoration plans before NT2 received World Bank approval (International Rivers Network, 2005; Lanza, 2005; Willing and Knoop, 2005), and many of these

Figure 4.3 *Nam Theun 2 resettlement homes*

problems have not been addressed. Villagers will be expected to grow cash crops on poor-quality land to sell in an as yet unidentified market. They were originally promised 10,000ha of production forest to be managed as a village 'business'; but the area has since been reduced by at least 40 per cent and is further threatened by illegal logging (McDowell et al, 2008). The forest area will be difficult for some villages to access, and important NTFPs such as bamboo will be flooded by the reservoir. Villagers were promised bountiful fish in the new reservoir, but it is likely to have initial poor water quality due, in part, to degrading and flooded vegetation, thereby threatening the development of a productive fishery (International Rivers, 2008a; McDowell et al, 2008, p18). While exact numbers are still unclear, as many as 2000 buffaloes may still need to be sold, which will weaken a critical livelihood safety net for villagers (NTPC, 2008b, p29).

In early 2007, the PoE warned that 'for a range of reasons, the forestry and agricultural livelihood programmes are unlikely to meet their originally planned targets before impoundment' (McDowell et al, 2007a, p9). A more recent PoE report indicates that after an initial boost from project jobs and other support, resettled villagers' living standards are likely to decline once the reservoir is flooded (McDowell et al, 2008, p11). Finally, the July 2008 World Bank–ADB update notes that while 'encouraging progress' is being made on the various livelihood programmes, significant challenges remain (World Bank and ADB, 2008, p6).

Villagers affected by construction activities

More than 10,000 people[1] (World Bank and ADB, 2007, pp23–24) have been affected by the construction of NT2's transmission lines, roads and project facilities, losing land, assets and access to natural resources. Households in Gnommalat District near the NT2 power station, the regulating pond and the downstream channel have been the most severely affected.

The downstream channel is 27km long and approximately 100m wide, cutting through paddy fields and other village land. The channel also blocks access to the forest and villagers' gardens and rice paddies on the other side. Villagers in the area have lost paddy land, houses, gardens, fruit trees, fisheries, irrigation water supply and other assets to varying degrees.

According to the NT2 Concession Agreement, villagers who lose less than 10 per cent of their productive assets are entitled to cash compensation, and those who lose more than 10 per cent are entitled to replacement land. Compensation payments only began in mid 2006, more than a year after NT2 construction activities had started to impact upon villagers' land and resources (International Rivers Network, 2006a), and it was not until mid 2008 that most compensation payments were finally made (World Bank and ADB, 2008, p12). As mentioned previously, the taking of land and assets before paying compensation is a violation of the Concession Agreement and of the World Bank's involuntary resettlement policy (International Rivers Network, 2007a).

There have also been problems with the assessment of entitlements and the delivery of compensation; nearly 400 grievances have been submitted to the District Grievance Committee (World Bank and ADB, 2008). It is unclear if the situation for the 200 households who are entitled to replacement land has been resolved. Only six households had received land-for-land replacement as of late 2007, and NTPC and project backers have said there is a shortage of available paddy land in the area (NTPC, 2008a).

In late 2007 and 2008, villagers indicated to International Rivers and to the PoE that replacement land was, indeed, available. Following more detailed investigations by the PoE, NTPC committed to work with villagers to identify and purchase adequate replacement land. While it was recommended that 'significant progress on land-for-land issues' should be made by the end of 2008, the PoE also noted that the Concession Agreement requirement to restore villagers' incomes within 18 months had not been complied with in many cases (McDowell et al, 2008, p16).

Downstream programmes

NT2 will affect more than 120,000 villagers (Shoemaker et al, 2001), or 75,000 people, according to NTPC's estimates, in the Xe Bang Fai River Basin. Villagers living along the river and some of its tributaries can expect more frequent and severe flooding, decimated fisheries and inundated riverbank gardens. About 40 villages that fish in the Theun River will also experience declines in fisheries and aquatic resources due to the reduced river flow downstream from the dam site.

In an attempt to mitigate NT2's impacts and compensate villagers in the Xe Bang Fai area, NTPC has developed a Downstream Livelihood and Asset Restoration Programme (Downstream Programme). This programme will be implemented in approximately 220 villages, including nearly 90 riparian villages. In breach of World Bank and ADB involuntary resettlement and information disclosure policies, the Downstream Programme Implementation Plan had still not been publicly disclosed as of this writing, although information has been provided to affected villagers.

The Downstream Programme focuses on microcredit funds to support agriculture, aquaculture and livestock projects. NTPC is also supporting water and sanitation improvements, and, in some villages, water-gate rehabilitation or mini-polder flood protection. In late 2007, NTPC said it intended to expand the programme to all the riparian villages in 2008 and to initiate activities in the remaining villages during late 2009 (NTPC, 2008a); but it now appears that the programme will reach only about half of the riparian villages by the end of 2008. As of mid 2008, projects had been initiated in less than 20 per cent of the total villages in the Xe Bang Fai Basin that are likely to be affected when NT2's operations begin (World Bank and ADB, 2008).

The World Bank, the ADB and the PoE agree that the Downstream Programme is behind schedule, and that continued delays pose risks to affected villagers. The PoE February 2008 report states:

> *While the formulation of the draft [Downstream Implementation] plan has gone on its leisurely way, the construction programme has forged ahead. As a result,* many impacts felt below the powerhouse will occur before the remedial or compensatory measures are in place. (McDowell et al, 2008, p26; emphasis in original)

The PoE also points to the Downstream Programme's short- and long-term funding gap, noting that the US$16 million budget 'was never going to be sufficient funds to complete the tasks envisaged' (McDowell et al, 2008, p35). Consultant fees will absorb US$1 million, leaving approximately US$200 per affected person[2] for mitigation and compensation measures. Although the budget is not being used to provide cash compensation to villagers, but rather to support infrastructure improvements and contributions to village savings funds, this figure highlights how inadequate the funding is to at least restore the livelihoods of affected people, as required by the Concession Agreement.

While NTPC has provided no additional funding, the World Bank approved a US$9 million Khammouane Development Project in June 2008 that will support irrigation development along the NT2 downstream channel and Lower Xe Bang Fai River. The PoE has also urged the ADB and AFD to commit additional resources to support irrigation and flood management (McDowell et al, 2008, p42).

Each of the affected Xe Bang Fai villages will receive approximately 2 million Lao kip per household (about US$200) from NTPC through a village savings fund. Villagers can borrow from these funds for various livelihood projects, ranging from fish ponds to pig-raising to tomato cultivation. However, villagers reported to International Rivers that they have to pay back the loans to the village savings fund, with monthly interest ranging from 1 to 3 per cent, *whether or not the projects succeed or fail.* Those people with unsuccessful projects have been forced to sell buffalo and other assets to repay debts to the village savings fund. Some villagers report that they have already stopped participating in the fund or will no longer borrow for livelihood projects. The reliance on a microcredit scheme to deliver compensation creates a cycle of debt if projects fail or if repayment terms are too demanding.

Flooding is another major concern for Xe Bang Fai villagers, some of whom lose rice crops and other assets every two to three years as a result. NT2 is expected to increase the frequency and the duration of floods in the Xe Bang Fai area, even if power production is stopped as promised when the river overflows its banks at Mahaxai town. At the nearby Theun-Hinboun Hydropower Project, recent research has shown that flooding along the Hinboun River has become increasingly severe over the past decade, leading to large-scale abandonment of rice paddy fields (FIVAS, 2007). While NT2 and Theun-Hinboun vary in some technical specifications, there are important lessons to be learned from the Theun-Hinboun experience. NTPC should prepare for a worst-case scenario where wet season rice production is no longer viable along sections of the Xe Bang Fai due to protracted annual flooding.

Most of the flood-prone villages visited by International Rivers in December 2007 said that they had requested flood protection works (such as dikes, mini-polders or water-gate rehabilitation) from NTPC, but in many cases were told that funding is not available. However, as of July 2008, 15 flood gates had been rehabilitated and work on an additional 4 gates was under consideration (World Bank and ADB, 2008). While it appears that the PoE's recommendation to allocate more resources to flood-protection works before power production starts has been followed (McDowell et al, 2008; World Bank and ADB, 2008), the total funds available will probably still be insufficient to mitigate NT2's flooding impacts.

Figure 4.4 *Woman fishing on the Xe Bang Fai*

Nakai-Nam Theun Protected Area

One of the selling points of the NT2 project was that NTPC would provide US$31.5 million to help protect one of the Mekong region's richest areas of biodiversity, the Nakai-Nam Theun National Protected Area (NPA). The 4000km^2 Nakai-Nam Theun NPA is the largest protected area in mainland Southeast Asia (World Bank, 2008a) and forms the watershed of the NT2 project. The funding provided by NTPC helped to establish the NT2 Watershed Management Protection Authority (WMPA) tasked with implementing an ambitious *Social and Environmental Framework and Operational Plan* (SEMFOP) in the Nakai-Nam Theun NPA. According to the PoE: 'without the NPA component, the NT2 Project as such would not exist' (McDowell et al, 2008, pp28–29).

The vision of the SEMFOP – and one of the key objectives of the WMPA – is to establish a balance between biodiversity conservation and development for the 6000 villagers living in the NPA. The PoE questioned the WMPA's progress on this challenging issue (McDowell et al, 2007b, pp27–28) and called for improvements in the WMPA's organization and staffing to more effectively meet the SEMFOP goals (McDowell et al, 2008, pp28, 47). A new director and a new chief technical adviser were reportedly appointed in mid 2008 (World Bank and ADB, 2008, p16).

Established as a biodiversity conversation area in 1993, the Nakai-Nam Theun NPA faces the threats of hunting and poaching, illegal logging and mining activities that are familiar to other protected areas in the region. In its February 2007 report, the PoE reported that the NPA was 'bleeding rosewood', referring to illegal logging of this valuable hardwood (McDowell et al, 2007a). One part of the problem was traced to the NANCY Company – tasked with clearing valuable timber from the NT2 reservoir area before it was flooded – for its alleged 'laundering' of illegal rosewood from the NPA during its salvage logging operation of the NT2 reservoir (McDowell et al, 2007a; EIA/Telepak, 2008, pp16–17). The PoE also exposed a significant mining operation that was threatening the integrity of the area, although that operation has reportedly since been shut down by the GOL. While the WMPA's efforts to control these threats have been commended, the PoE notes that 'to date, the patrolling results are still inadequate' (McDowell et al, 2008, p28). Once the NT2 reservoir is filled, it could increase access to the NPA and make illegal logging and poaching even more difficult to control.

Environmental management during construction

The NT2 project was supposed to demonstrate best practice in terms of environmental management. However, the construction companies involved – which include NTPC shareholders Electricité de France as the head construction contractor and Ital-Thai Development as the principal civil works contractor – have repeatedly been cited for environmental infractions by the Lenders' Engineers and the PoE (McDowell et al, 2006b; PB Power, 2007). These violations include road-building negligence, leading to excessive deforestation, and a failure to control dust, erosion and sedimentation, leading to water quality problems and respiratory difficulties for villagers.

A leaked copy of a Lenders' Engineer report summarizes NT2's environmental management shortcomings:

> *As noted in our previous report, for a project which is intended to set a benchmark of world's best practice against which future projects can be assessed, the environmental performance still falls significantly short of this benchmark in many areas and, in some aspects, still barely meets 'business as usual' levels.* (PB Power, 2007, p9)

The PoE, the World Bank and the ADB continued to report on problems with road-building, wastewater treatment and revegetation, in particular (McDowell et al, 2007b, p20; World Bank and ADB, 2008). It seems that no penalties or fines have been levied for these breaches despite the fact that the head construction

Figure 4.5 *Excessive road clearance*

contractor's compliance with the *Environmental Mitigation and Management Plan* 'in all material respects' is a requirement of the World Bank's agreement with NTPC and the GOL (World Bank, 2005d).

Revenue management framework

NT2's development justification – and the justification for its World Bank and ADB support – hinges on the GOL using project revenues to help the poor. The establishment of the NT2 revenue management framework is being supported by a number of World Bank- and ADB-backed public expenditure and public financial management programmes. The goal of these loans, grants and technical assistance initiatives is to support improvements in public financial management, broadly speaking, and to ensure that NT2 revenues, once generated in early 2010, are directed to poverty reduction programmes as required.

The most recent Public Expenditure Review–Integrated Fiduciary Assessment for Laos points to a number of shortcomings. These include poor budget planning, 'weak authority and capacity of the Treasury and deficiencies in the systems for financial control, accounting and budget reporting', and poor fiscal transparency (World Bank et al, 2007, p28). The same review acknowledges 'potential issues' with regard to NT2 revenues: 'in the short term, NT2 revenues are likely to be used primarily to increase the timeliness of wage and salary payments to [Ministry of Agriculture and Forestry] staff and to increase the wages of personnel stationed in remote rural areas' (World Bank et al, 2007, pp72–73). This would seem to run counter to the GOL's commitments to direct NT2 revenues initially to health, education and rural infrastructure. The World Bank bemoans the fact that 'despite sizeable donor contributions', Laos still spends much less on health and education than other low-income countries (World Bank et al, 2007, p25).

The implementation of NT2 revenue arrangements has reportedly progressed, with the identification of NT2-eligible expenditures, such as health, education and rural roads expected to occur before the end of 2008 (World Bank and ADB, 2008, p2). Baseline allocations for these programmes will reportedly be assessed in the 2008/2009 financial year so that the 'additionality' of NT2 revenues in 2010 can be assessed (World Bank and ADB, 2008). But after more than a decade of World Bank and ADB support to improve public financial management in Laos, it is clear that significant weaknesses remain (World Bank et al, 2007; World Bank, 2008b). If the GOL loses interest in these reforms, there will be no external controls to ensure that NT2 revenues are not misdirected.

Lessons and implications for Laos

The fact that NT2's social and environmental programmes have run into such major difficulties raises concerns about the commitment of NTPC and the GOL to deliver on earlier promises, as well as the feasibility of the promises made to justify the project in the first place. If the World Bank, ADB and other project lenders are unable to ensure compliance with key social and environmental obligations while the project is still under construction, it is difficult to see what leverage they will have once the project has been completed and electricity is being generated. At that stage, NTPC will have even fewer incentives to meet its commitments to affected communities and the environment.

Furthermore, the World Bank's promises that NT2 would provide a model for more sustainable hydropower development in Laos are largely ringing hollow. The GOL's National Policy on the Environmental and Social Sustainability of the Hydropower Sector in Laos, enacted at the time of NT2 project approval with World Bank support, is still not being implemented (International Rivers Network 2006b, 2007b). Even a basic provision of the policy, that environmental impact assessments for hydropower projects should be disclosed, is not being followed.

Construction on new hydropower projects, such as Nam Theun 1, Nam Ngum 5 and Xekaman 3, is proceeding before environmental licences have been granted. Resettlement guidelines are not being followed (Vattenfall et al, 2008). The GOL's Water Resources and Environment Agency (WREA) lacks the capacity and the authority to assess, approve and monitor these new dam developments effectively.

There is also no discussion of earmarking revenue from proposed hydro projects for poverty reduction expenditures, as with NT2.

Why have problems occurred with NT2?

The problems inherent in the planning and implementation of NT2 point to broader challenges with the process for deciding upon and implementing dam projects in Laos and, more broadly, the Mekong region. First, the preponderance of non-democratic regimes in the region makes it difficult for meaningful participatory processes to be undertaken due to the lack of true freedoms of speech and assembly in countries such as Laos, Vietnam and Myanmar/Burma. Because of the political situation, it is difficult for affected communities to access independent information about the risks of particular projects or to organize against projects if they feel that they will not benefit from them.

These countries also have rudimentary legal systems and few laws granting basic rights to citizens. This lack of the rule of law makes it impossible for affected communities to seek redress when their rights are violated, or to challenge decisions taken by the government on whether or not to proceed with a particular project. They also make it difficult to ensure that project agreements and commitments to villagers and the environment are upheld. This leaves affected communities vulnerable to exploitation, allows the government and the private developers to avoid covering the true social and environmental costs of their project, and leads to poor outcomes for both people and the environment. While NT2 has gone farther than most projects in terms of public reporting and participation, its efforts have inevitably been constrained by the political environment in Laos.

Second, although numerous studies were conducted for NT2, some of them were of poor quality, containing unrealistic assumptions and insufficient data (International Rivers Network, 2005; Lanza, 2005; Willing and Knoop, 2005; McDowell et al, 2007b). This illustrates a more fundamental problem inherent in hydropower planning around the world: it is up to dam developers to commission and endorse the environmental impact assessment and other studies, and to agree on the funding for compensation and mitigation measures. By the time the EIAs are submitted to the GOL's environmental regulatory agency, for example, the dam has been designed, costs have been determined, and often the Project Development Agreement has been signed. Even if the GOL's WREA had sufficient capacity and resources to review these documents, their ability to stop or change a project at that stage is limited.

Additionally, environmental and social assessments are done by consultancy companies who often have vested interests in presenting all environmental impacts as 'manageable' and seeing that the hydropower project gets built by minimizing the social and environmental problems that it will pose. This is one way of ensuring that they receive more contracts from dam developers. The author is aware of a number of cases of a consultant's work being replaced by that of another firm – or strong indications that this would be the case – if the consultant's predictions of likely impacts and mitigation costs are too high (International Rivers, 2008c). This situation is certainly not unique to Laos.

Third, NT2 has suffered from a lack of qualified staff, resources and a high-level commitment from NTPC to the project's social and environmental objectives. The PoE highlights several examples of Concession Agreement violations that occurred because NTPC preferred to cut costs of provisions for irrigation and adequate compensation for fisheries losses (McDowell et al, 2008, p19). The PoE also cites NTPC's failure to adequately account for the staffing levels and budget that would be required to meet the obligations of the Concession Agreement (McDowell et al, 2008, p11), although that situation has reportedly improved as of mid 2008 (McDowell et al, 2008, p43; World Bank and ADB, 2008). To improve the implementation of resettlement and livelihood programmes, a more realistic assessment of local capacity and the resources and staffing level required to meet NT2's social and environmental requirements should have been conducted before the project was initiated and acted upon.

Finally, NT2 points to the limited leverage that international institutions such as the World Bank and ADB have once a project is under construction. Short of withdrawing financial assistance from the country or holding up project loans and guarantees, which they are notoriously reluctant to do, the MDBs have difficulty in ensuring compliance once funds are disbursed and construction has begun, since their supervision of the project decreases. This issue, is, however, becoming increasingly irrelevant as the MDBs are displaced by other dam funders who may be less concerned with the environmental and social risks of their investments.

Conclusions and recommendations

The problems outlined above highlight the difficulties of managing the social and environmental impacts of large dams, particularly in a country with a repressive political climate and without a well-developed legal system through which affected communities' rights can be defended and environmental laws upheld. They also show that time and money are needed to do these projects right. When developers cut corners, the GOL and the Lao people ultimately bear the high costs of damaged livelihoods, lost fisheries and degraded agricultural land. On the other hand, the added costs of genuine mitigation and compensation measures may only be a fraction of the profits that dam developers stand to gain.

Increasingly, though, the dam playing field in Laos is not being shaped by the World Bank or Western hydropower developers, but by companies from Thailand, China, Vietnam, Russia and Malaysia. If the GOL does not quickly take the reins of the hydropower sector, these private developers and their regional backers will do it for them. It is time for the GOL to learn from NT2's mistakes and build on some of the things that NT2 has done better. The flood of interest in the hydropower sector in Laos means there is room for higher standards, especially if it involves weeding out the most irresponsible dam developers.

First, the GOL and donors should comprehensively assess, through a broad-based participatory process, all the poverty reduction and revenue generation options for Laos in a way that honestly evaluates the trade-offs and costs of hydropower development. Laos needs a development strategy that does not destroy the rivers and resources upon which the majority of the population depends. Second, the GOL should recognize that the costs of some dams, including in economic, social, environmental and cultural terms, will be too high to merit proceeding. Projects proposed for the Mekong River mainstream would probably fall into this category. In particular, their non-financial costs need to be more carefully studied and subject to independent review.

Dam projects in Laos need to be selected based on strategic environmental assessments and basin-wide planning, not on an *ad hoc* basis driven by the interests of dam developers or select government officials. For dam projects that do go forward, social and environmental obligations must be considered a core part of the project, accompanied by adequate budgetary resources to both implement mitigation programmes and compensate affected people. The GOL should ensure that quality assessments are submitted which include viable mitigation and livelihood restoration plans. Then the GOL must hold developers accountable to these plans. Failure to meet social and environmental commitments should carry the same penalties as failure to meet engineering deadlines.

In order to evaluate and enforce developers' environmental and social commitments, as well as ensure compliance with the laws and regulations of Laos, the GOL urgently needs to strengthen the capacity and authority of WREA and move its project review process upstream. Working with donors, the GOL should establish a well-resourced WREA unit that is responsible for conducting regular project monitoring during the construction and operation phases. The GOL's ability to negotiate favourable agreements with power purchasers should also be strengthened. Until this capacity has been built, the GOL should consider a moratorium on the signing of new Concession Agreements and PPAs, as well as increasing the use of independent experts and external monitors to assess dam projects.

Finally, transparency and accountability to affected people needs to be dramatically improved. The rural villagers who bear the costs of these dam projects should have more of a say in whether or not they are developed. They also need a better understanding of their rights and entitlements and the space to ask that those rights be respected. In addition to receiving full compensation for their losses, all communities affected by a hydropower project should be provided with a percentage of the dam's revenues – for the life of the project – to support their own community development initiatives.

NT2 may have improved upon the performance of past dams in Laos, but it is still a long way away from 'doing dams right'. Laos will have only one chance to manage this hydropower boom to minimize costs and maximize benefits for its people. Urgent action is needed to both address the problems with NT2 and avoid a repeat of past dam mistakes in Laos.

Notes

1 According to the World Bank in June 2008, the numbers for significantly affected households have been revised downward to approximately 200. Therefore, 2200 households multiplied by the average household size in Laos (6) means that an estimated 13,200 villagers have been affected by construction activities.
2 Using NTPC's numbers of only 75,000 affected people.

References

ADB (Asian Development Bank) (2003) *Technical Assistance to the Lao PDR for Preparing the Greater Mekong Subregion: Nam Theun 2 Hydropower Development Project*, ADB, Manila

ADB (2004) *Summary Environmental and Social Impact Assessment, Nam Theun 2 Hydroelectric Project in Lao PDR*, ADB, Manila

Amnesty International (2003) *Amnesty International Report 2003*, Amnesty International, Lao PDR

AusAid (2005) *Review of Nam Theun 2 Hydroelectric Dam, Lao PDR*, Final report to AusAID, Vientiane

duPont, P. (2005) *Nam Theun 2 Hydropower Project: Impact of Energy Conservation, DSM and Renewable Energy Generation on EGAT's Power Development Plan*, Danish Energy Management A/S for the World Bank, Bangkok

EIA/Telapak (2008) *Borderlines: Vietnam's Booming Furniture Industry and Timber Smuggling in the Mekong Region*, www.eia-international.org/files/reports160-1.pdf, accessed 10 July 2008

FIVAS (2007) *Ruined Rivers, Damaged Lives: The Impacts of the Theun-Hinboun Hydropower Project on Downstream Communities in Lao PDR*, Printhouse, Oslo, Norway

Friends of the Earth–Japan, JACSES and Mekong Watch (2005) 'Letter to World Bank President James Wolfensohn endorsed by 153 civil society organizations citing concerns about Nam Theun 2', 14 March

Gasparini, N. and Rex, W. (World Bank) (2008) Pers comm in a meeting with S. Lawrence from International Rivers, Vientiane, Laos, 17 June

GOL (Government of Laos) (2002) *NT2 Project – Lao PDR*, Donor Roundtable Presentation, July

Greacen, C. and Sukkamnoed, D. (2005) 'Did the World Bank fudge figures to justify Nam Theun 2', unpublished paper, 23 July, www.palangthai.org/docs/NT2EconMalfeasRefs.pdf

Imhof, A. and Lawrence, S. (2005) *An Analysis of Nam Theun 2 Compliance with World Commission on Dams Strategic Priorities*, International Rivers Network and Environmental Defense, Berkeley, CA and Washington DC

International Rivers (2008a) *Nam Theun 2 Trip Report and Project Update*, May, International Rivers, Berkeley, CA

International Rivers (2008b) N*am Theun 2 Hydropower Project: Risky Business for Laos*, International Rivers, Berkeley, CA

International Rivers (2008c) *Power Surge: The Impact of Rapid Dam Development in Laos*, International Rivers, Berkeley, CA

International Rivers Network (2004) *The Legacy of Hydro in Laos*, International Rivers Network, Berkeley, CA

International Rivers Network (2005) *Review of Agriculture and Livestock Development Plan for the Nam Theun 2 Hydropower Project*, International Rivers Network, Berkeley, CA

International Rivers Network (2006a) *IRN Visit to Nam Theun 2 Hydropower Project in Laos June 2006*, International Rivers Network, Berkeley, CA

International Rivers Network (2006b) 'International Rivers Network letter to WREA', 11 September, www.internationalrivers.org/files/Xaypaseuth%20letter%20Aug%2006.pdf, accessed 10 June 2008

International Rivers Network (2007a) *Nam Theun 2 Trip Report and Project Update*, February, International Rivers Network, Berkeley, CA

International Rivers Network (2007b) 'International Rivers' letter to WREA', 9 November, www.internationalrivers.org/files/Letter%20to%20WREA_MEM%20Nov07.pdf, accessed 10 June 2008

International Rivers Network, Environmental Defense Fund, Centre for International Environmental Law and National Wildlife Federation (1997) 'Letter to US executive director to the World Bank citing concerns about NT2'

Kuankachorn, S. (2005) 'Paying the bill for destruction: Thai society and the Nam Theun 2 Dam in Lao PDR', *Watershed*, vol 10, no 2, pp56–60

Lang, C. (2003) 'Laos: French company pulls the plug on Nam Theun 2', *World Rainforest Movement Bulletin*, no 72, pp10–11

Lanza, G. (2005) *Review of the Water Quality Assessment (Environmental Assessment and Management Plan) for the Proposed Nam Theun 2 Hydroelectric Project*, University of Massachusetts, MA

Les Amis de la Terre, CRBM (Campagna per la Riforma della Banca Mondiale), BIC (Bank Information Center) and Environmental Defense (2003) *NGO Visit to Nam Theun 2 Hydroelectric Project in Laos*, Les Amis de la Terre, CRBM, BIC and Environmental Defense, Paris, Rome and Washington DC

Manolom, S. (2002) 'Nam Theun 2: Window on the future', Symposium presentation at 'Nam Theun2 Project', Laos

McDowell, D., Scudder, T. and Talbott, L. (2006a) *Lao PDR Nam Theun 2 Multipurpose Project*, Ninth Report of the International Environmental and Social Panel of Experts, 15 February

McDowell, D., Scudder, T. and Talbott, L. (2006b) *Lao PDR Nam Theun 2 Multipurpose Project*, Tenth Report of the International Environmental and Social Panel of Experts, 30 October

McDowell, D., Scudder, T. and Talbott, L. (2007a) *Lao PDR Nam Theun 2 Multipurpose Project*, Eleventh Report of the International Environmental and Social Panel of Experts, 23 February

McDowell, D., Scudder, T. and Talbott, L. (2007b) *Lao PDR Nam Theun 2 Multipurpose Project*, Twelfth Report of the International Environmental and Social Panel of Experts, 29 September

McDowell, D., Scudder, T. and Talbott, L. (2008) *Lao PDR Nam Theun 2 Multipurpose Project*, Thirteenth and Fourteenth Reports of the International Environmental and Social Panel of Experts, February–April

Mekong River Commission (2007) *Review of the Hydraulic Study for Discharges from the NT2 Regulating Pond and Impacts on the Xe Bang Fai*, Mekong River Commission, Vientiane

Nette, A. (1997) 'Nam Theun 2 stranded with no buyer for its power', *World Rivers Review*, vol 12, no 2, pp7–9.

NTPC (Nam Theun 2 Power Company) (2005a) *Nam Theun 2 Social Development Plan: Nakai Plateau and EMDP and RAP*, NTPC, vol 2, Vientiane

NTPC (2005b) *Nam Theun 2 Social Development Plan: Downstream Areas – EMDP, Resettlement and Livelihood Restoration*, NTPC, vol 3, Vientiane.

NTPC (2005c) *Nam Theun 2 Social Development Plan: Project Lands Resettlement (Acquisition and Compensation) Plan*, NTPC, vol 4, Vientiane.

NTPC (2005d) 'NTPC signs US$1 billion loan agreements', Press release, 3 May, www.namtheun2.com/Finance/Press%20_Release_%20Financing.pdf, accessed 13 June 2008

NTPC (2005e) *Nam Theun 2 Hydropower Electric Project: Concession Agreement between the Government of Lao PDR and the Nam Theun 2 Power Company, Schedule 4*, NTPC, Vientiane

NTPC (2005f) *Frequently Asked Questions (FAQ) – Environmental*, www.namtheun2.com/faq/faq_env.htm, accessed 25 June 2008

NTPC (2008a) *NTPC Response to International Rivers*, www.internationalrivers.org/files/IR_NT2ques_NTPCresp_108.pdf, accessed 20 May 2008

NTPC (2008b) *Specific E&S Action Plans for Reservoir Impoundment*, NTPC, Vientiane, disclosed April 2008

PB Power (2007) *Lenders' Engineer Quarterly Site Visit Report No 8, Part C*, PB Power, Bangkok

Permpongsacharoen, W. (2004) 'An alternative to Thailand's *Power Development Plan*', *Watershed*, vol 10, no 1, pp30–37

Praiwan, Y. (2003) 'Nam Theun 2 draws interest from overseas', *Bangkok Post*, 10 September

Reuters (2003) 'Thai firms agree deadline for Laos dam investor', Reuters, 23 July

Salignat, O. (NTPC) (2007) Pers comm in meeting with S. Lawrence, International Rivers Network, Vientiane, Laos, March

Scudder, T. (2005) *Laos Nam Theun 2 Dam*, Unpublished manuscript, www.hss.caltech.edu/~tzs/Lesotho%20&%20Laos%20Cases.pdf, accessed 15 August 2008

Shoemaker, B., Baird, I. and Baird, M. (2001) *The People and their River: A Survey of River-Based Livelihoods in the Xe Bang Fai River Basin in Central Lao PDR*, Lao PDR/Canada Fund for Local Initiatives, Vientiane, Laos

Transparency International (2008) *Corruption Perceptions Index*, Transparency International, www.transparency.org/policy_research/surveys_indices/cpi/2008, accessed 1 November 2008

United Nations Committee on Elimination of Racial Discrimination (2003) 'Decision on the situation in Lao PDR', 21 August

US State Department (2005) *Country Reports on Human Rights Practices – 2004*, Released by the Bureau of Democracy, Human Rights, and Labor, Washington DC, 28 February

US State Department (2008) *Country Reports on Human Rights Practices – 2007*, Released by the Bureau of Democracy, Human Rights, and Labor, Washington DC, 11 March

US Treasury Department (2005) *Official Remarks from US Treasury Department on Abstention on Nam Theun 2 Hydropower Project*, 1 April

Vattenfall Consultants AB, Ramboll Natura AB and Earth Systems (2008) *Lao People's Democratic Republic: Preparing the Cumulative Impact Assessment for the Nam Ngum 3 Hydropower Project*, Technical Assistance Consultant's Report to the Asian Development Bank, Linköping, Sweden

Willing, P. and Knoop, K. (2005) *Review of Hydrology Component of the Environmental Assessment and Management Plan for the Proposed Nam Theun 2 Hydropower Project*, International Rivers Network and Environmental Defense, Berkeley, CA and Washington DC

World Bank (2002) *Decision Framework for Processing the Proposed NT2 Project*, World Bank, Washington DC

World Bank (2004a) *Project Performance Re-assessment Report, Lao PDR, Second Structural Adjustment Credit*, Operations Evaluation Department, World Bank, Washington DC, 24 June

World Bank (2004b) *Lao PDR Country Economic Memorandum: Realizing the Development Potential of Lao PDR*, World Bank, Washington DC, August

World Bank (2005a) *Lao PDR Nam Theun 2 Hydroelectric Project: Project Economic Analysis*, World Bank, Washington DC

World Bank (2005b) *Proposed Nam Theun 2 Hydroelectric Project in Laos World Bank Responses to IRN-EDF Technical Reviews*, World Bank, Washington DC

World Bank (2005c) *Revenue and Expenditure Management Technical Brief, Nam Theun 2 Hydroelectric Project*, World Bank, Washington DC

World Bank (2005d) *Project Appraisal Document for the Nam Theun 2 Hydroelectric Project*, World Bank, Washington DC, March

World Bank (2008a) 'Nam Theun 2: A project in the heart of Laos', PowerPoint presentation

World Bank (2008b) *Program Document for a Proposed Grant in the Amount of SDR 6.1 million to the Lao People's Democratic Republic for a Fourth Poverty Reduction Support Operation*, World Bank, Washington DC, May

World Bank and ADB (Asian Development Bank) (2007) *Update on the Lao PDR: Nam Theun 2 (NT2) Hydroelectric Project*, World Bank and ADB, Washington DC and Manila, December

World Bank and ADB (2008) *Update on the Lao PDR: Nam Theun 2 (NT2) Hydroelectric Project*, World Bank and ADB, Washington DC and Manila, July

World Bank, IMF (International Monetary Fund), ADB (Asian Development Bank) and the European Commission (2007) *Lao PDR Public Expenditure Review and Integrated Fiduciary Assessment*, World Bank, IMF, ADB and EC, Washington DC, Manila and Brussels

Appendix

Nam Theun 2 and Its Impact upon Hydropower Development in Laos[1]

Patchamuthu Illangovan, World Bank country manager for Laos

Laos, a landlocked country in Southeast Asia, is one of the poorest countries in the region. During the early 2000s, the country was characterized by weak infrastructure, low capacity in its institutions, lack of industries, extremely low foreign investment, high poverty and few opportunities for growth. However, Laos had a goal: to exit from poverty and join the ranks of middle-income countries by 2020.

The development challenges of Laos were, and are, abundant; but the country is working hard to overcome them. More children need to attend school, healthcare facilities need significant improvements, roads are needed to give access to communities, agriculture systems need to be upgraded and, crucially, the government needs revenues to invest in poverty reduction programmes. Moreover, these revenues must be well managed so that they can be utilized effectively.

The case for hydropower

In order to tackle these challenges, the Government of Laos (GOL) and the international community have recognized a need to implement far-reaching reforms in the public financial management sector. Moreover, the government has highlighted the imperative of developing what is becoming its regional comparative advantage: hydropower.

Surrounded by neighbours in need of electricity, Laos is blessed with abundant water resources that it can tap and sell, thereby generating much needed revenues. But hydropower also brings social and environmental impacts that must be well assessed and managed. The sustainable management of the country's natural resources has the potential to generate revenues, reduce poverty, improve people's life and protect the environment. The key is to balance water resources and potential developments in Laos so that the country and its people benefit.

Back in 2002, the Nam Theun 2 Hydroelectric Project (NT2) was an opportunity to undertake necessary reforms and to begin to strategically develop this important sector of the economy of Laos, ensuring that local communities benefited. The GOL reached out to the international community for support, including the World Bank, Asian Development Bank, Nordic Investment Bank, European Investment Bank and the French Development Agency (AFD). Together, the institutions, GOL and project developers designed the preparation for a project that would have impacts across the board.

UNDERTAKING IMPORTANT REFORMS

The basic premise behind NT2 was that it would help to generate much needed revenues that the GOL could invest in poverty reduction projects. However, to do this, the GOL had to strengthen its public financial management capabilities and its budgetary framework.

With the support of the development community, including several World Bank projects, the GOL has been undertaking a considerable number of important reforms in the last four years that are helping to establish a more rational and consistent public financial management system. From acquiring and storing data in a more comprehensive manner to implementing a strengthened and new budgetary law, the GOL has been upgrading and enhancing the way in which it manages the country's finances.

These reforms, complementary to the NT2 project, are having far-reaching implications for the future of Laos. The reforms that are being undertaken, the institutions that are being put in place and the capacity that is being built, among others, will not only be useful to manage NT2 revenues, but all revenues in the country and their allocation. This is a very significant impact that the project has had upon the country.

NAM THEUN 2

NT2 has brought with it the most rigorous social and environmental mitigation programmes that any project of its kind has had in Laos. These include detailed assessments of project impacts; a thorough consultation process with communities and national and international stakeholders; resettlement standards that led to the creation of a country-wide resettlement policy; the conservation of a 4000km^2 protection area that is helping to shape the country's thinking about conservation initiatives; appropriate compensation policies for people losing land to project areas; and mitigation measures and a programme for those affected downstream.

The social and environmental programmes in NT2 are helping to improve the lives of those affected by the project. In the case of the resettled villagers, the project is helping the communities to overcome poverty by raising their living standards above the country poverty line five years after relocation. In the downstream, the livelihood restoration programme is working to ensure that people's lives are not degraded as a result of the project and that they can maintain their living standards.

Moreover, the NT2 preparation process and its current implementation are raising the standard of hydropower development in Laos, as well as building significant capacity across a range of ministries and the project area. NT2 is also a catalyst in framing the discussions about hydropower in the country.

THINKING SUSTAINABLY ABOUT HYDROPOWER

The discussions on hydropower taking place in Laos today involve communities who are affected, national and international civil society, international donors, academia, developers and GOL officials. Many of these discussions centre on the aim of not only developing projects and generating revenues, but developing them sustainably, ensuring that communities benefit and that the environment is protected.

Discussions also centre on the need to attract investments from responsible private-sector partners, as exemplified in discussions during late 2007 at the first Lao–Thai Hydropower Forum, as well as at various other forums. Laos understands the importance of engaging with a private-sector partner who can help to improve the lives of affected people and the environment.

THE SUSTAINABLE DEVELOPMENT OF HYDROPOWER IS AT THE TOP OF LAOS'S AGENDA

Moving forward

The preparation and implementation of the NT2 project in Laos has triggered a number of reforms of economic and social processes, which are slowly transforming various areas in Laos. The impact of the project goes beyond the actual development of the dam and its social and environmental mitigation programme to encompass the broader effects of the project on Laos's public financial management system and natural resource extraction.

Laos's natural resources have the potential to help its people to overcome poverty and therefore need to be sustainably managed. This will require the involvement of all stakeholders to ensure that this potential is met and that the Lao people lead better lives.

Going forward, Laos needs to carefully consider and balance its abundant resources, evaluating the potential to develop them, and the impacts and benefits to all of its population and future generations. Laos, thus, will need to ensure that potential projects are well assessed, that they are carefully reflected upon, that social and environmental programmes are well implemented, that local populations benefit from the projects, that investments are sound, and that the country generates the revenues which it needs, allocating funds to poverty reduction projects.

NOTE

1 For more information about Laos, the World Bank's programme in Laos and the Nam Theun 2 Project, visit www.worldbank.org/lao and www.worldbank.org/laont2.

5

Damming the Salween River

Darrin Magee and Shawn Kelley

Introduction

The Nu-Salween River[1] is one of Asia's principal rivers, the source of livelihood for an estimated 6 million people in China, Myanmar/Burma and Thailand (IUCN et al, 2003). The mountains and valleys of the watershed are home to some of the most culturally and biologically diverse areas of the world. Over its 2800km course the river drops some 5000m, much of that in steep gorges, making the Nu-Salween extremely attractive from a hydropower development perspective (Magee, 2006b). Until recently, the remoteness and lack of basic infrastructure throughout much of its watershed made such development technically and economically infeasible.

All of this has changed, though, as regional economies have grown and electric power shortages have become acute, especially since 2003. China's plans to construct hydropower installations on the Yunnan portion of the Nu originally emerged during the early 1990s. Proponents of large hydropower development on the Nu argue that such development would ease the country's energy crunch while providing revenues to areas highly dependent upon central government subsidies for local governmental operations. Yet, whereas the dams on the neighbouring Lancang seem largely unalterable, the Nu cascade has seen a much greater tide of international and domestic criticism, which doubtless played a role in the central government's decision to suspend the projects in early 2004.

Further downstream, Thailand's plans to build dams in neighbouring countries were first proposed 30 years ago, but gained renewed momentum amid Thailand's foreign investment and liquidity boom of the late 1980s, when the then government of Prime Minister Chatichai Choonhaven first aired the idea to turn the Mekong region 'battlefields into marketplaces'. As Thai capital moved abroad seeking opportunities in Cambodia, Laos and Myanmar, opposition movements within

Thailand demanding greater accountability and transparency in infrastructure development became a serious force posing a nuisance for developers at home. By the late 1990s, demonstrators had successfully blocked the construction of various dams and other industrial projects across the country, even as Thailand's neighbours were showing greater enthusiasm for foreign hydropower development.

Like their counterparts in Thailand, but only later, Chinese dam development companies are now poised to export not only electricity, but also dam-building expertise and capital throughout mainland Southeast Asia and even further afield to Africa and the Middle East. The most influential of these companies, once part of the Chinese government's Ministry of Electric Power, now raise capital on foreign financial markets and bring 'made in China' technologies to projects over which other lenders may bulk for technical, political or economic reasons. Myanmar, as we discuss below, is home to several of those projects, and Chinese companies are making inroads there and throughout the Lower Mekong watershed.

We begin with an overview of the Nu-Salween watershed and then provide details of the projects planned for the Chinese and Myanmar stretches of the river, as well as of the principal actors in the three countries involved in surveying, designing, financing, constructing and operating the dams. We then sketch the decision-making contexts in which the dams are situated. Finally, we conclude with an assessment of leverage points in decision-making processes and modest recommendations for reducing ecological and socio-economic impacts while striving to meet regional energy needs.

Overview of the Nu-Salween watershed

In China, where the river has its source, the Salween is known as the Nu Jiang, or the 'Angry River'. From its headwaters on the Qinghai-Tibet Plateau at an elevation of nearly 5000m, the river tumbles southward between steep gorges through Yunnan Province before entering Myanmar, where it forms the border with Thailand for some 800km and finally empties into the Andaman Sea. Having begun its journey as a trickle of glacial melt 2800km upstream, the Nu-Salween swells to a muddy brown river several kilometres wide at its mouth, discharging an annual average of $1650m^3/s$ into the sea. Along the way, it drains a basin of approximately $271,914km^2$ in area (IUCN et al, 2003).

Given its remoteness and limited infrastructure, the socio-economic situation of much of the Nu-Salween Basin is rather poor. In China, both the Tibet Autonomous Region (TAR) and Yunnan Province have been targeted since 1999 by the state's Western Development Campaign, which seeks to address gaps in economic development between China's western interior and its eastern seaboard. Many of the campaign's initiatives involve basic infrastructure construction, so-called ecological construction to re-engineer previously engineered environmental degradation, education and social development. All four counties of Nujiang prefecture in north-western Yunnan are designated as national-level poverty counties, the governments of which derive the bulk of their revenues from central government subsidies. Aside from sporadic non-ferrous mining operations in the region, most economic activity is agricultural. Principal crops include maize, rice, wheat, buckwheat, sorghum and beans, as well as rapeseed (canola) and Tibetan barley.

The Nu Valley is one of the most ethnically diverse areas of China. Yunnan, as a whole, is home to significant populations of Yi, Naxi, Bai, Zang (Tibetan), Dai and a number of other ethnic groups (Magee, 2006a; McDonald, 2007). In 2003, the United Nations Educational, Scientific and Cultural Organization (UNESCO) added the Three Parallel Rivers area (including portions of the Nu watershed) to its World Heritage list (UNESCO, 1992–2008). Within Yunnan, the watershed is home to approximately 5 million people, many of whom are subsistence farmers. The area is also extremely rich in biodiversity, with an estimated more than 12,000 species of plants alone, some 3500 of which are endemic. Conservation International includes the entire Nu Valley as a part of its Mountains of Southwest China Biodiversity Hotspot (Conservation International, 2007). One study (Xu and Wilkes, 2004) identified livelihood activities such as fuelwood collection, agriculture and livestock grazing as primary threats to biodiversity in the area.

For the first 1400km of its journey, the shallow and braided Nu winds its way through high mountains and plateaus, with wide valleys in southern Qinghai and eastern Tibet, narrowing and deepening as it approaches north-western Yunnan. Over its 621km course in Yunnan, the Nu drops 1116m, making it extremely attractive for hydropower development.

Exiting Yunnan, the Nu (now Salween) enters the Shan State of Myanmar before continuing on through Karen (Kayan) and Mon. Here, as in China, the river traverses remote regions populated principally by ethnic minorities, many of whom are subsistence farmers who depend for a large portion of their livelihoods upon the Salween River and its related ecosystems. As detailed below, armed militias in many of these areas are openly hostile to Myanmar's ruling military junta, a situation that further strengthens the junta's resolve to pacify (at least partially) the region through large hydropower projects that will flood much of the bottomland areas and disrupt the lives and livelihoods of ethnic communities.

DEVELOPMENT PLANS

Upstream (China)

Construction of large hydropower in western China is a central component of national-scale discourses of development, most importantly the Great Western Development Campaign. Supportive policies such as Send Western Electricity East (*xidian dongsong*) and Send Yunnan Electricity Out (*Yundian waisong*), as well as the creation of entirely new regional constructions such as the Pan Pearl River Delta (stretching from Shanghai to Yunnan) help to legitimize and even *naturalize* large-scale power generation and transmission infrastructure. Discourses of power – and the power of those discourses – resonate loudly internationally as well. The Greater Mekong Sub-Region (GMS), which did not exist as a geographic entity before the Asian Development Bank (ADB) created it in 1992, has now become the backdrop for Mekong region-wide infrastructure development, such as highways and power grids.

Initial calls for a hydropower cascade on the Nu came as early as 1995 (Wei, 2005); yet serious planning and surveying did not begin until 2001. As plans began to coalesce, construction on a similar hydropower cascade on the neighbouring Lancang (Upper Mekong) was already under way. Hydrolancang, the Lancang developer, negotiated an agreement in 1998 for Thai co-financing on the Jinghong Dam, at the time planned as a 1500MW project.[2] In exchange for a 70 per cent stake, Thailand would receive all electricity produced for the first two years of the dam's operation, then projected to be 2013 to 2015. Interviews with officials in China in 2005, however, revealed that the dam is now being built without Thai financing due to an accelerated development timeline; the first turbine of Jinghong, now designed for a total 1750MW, came online in June 2008.

Thus far, the joint venture model has not been openly discussed as an option for the Nu dams within China. On the Lancang, the developer's model of 'rolling development', where power and revenue generated by one dam are used to build the next, has been supplemented heavily by Chinese central bank loans, which provide up to 75 to 80 per cent of the capital. It is likely that the controversial nature of the Nu dams will make foreign investors, already deterred by long construction periods, high start-up costs and delayed returns on investment, even less interested in investing. Moreover, given the Nu's distance from key load centres such as Guangdong, significant start-up costs and delays will probably result from solidifying grid infrastructure to efficiently and safely transmit power over such long distances. To this end, China has become a world leader in ultra-high voltage (800kV) direct current (DC) transmission lines.

Central authorities delegated survey and design work for the Nu cascade to the Beijing Institute of Hydropower Survey and Design and the East China Institute of Hydropower Survey and Design. Plans were submitted in July 2003 as the *Middle and Lower Nu River Hydropower Planning Report*. Supporters cited practical advantages of developing large-scale hydropower on the Nu, including the river's steepness and the relatively small number of people who would have to be resettled, estimated at some 50,000 (He and Feng, 2004). Development costs are also expected to be low relative to other large hydropower projects, which will presumably result in low prices for electricity sold to the grid and, in turn, to end users.

Environmentalists and cultural preservationists have criticized the dam plans as threatening to the cultural and biological diversity of the area, and have repeatedly made appeals to save one of China's last 'undammed' or 'virgin' rivers (even though there are already two completed dams and a third under way on the upper reaches in Tibet). Semantic questions notwithstanding, the most substantive critiques of the projects have questioned the decision-making processes leading to their apparent initial approval, as well as the gaps in those processes that seem to allow developers to skirt laws regarding environmental impact assessments and public input, and to avoid oversight from relevant watershed authorities.

In response to domestic and international outcry, the Nu projects were suspended by Premier Wen Jiabao in 2004, officially for failure to comply with environmental reporting requirements. Over two years later, then Minister of Water Resources Wang Shucheng referred to the 13-dam cascade as a case of 'predatory development'[3] in a speech in Hong Kong (Xiang Gang Shangbao, 2006). Due to their controversial nature, there is limited publicly available information regarding the Nu dams. We provide here only a preliminary sketch of the projects, based on close examination of Chinese and Western sources, recognizing that details such as capacity, location and timelines may change. Figure 5.1 indicates the approximate locations of the Nu-Salween hydropower projects; Table 5.1 provides basic information about each dam, including map abbreviations. The final status of the cascade is still uncertain, so our grouping of certain dams as more or less likely may be inaccurate or premature.

The original two-reservoir, 13-step (*liang ku shisan ji*) plan called for two dams with major reservoirs in a cascade totalling 13 dams. Large reservoirs provide multi-seasonal regulation (storage capacity), enabling more consistent power generation even in the dry season. Two of the Nu dams, Songta and Maji, were designed with reservoirs of 6.3 and 4.7 billion cubic metres in capacity, respectively. Preliminary work has already begun at the two sites. A report in late 2004 claimed that Songta would probably be one of the projects approved in a 'slimmed-down version' of the Nu development plan (Cheung, 2004). Once Songta and Maji are built, the economic logic of filling in the gaps by building the smaller projects downstream becomes more compelling.

Yabiluo (1800 MW) and Maji (4200 MW) made early headway toward central government approval. According to the 2003 plan, those projects, along with Bijiang, Lushui and Yangsangshu, were to be completed between 2015 and 2020 (He and Feng, 2004). Such expectations were later scaled back; in 2005, Minister of Water Resources Wang Shucheng suggested one or two dams were likely to be approved in the short term (Ma, 2005). Most observers understood that to include the cascade's smallest dam, Liuku, work on which has been under way in conditions of questionable legality since 2006. At 180MW of installed capacity, Liuku ranks as a medium-sized dam in China, and will almost certainly supply power locally. According to recent media reports, the resettlement sites for housing villages moved from near the dam site are already constructed. The same source reported that many villagers protested the terms of resettlement (Shi, 2008). Meanwhile, the Yunnan subsidiary charged with developing the Nu met in 2006 with the Beijing-based

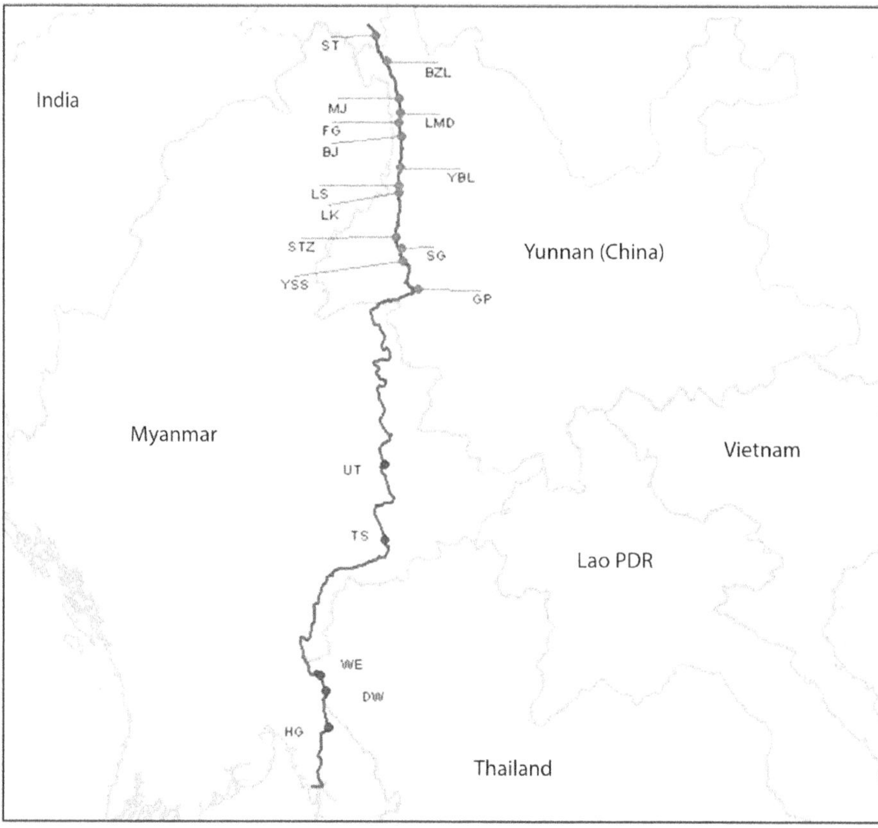

Figure 5.1 *Approximate locations of the Nu-Salween hydropower projects (including proposed and ongoing projects)*

Source: chapter authors, based on Magee (2006b)

Institute for Water Resources and Hydropower Research to discuss turbine design for the 1000MW Saige Dam (IWHR Office, 2006). Saige and Liuku were both cited in the *11th Five-Year Plan (2006–2010)* for renewable energy development as key projects to be pushed forward by 2010 (Yundian Xinwen, 2008).

Given the controversy surrounding the original design, a smaller one-reservoir/four-step design (*yiku siji*) has been suggested as a compromise. This would comprise Liuku, Yabiluo, Saige and Maji, with Maji the major upstream reservoir for the other three dams. Aside from the four projects that seem most likely to proceed in the near term, and with the possible exception of Songta on the Tibetan side of the Yunnan–Tibet border, nine dams remain on the drawing board. Of those, all but three have planned installed capacities greater than 1000MW, meaning they will play an important role in electricity transfers out of Yunnan over the coming

Table 5.1 *Basic information about the Nu-Salween projects (including map abbreviations)*

Dam name	Map abbreviation	Projected installed capacity (MW)
Songta	ST	4200
Bingzhongluo	BZL	1600
Maji	MJ	4200
Lumadeng	LMD	2000
Fugong	FG	400
Bijiang	BJ	1500
Yabiluo	YBL	1800
Lushui	LS	2400
Liuku	LK	180
Shitouzhai	STZ	440
Saige	SG	1000
Yangsangshu	YSS	100
Guangpo	GP	600
Upper Thanlwin	UT	2400–3000
Tasang	TS	7000
Weigyi	WE	4540–5600
Dagwin	DW	500–900
Hutgyi	HG	1190

Source: Nujiang Lisu Authonomous Prefecture Government (2005)

decades. In an interview, one development executive noted that he expected a small number of dams would first be approved, with the others probably following.

One of the principal arguments against Nu hydropower development is that a number of the dams lie adjacent to the Three Parallel Rivers UNESCO World Cultural Heritage preserve. Activists and academics are concerned that some of the reservoirs would threaten the preserve, suspected to be one of the greatest concentrations of biodiversity in the world (Fan, 2005). Others, however, counter that the elevation of the reservoirs lies below that of the preserve and therefore the impact will be limited (He and Feng, 2004).

Downstream (Myanmar, Thailand)

As early as 1981, the Electricity Generating Authority of Thailand (EGAT) had studied potential hydropower development on the Salween (EGAT, 1981; TERRA, 2006), and Thailand's economic boom that followed a few years later gave the idea stronger impetus. Preliminary studies commissioned by Thailand and Myanmar and conducted by Japan's Electric Power Development Company during the early 1990s identified about ten potential dam sites on the Salween. But the economic crisis in 1997 sidetracked those plans by bankrupting Thai developers and raising new questions about the viability of investing massive amounts of public and private funds in foreign megaprojects (Greacen and Palettu, 2007). At that time, Thailand's relations with Myanmar's State Peace and Development Council (SPDC) government were growing increasingly strained over a number of security issues: the SPDC's alleged complicity in the production and distribution of the methamphetamines and heroin that enters Thailand, boundary disputes, the ethnic insurgencies and refugees. The resulting tensions led to occasional armed clashes on the border, and at least one Thai military offensive well inside Myanmar (Pathan, 2005).

While some Thai-built dams in Laos are already in operation, building on the Salween is proving more problematic. One reason is that the sites will be in outlying areas that the central Myanmar government has never fully controlled. Myanmar's key load centres are concentrated in the central and lower parts of the country; thus, hydropower development has mainly been in these regions close to the national grid (Myanmar Department of Hydropower Planning, 2006). But an estimated 60 per cent of Myanmar's hydropower potential, including the Salween dams, lies in the more remote central and eastern hills region, mostly in the Karen and Shan states (Bartle, 2005), home to an array of organized ethnic insurgents, pro-government militias and smaller private guerrilla units. As Myanmar's army gradually asserts control over its hinterland and weakens its political opponents, however, and as its ambitious hydropower development programme gathers pace, plans to dam the Salween River look more realistic. Political stability remains a concern; but construction of the dams and the resulting flooding and dislocation of people around them would probably deliver a crushing blow to the ethnic insurgencies.

Myanmar's limited financial and technical capacity has also hampered progress and underlines the necessity of foreign assistance. Japanese war reparations financed construction in 1960 of Myanmar's first major hydropower station in Karen State on the Baluchaung River, a tributary of the Salween (Japan Ministry of Foreign Affairs, 2002). Japanese loans and emergency assistance have since covered critical repairs and maintenance costs. During Myanmar's socialist period and self-imposed isolation beginning in 1962, the government did not build another major dam until it completed the Kinda Dam in 1985, with Japanese public funding, and the Sedawgyi Dam, completed in 1989 with ADB loans (ADB, 1989; Myanmar Department of Hydropower Planning 2006), both in Mandalay Division. According to one member of a survey team from MDX Group, a privately owned Thai developer, MDX was among the first callers to explore potential large-scale hydropower development opportunities in Myanmar following the Myanmar government's decision to open its door to foreign investment in 1988.

Financing large dams, however, proved problematic. Thai firms lacked the funds and access to capital required for these big-ticket projects. And unlike in Laos, where the World Bank and ADB have backed various dam projects with grants, loans and technical assistance (ADB, 2008a), Myanmar is not eligible for similar assistance (World Bank, 2008). The ADB has not extended loans to the country since 1986, and bilateral technical assistance ended in 1987 (ADB, 2008b). Similarly, the World Bank has approved no new lending for Myanmar since 1987, citing defaulted payments and lack of reforms (World Bank, 2008).

Moreover, as part of its economic sanctions package against Myanmar's military rulers and their associates in response to the large-scale repression and violence directed against their political opponents, the US government is required to 'vote against the extension of any financial assistance to Burma by international financial institutions' (US Government, 2003; Niksch and Weiss, 2008). First imposed in 1997 and tightened in subsequent years, restrictions also include a ban on imports from Myanmar and the prohibition of investment by US companies there, as well as a freeze on assets of companies and individuals linked to the junta. Fearful of a backlash from US equity markets, some banks in Singapore and China have recently ceased dealing with some firms and banks linked to the military (Lwin, 2006; Levett, 2007). Additionally, the European Union has adopted similar, if weaker, restrictions against Myanmar (Council of the European Union, 2007).

Given these legal complexities and political sensitivities for Western firms and lending institutions, the entry of Thai and Chinese developers and financiers has given Myanmar's hydropower regime a needed boost. Thai energy planners, for instance, recently identified the Salween as the 'most favourable' location for transboundary hydropower development, notwithstanding security and political concerns (EGAT, 2003). In August 2003, China approved a US$200 million loan for the 790MW Yeywa Dam project (Myanmar Ministry of Foreign Affairs, 2003; Bosshard, 2004), currently Myanmar's largest hydropower facility. Yeywa is being built by a consortium of Chinese companies that includes China's Gezhouba, which also reportedly is contracted for part of the construction work at Tasang (*International Water Power and Dam Construction*, 2007a).

The Salween dams will generate electricity for export and for the domestic market, where the country's notoriously unreliable power supply causes daily power outages even in its largest cities. The regime estimates that it has so far tapped only 1 per cent of its total hydropower potential (Myanmar Department of Hydropower Planning, 2006), which currently produces roughly one third of the country's entire electricity output.

In order to tap the country's hydropower potential, the Myanmar regime, in 2002, restructured its Ministry of Electric Power and its Department of Hydropower (see 'Principal actors' section below), and also signed the Inter-Government Agreement on Regional Power Trade in the Greater Mekong Sub-Region Countries,[4] which the regime hoped could allow it to export power generated from the Tasang Dam and other planned projects to other GMS countries through the Asia Power Grid (Bartle, 2005). National development strategies, meanwhile, have placed hydropower at their centre. The country's first two five-year development plans (2001–2005 and 2006–2010) focus on hydropower growth to feed the domestic market. Its third five-year plan (2011–2015) outlines strategies to ramp up power trade with neighbouring GMS countries to the east and with India and Bangladesh (Myanmar Department of Hydropower Planning, 2006). Foreign companies are being invited in to form joint ventures with local partners to help finance and construct dams, typically with the condition that Myanmar is entitled to '10 to 15 per cent of annual electricity generation from the power stations free of charge' (Myanmar Department of Hydropower Planning, 2006). In December 2007, the completion of 13 hydropower projects in Myanmar was given priority over all other projects, including those in the increasingly significant oil and gas sector (Thu, 2007).

The Salween dams, however, were not included in that list as they are Chinese- and Thai-led projects, the output of which will be diverted abroad. But the dams are included in the roster of more than 40 projects that the regime hopes to commission in the coming years (Myanmar Department of Hydropower Planning, 2006). Owing to their political sensitivity, the five dams planned for the Salween have proceeded under a high degree of secrecy. The lack of a clear regulatory framework for hydropower development on the river, allegations of human rights violations conducted in preparation for the dams, and the potential environmental destruction that may result from their construction all raise further questions about the viability of the Salween projects.

Hutgyi

The first dam on the Salween targeted for construction is the Hutgyi Dam (sometimes also spelled Hatgyi, Hutgi or Hatkyi), a US$1 billion run-of-river power plant located in Karen State, some 33km downstream from the confluence of the Moei River at the Thai border. It is a joint venture project between Myanmar's Ministry of Electric Power No 1, the Electricity Generating Authority of Thailand and China's Sinohydro Corp (Thu, 2006). An initial feasibility study in 1999 conducted by NEWJEC (formerly known as New Japan Engineering Consultants, Inc.), a Japanese development consultant, recommended a low-height, run-of-river dam with a capacity of 300MW (Vatcharasinthu and Babel, 1999); but Thai official figures list the dam at 1190 MW, with some 75 per cent of the output to be delivered to Thailand starting in 2019 (EGAT, 2008).

In December 2007, Russian manufacturer Power Machines Company, which makes equipment for thermal, nuclear, hydraulic and gas-turbine power plants (Power Machines Company, 2006), entered the picture when it announced that its joint venture with Chinese firm Zhejiang Fuchunjiang Hydropower Equipment would deliver eight turbine units – seven at 170MW and one at 132MW – to the Hutgyi plant (*International Water Power and Dam Construction*, 2007b).

Thailand and Myanmar signed a memorandum of agreement in December 2005, which stated that EGAT would begin construction of Hutgyi in late 2007, and Sinohydro signed a memorandum of understanding (MoU) with EGAT and Myanmar's Hydropower Implementation Department (HPID) in 2006 for joint investment in the project (SHAN, 2006; Thu, 2006). But the parties hadn't worked out an investment model at that point and no other details were made public according to the terms of an earlier 2005 MoU, which states that 'each party shall strictly keep confidential any and all technical, legal and commercial data and information'. An environmental impact assessment (EIA) was conducted by the Environment Research Institute at Chulalongkorn University in Bangkok. The EIA, not required by Thai law, was criticized for downplaying the environmental and human impact and for making dubious claims about the extent of the opposition to the project by the local ethnic Karen. The institute was preparing a revised EIA before EGAT halted all work on the project in late 2007, after two of its staff were killed in just over a year. The deaths, allegedly by a landmine explosion and artillery ambush near the project site (*Bangkok Post*, 2007), prompted EGAT Governor Kraisi Karnasuta to shelve the project 'indefinitely' and then Energy Minister Piyasavasti Amaranand to urge Thai officials to expedite power development plans in Laos instead (Energy for Environment Foundation, 2006). The Myanmar government blamed the Karen National Union (KNU) for the attack, which it denied, pointing out that the Hutgyi site is located in territory controlled by its rival, the pro-junta Democratic Karen Buddhist Army (DKBA). In February 2008, KNU Secretary-General Mahn Sha, a staunch opponent of the dam, was gunned down in his home by unidentified assailants (Associated Press, 2008). The motives and the culprits behind these killings remain publicly unknown, while the episodes highlight the security concerns associated with the Salween projects.

Tasang

Tasang (sometimes spelled Tarhsan, or Tar-hsan) would be the largest dam in Southeast Asia, with a total capacity of 7000MW (EGAT, 2008) and annual generation of 35,446 million kilowatt hours (Xinhua News Agency, 2007). At a cost of US$6 billion, it would be the single largest investment ever in Myanmar. The dam site is located in southern Shan State some 130km northwest of the Thai border pass at Baan Arunothai/Nong Ook. Thailand is expected to purchase at least 85 per cent of the annual production generated by the plant; but no power purchasing agreement has been signed thus far.

Construction works on the project will include an 876m long, 230m high concrete dam, and two 8m diameter tunnels, the longest of which will stretch 1.2km (Bartle, 2005). The project's initial investors were Myanmar's Department of Hydropower Planning (DHP) and MDX Group at 15 per cent and 85 per cent, respectively (MDX PCL, 2007b). Myanmar's semi-official state press reported recently that China Gezhouba Water and Power Group bought a controlling 51 per cent stake in Tasang (Thu, 2007). But according to a senior executive of MDX Group and the company's filings to the Thai Stock Exchange, MDX still holds an 82.88 per cent stake, while DHP holds the remaining 17.12 per cent in the Tasang Hydropower Company Ltd, the operating company for the project.[5]

After several early studies of Tasang in April 2006, the Myanmar Ministry of Electric Power signed a development agreement for the project with MDX, with completion scheduled around 2020. A year later the official Myanmar media reported that implementation was under way for Tasang, now listed at 7110MW (Xinhua News Agency, 2007). Tasang was officially inaugurated on 30 March 2007, when officials from MDX and other involved parties cut the ribbon at the groundbreaking ceremony (New Light of Myanmar, 2007b); but heavy rains halted construction soon thereafter (Thu, 2007).

Tasang has also been constantly surrounded by accusations of human rights abuses and widespread environmental damage. A Shan advocacy group has said that over the past ten years, the Myanmar army has relocated more than 60,000 villagers from areas adjoining the dam site and the projected flood zone (SSEO, 2006). Other human rights groups have said the project would displace tens of thousands more from their homes in the Shan, Karenni and Karen states in Myanmar, as well as from Mae Hong Son Province in Thailand, and that others have been press-ganged into forced labour, raped and killed in preparations for construction of the dam. Preliminary feasibility studies required an increased military presence near the dam site (EarthRights International, 2005).

In 2002, the ADB studied the Tasang Dam as part of a master plan for a regional power grid, but backed away, citing 'serious socio-environmental concerns'. Rajat Nag, who heads the ADB's Mekong Department, told the Associated Press:

> *It didn't pass our first filter. The dam would have a profound impact on the Salween River. The project would fragment a fragile river ecosystem, reduce the flow of nutrients and water downstream and reduce the biodiversity. Deforestation is likely and would lead to soil erosion in the rainy season, which would exacerbate flood damage.* (Gray, 2006)

Upper Thanlwin

Myanmar's Hydropower Implementation Department signed an MoU in 2007 with Farsighted Investment Group Co Ltd, now Hanergy Holdings Group Company Ltd, and Gold Water Resources Co Ltd of China to develop the Upper Thanlwin Dam in northern Shan State, which will reportedly have an installed capacity of between 2400MW and 3000MW (New light of Myanmar, 2007a; Siripol, 2007). Its precise location is undisclosed. Also signatory to the agreement was Tun Myint Naing, managing director of Asia World; both the company and Tun Myint Naing have been barred from doing business with individuals or business from the US (US Department of the Treasury, 2008).

Weigyi

The Weigyi Dam will be located on the border in Papun district in Karen State, on the Myanmar side, and in the Salween Wildlife Sanctuary on the Thai side, with the access road cutting through the adjacent Salween National Park. The dam has a proposed height of 168m, an estimated power capacity of between 4540 and 5600MW (KDRG, 2006) and a price tag of US$3 billion to US$6 billion (Foundation for Ecological Recovery, 2003). The dam could create a reservoir, mostly in Karenni State, of between 640km^2 to 1000km^2 of forest, river and farmland, roughly the size of Singapore, affecting an estimated 30,250 people living in flood zones (Foundation for Ecological Recovery, 2003; KDRG, 2006). The status of this dam is unknown; but it is likely to be the third construction project, after Tasang and Hutgyi.

Dagwin

The Dagwin Dam site is also located on the border, just south of the Weigyi site, near Tha Ta Fang village, Mae Hong Son Province. The dam's projected capacity is variously given as 500, 792 or 900MW (Foundation for Ecological Recovery, 2003); but its main purpose would be to trap and regulate large amounts of water released by the Weigyi Dam during peak hours. It would use off-peak power to pump water back up into the upper dam. The estimated US$900 million cost and the fact that it has no practical water diversion route make this dam exceptionally impractical. Both the Dagwin and Weigyi dams appeared in EGAT's 2004 *Power Development Plan* (PDP) (EGAT, 2005) but not in its 2007 PDP (EGAT, 2008).

PRINCIPAL ACTORS

China

China's principal developers and exporters of hydropower expertise, capital and technologies were carved off the former Ministry of Electric Power (MEP). Some have referred to the 'privatization' of the former ministry and its subsequent state-owned enterprise; yet since most of the stock in these companies is still controlled by the central government's State Assets Supervision and Administration Commission, referring to the companies as 'private' seems premature. The power sector has seen extensive reforms since the mid 1990s, aimed at promoting better governance, increased competition, improved technologies and lowered tariffs (Xu, 2002; Yeh and Lewis; 2004; Magee, 2006a). One specific objective was the separation of generation and transmission facilities, all of which had, before 2002, been part of the State Power Corporation of China and its predecessor, the MEP. The 2002 reforms divided the generation assets of the State Power Corporation among five national-level generation companies. The National Development and Reform Commission (NDRC) then apportioned development rights on the country's rivers to those companies. Rights to the Nu went to Huadian. Yunnan Huadian Nujiang Hydropower Development Company Ltd, Huadian's subsidiary responsible for the Nu cascade, was established in June 2003 through joint investment from China

Huadian (51 per cent), Yunnan Development Investment Corporation (20 per cent), Yunnan Electric Power Group (19 per cent) and Yunnan Nujiang Electric Power Group (10 per cent) (Zhou, 2003).

Sub-national grid infrastructure was divided between two national-level grid companies, State Grid Corporation of China and China Southern Grid Corporation. Finally, four other national-level companies devoted to design, technological development, consulting and construction were created out of the restructuring. Sinohydro, a construction company, traces its lineage to the China National Water Resources and Hydropower Development Authority, founded after the establishment of the People's Republic of China (PRC) in 1949. The company has led the development and construction of some 80 per cent of the large- and medium-scale hydropower projects in China built since then, and is involved in an increasing number of international projects, including several in Myanmar. China Gezhouba Group, a design and construction company, derives its name and reputation from the first dam on the Yangtze and plays a significant role in building dams overseas. The corporation has spearheaded projects in over 30 countries in Asia and Africa, including the Tekeze Dam in Ethiopia and the Yeywa Dam in central Burma.

Thailand

The Electricity Generating Authority of Thailand (EGAT) was established in 1969 when three regional state-owned generating enterprises were consolidated as a single state enterprise under the Office of the Prime Minister, and is now under the Ministry of Energy (EGAT, undated). Responsible for electricity generation and transmission, EGAT builds, owns and operates thermal, hydropower and alternative energy power plants and operates the national grid. It also purchases electricity from private power companies and from neighbouring countries (EGAT, 2008), including two dams in Laos. Plans to privatize EGAT faltered in 2005; but some subsidiary companies have been spun off to the private sector, such as the Ratchaburi Electricity Generating Holding Company Ltd, although EGAT retains an approximately 45 per cent interest in the company (EGAT, undated). EGAT would be the main purchaser of electricity generated at Hutgyi and Tasang; but in the absence of power purchasing agreements for the two dams, its role in Myanmar is not clearly determined (EGAT, 2008).

Unlike at Hutgyi, where EGAT may act as the lead investor, the lead entity at Tasang is Thai developer MDX, established in 1988 'to invest in hydropower generating dam projects in the Greater Mekong Sub-Region' (MDX PCL, 2007b). It also expanded into public infrastructure works, industrial development parks and extensive real estate holdings, and was listed on the Stock Exchange of Thailand in March 1992.

The company is steered by its honorary adviser, Subin Pinkayan (pers comm, 23 January 2007), former minister of foreign affairs and minister of commerce, and a key architect of plans to open neighbouring markets to Thai companies. In 1989, as commerce minister, he was part of the government that announced it wanted to turn the Southeast Asian mainland into Suwarnabhumi, or a 'golden land', with Thailand as the regional centre of trade and finance. In 1997, he was ordered by the Supreme Court to pay US$10 million in back taxes on income that the court deemed he earned unlawfully as a minister between 1988 and 1990 (Boonlom, 1997). An engineer by training, he now promotes his regional vision as a private construction consultant and university lecturer. Other notable directors and shareholders of MDX-controlled companies include Subin's relatives, the editor of a major Thai daily newspaper, the relatives of the former head of the Royal Thai Third Army, which is responsible for northern Thailand, including a large portion of its border with Myanmar, and the relatives of a former Thai ambassador to Myanmar (MDX PCL, 2007b). Some American and European banks and investment funds also hold shares.

The company's troubled financial past set back its Tasang plans several years and raised questions about its ability to raise the necessary capital for the project through debt or equity financing. In 1996, the company defaulted in payment on US$100 million worth of dollar-denominated convertible debentures (MDX PCL, 2007a). By the end of 1997, amid the economic slowdown, the Stock Exchange of Thailand suspended trading of MDX. In 2004, the Central Bankruptcy Court ordered the company into rehabilitation. MDX then restructured its capital and debt and resumed trading on the exchange in August 2007, with the explicit aim of building the Tasang Dam (MDX PCL, 2007a).[6]

Myanmar

Myanmar's military regime established the Ministry of Electric Power in November 1997 and in May 2006 split the agency into two parts: the Ministry of Electric Power No 1, responsible for generation of electricity and hydroelectric power implementation, and the Ministry of Electric Power No 2, responsible primarily for transmission and distribution and gas-fired power implementation (Myanmar Department of Hydropower Planning, 2006). MEP No 2 is also tasked with restoring the national power grid and preparing it for the opening of the 790MW Yeywa plant, perhaps as early as 2010 (New Light of Myanmar, 2007a).

Under the MEP No 1, the former Department of Hydroelectric Power was renamed the Hydropower Implementation Department (HPID), and is tasked with planning, designing and constructing hydropower projects (Myanmar Department of Hydropower Planning, 2006). It also signs memoranda of understanding and of agreement, and joint venture agreements with foreign companies to develop new hydropower projects (New Light of Myanmar, 2007a). A second new unit, the Department of Hydropower Planning (DHP), manages the internal affairs of the ministry. A third unit, the Hydropower Generation Enterprise, has taken over operation of the existing network of larger hydro plants from the Myanmar Electric

Power Enterprise and is responsible for the installation and maintenance of power-generating equipment at hydropower stations (New Light of Myanmar, 2007a).

It is the Electric Power Development Project Lead Committee (or the Leading Committee on National Electricity Development), however, which has the ultimate authority over hydropower development. Under the direction of junta chief General Than Shwe and staffed with other high-level authorities, the committee coordinates dam construction with the line agencies and, importantly, controls the allotment of state funds (IED, 2007; Myanmar Department of Hydropower Planning, 2006). Strong centralized control has been a hallmark of the military regime and government contracts are often awarded to firms close to the country's ruling generals (Lintner, 2007; *The Economist*, 2008), including Asia World and Hongpang, which have both expressed interest in the Salween projects.

Myanmar's largest construction company, Asia World Co, was founded in 1992 by Lo Hsing Han, a Kokang Chinese from the opium-producing region of Myanmar's Golden Triangle who controls one of the largest armed drug trafficking gangs in Southeast Asia. The company has received numerous government construction concessions and was one of the two major contractors to build the new capital at Naypyidaw (Lintner, 2007). In April 2007, its managing director and Lo's son, Tun Myint Naing, signed an MoU on the implementation of the 2400 MW Upper Thanlwin Project with Farsighted Investment (now Hanergy Group), Gold Water Resources of China, and the HPID director general (New Light of Myanmar, 2007a). Washington has accused both Lo and Tun of 'having a history of illicit activities that supported Myanmar's junta' and banned Americans from doing business with Asia World and ten Singapore-based companies owned by Tun's wife (US Department of the Treasury, 2008).

Hongpang General Trading Co Ltd is similarly blacklisted by Washington for its close association with a United Wa State Army commander, Wei Hseuhkang, the reputed founder of the company, who was indicted, along with seven other Wa leaders, by a US court in 2005 on heroin and methamphetamine trafficking charges (US Drug Enforcement Agency, 2005). Founded in 1998, Hongpang is involved in a range of activities, including manufacturing, agriculture, gem mining and highway construction (SHAN, 2005).

Civil society

In addition to state and business actors, a number of what might be called civil society actors within China and Thailand have also become involved in the Nu and Salween dams debates. These include several 'civil society' organizations, segments of the media, and a number of academics. Public discussion of the projects barely exists in Myanmar, so most civil society actors from there work on the Salween with local and international organizations in Thailand. Broadly speaking, these individuals and groups seek to raise public awareness of river conservation, cultural and biological diversity protection, and socially and environmentally responsible energy development, aiming to affect the direction, magnitude and pace of energy resources development on the Nu-Salween.

Perhaps the greatest hurdle that these actors face lies in promoting their messages of conservation, preservation and socio-environmental responsibility without being seen as opposed to economic development in areas where conditions of extreme poverty frequently prevail. Hydropower developers in China have made a rock-solid connection, through media and governmental channels, between large dam development and poverty alleviation. For the Salween, the correlation has been less clearly or strenuously expressed. The river for most Thais invokes notions of a remote and dangerous frontier and few pay it any heed. Thus far, civil society organizations have failed to successfully (or convincingly) articulate compelling alternatives, with the exception, perhaps, of ecotourism, that would provide comparable economic development benefits without compromising environmental integrity or biological or cultural diversity. In a 2005 interview conducted by Magee, a Chinese hydropower development official argued that the infrastructure improvement required for ecotourism development – namely, in roads, bridges, water, electricity, waste management and lodging – would be greater than those required for dam development, with financial returns far lower.

Governance

China

The history of hydropower development leaves little reason to believe that decision-making processes about dams are always (or even usually) rules based. Yet, understanding how decision-making processes have shaped the trajectory of hydropower development on the Nu River, in a context of enterprise restructuring, industry reforms, loosening of political controls, and increasing engagement of China with its neighbours on resource development projects, paves the way for identifying leverage points in those processes. In this section we outline the overall contours of decision processes, recognizing that we have surely overlooked numerous subtleties and cannot hope to capture all the nuances, personal relations and backdoor deals that help to move projects from the drawing board to the river.

The Nu case has been characterized by sustained debate, influenced by past experiences with the Three Gorges Dam and, more recently, the Lancang River hydropower cascade and the Dujiangyan–Zipingpu–Yangliuhu case in Sichuan (see Mertha and Lowry, 2006). Several dynamics have complicated decision-making processes. First, recent changes in the Chinese legal system have given greater voice to social organizations to challenge development projects. Such challenges increasingly rely on new Chinese laws regarding environmental impact assessments, pollution, resource extraction or resettlement compensation. At the same time, these apparent gains in transparency and pluralism have been offset by the Chinese government's sporadic tightening of restrictions for reasons of social stability and national security. Thus, despite new regulations in 2003 calling for public input on EIAs, authorities insist that the Nu's status as a transboundary river means that detailed hydrological data on the river are a national security concern, and therefore that the EIA cannot be made public.

National security arguments notwithstanding, major river development projects in China should theoretically be subject to a fairly straightforward approval process that begins with one of seven river basin commissions and ends either with the same commission or, in the case of 'major' or transboundary rivers, with the NDRC and the State Council. In the case of the Nu, the Changjiang (Yangtze River) Water Resources Commission (CWRC), which holds authority over all rivers in south-western China, would first develop a comprehensive plan for the basin, covering everything from shipping and transportation to hydropower and forestry. Next, the developer, in conjunction with design institutes, conducts a pre-feasibility study and submits the results to the basin commission to check for compliance with basin-wide priorities. The plan then proceeds to the design stage, with the resulting full feasibility study and detailed design report submitted to the basin commission. Final approval from other authorities, including the provincial government and even the NDRC and State Council, may be necessary in certain cases.

In interviews, CWRC officials lamented that for rivers such as the Nu, hydropower planning often leads comprehensive planning, rather than the opposite. Moreover, the commissions lack oversight on projects outside China, even those on rivers that have part of their watersheds in China. Instead, relevant foreign affairs bureaucracies (such as the Ministry of Commerce and the Ministry of Foreign Affairs) and the NDRC take the lead in approving or denying projects, based primarily on political and economic considerations. The practice by which Chinese companies form consortia expressly for bidding on specific foreign projects reinforces this since the transactions would have to be approved by a number of central government departments, especially since companies such as Sinohydro and Gezhouba still have a majority of their stock owned by the central government's State-owned Assets Supervision and Administration Commission.

Thailand

The Salween projects were planned during the period of Thailand's rapid economic growth to meet EGAT's forecasts of rising energy demand and to diversify energy sources away from imported natural gas. But given EGAT's status as a self-regulating monopoly utility, it has an incentive to overestimate demand, while its planning process is susceptible to political intervention and other conflicts of interest (Greacen and Palettu, 2007). Critics of EGAT's decision processes also say that it is highly centralized and lacking public participation, and that it neglects alternative energy investment in favour of unnecessary and inefficient mega-power plants (Greacen, 2006).

Thai energy officials have already signed various agreements with their Myanmar counterparts for the Salween projects, and have conducted feasibility studies and discussed transmission systems between the two countries. The dams are included in EGAT's PDP for 2003–2016, which plans for Hutgyi to come online in 2012. But the dams continue to encounter financial and political setbacks. Work sites and roads are being built at Tasang and Hutgyi, although further construction is unlikely in the absence of a power purchase agreement.

Adequate impact studies are also lacking. Although the Thai Ministry of Natural Resources and Environment requires environmental impact studies for domestic hydropower development, cross-border projects are subject to the laws of the host country. Environmental regulations in Myanmar are undefined, and the lead agency for impact studies, the National Commission for Environmental Affairs, lacks a clear institutional framework or the political muscle for environmental management (Habito and Antonio, 2007). Formal regulation of hydropower through the GMS initiatives is similarly absent.

Opponents of the dam say that Thailand's laws require public disclosure of project details and stakeholder input at public hearings, and call for work on the Salween to cease until these conditions are met. Further development on the Salween, say military officials interviewed on the border, could inflict insuperable ecological and population pressures on large swathes of northern Thailand. Until the questions surrounding the viability of the projects are addressed publicly, the Salween dams will continue to be dogged by harsh criticism and sustained debate.

Conclusion: Leverage points

Following the Asian financial crisis during the late 1990s, and with increased public outcry about hydropower externalities and performance versus predictions, large hydropower came under intense international scrutiny. Major international investors and risk insurance providers such as the World Bank were accused of prioritizing macro-economic development goals over more targeted projects that were sensitive to local socio-economic, cultural and ecological conditions. The World Commission on Dams (WCD) report in 2000 provided a harsh indictment of many of the world's large dams, indicating that most fell far short of their power and revenue generation targets, including the Pak Mun Dam in northeast Thailand (see Chapter 3). In Southeast Asia, the ADB openly refrained from investing in large hydropower projects due to their controversial nature (although ADB did provide funding for several related projects, such as transmission lines and feasibility studies). More recently, both multilateral banks have signalled their intention to re-engage in hydropower.

Now, however, with demand for electric power in China and throughout the GMS on the rise, and increasing concern about fossil fuel-based generation systems, large hydropower is enjoying a comeback with or without the traditional lenders. This is partly eased by the perception of hydropower as a clean and renewable energy source that stands to be reinforced as global prices for fossil fuels rise. Chinese developers equipped with technology, expertise and growing clout within China and abroad are spearheading hydropower developments in Africa, Asia and the Middle East. Hydropower deals are often part and parcel of 'development packages' that guarantee Chinese access to natural resources vital to China's continued economic development. Much of this is similar to the zeal with which the United States promoted large dams around the world following World War II. To be sure, consultants, developers and funders from Europe and elsewhere in Asia are involved in hydropower development in the GMS region; but China is emerging as the major player.

Understanding processes is the first step in influencing them. This is the primary motivation for our research into the Nu and Salween dams. While ours is but a preliminary sketch about the projects, major actors involved and processes through which the projects are designed, funded and approved, we conclude with several observations about potential leverage points in those processes.

Ironically, the greatest potential for exerting constructive influence may lie in China. The majority of the Nu projects are currently stalled, and while it is likely that some will go forward, there is also a chance that others will be shelved. Significant efforts are currently under way to rethink development and operation of large dams on China's major rivers, especially since many of the flood control objectives can be met through wetland preservation/rehabilitation and spillway management. Similarly, many of the power provision objectives may be alternatively met through end-use conservation. For those projects that do go forward, new laws (e.g. the EIA law and its public participation requirements, resettlement laws, etc.), greater sophistication in the legal profession, and gradually increasing transparency in decision processes, may help to maximize benefits and reduce the negative social and ecological impacts of the dams. Additionally, the past four years of debate have opened the door for increased consideration of scientifically informed development alternatives for western Yunnan, as well as for sustained public pressure for procedural justice (i.e. adherence to EIA processes and public participation requirements).

In our estimation, two things are crucial to the success of any alternative proposal for development in south-western China. First, the notions of 'sustainable' (*kechixu de*) and 'scientific' (*kexue de*) must be decoupled. These terms are frequently conflated in China; yet the latter in no way implies the former. 'Scientific development' figures prominently in China's *11th Five-Year Plan* (2006–2010), and projects deemed 'scientifically' (technically) sound are frequently assumed to be sustainable as well; this is especially true given hydropower's 'green' reputation. Second, evidence must be provided showing that alternative development projects will match or exceed hydropower projects' ability to achieve poverty alleviation goals. Admittedly, amassing such evidence from experience elsewhere is difficult. As noted above, though, the link between large hydropower development and poverty alleviation has been cemented in development discourse in China, and any alternative proposal must clearly demonstrate the frailty of that link.

In Thailand, civil society groups have long called on EGAT and other major power producers to improve standards of transparency and accountability, but with limited success. For instance, while public pressure forced EGAT to shelve its privatization efforts in 2006, it has failed to compel the company to decommission controversial dams and undertake other reforms. As for the Salween dams, Thai energy planners have been urged to reconsider large-scale hydropower development, in general, and to open the energy sector to more small and independent power producers. Decoupling generation from transmission and improving the regulatory environment would also help to depoliticize energy planning and encourage greater efficiency and conservation, thus rendering the output from the Salween dams unnecessary. Improving standards of public disclosure and stakeholder input must first be improved to foster informed dialogue among concerned parties. The Chinese government, for its part, has long been one of the few allies of the Myanmar junta, and arguably wields the greatest degree of influence with the reclusive military regime. Yet, trade and development practice, along with sovereignty concerns, make it unlikely that the Chinese government would require, or that Myanmar would accept, environmental impact assessments based on Chinese standards for projects on foreign soil. That said, most of the dams discussed here are in areas where even the Myanmar government lacks firm control; indeed, its hydropower pursuits are part of its broader nation-building programme. If the political and economic costs become too high, however, it is conceivable that the Chinese developers might pack their bags and head for friendlier sites in Africa or the Middle East.

NOTES

1 In the Burmese language, the river is known as the Thanlwin. Here we refer to the Chinese (including the Tibetan) portion of the river as the Nu, and the Myanmar/Thai section as the Salween.
2 For more on the Lancang cascade, see Magee (2006a, 2006b).
3 Technically, Wang was speaking in his personal (not official) capacity; but his comments probably evince some frustration about the Ministry of Water Resources' relative lack of influence on the direction of large-scale hydropower development (vis-à-vis the National Development and Reform Commission and the development companies).
4 The GMS includes Cambodia, China, Laos, Myanmar, Thailand and Vietnam.
5 Myanmar's Hongpang General Trading Company Ltd has also expressed interest in participating in the project (Thu, 2007), presumably to build roads leading to the site.

6 According to a company filing to the Stock Exchange of Thailand, MDX shall first invest in 'the Tasang Hydro-Power Project in the Union of Myanmar, which shall be run by a joint-venture company under establishing [sic] in the Union of Myanmar, with registered capital of US$250 million, comprising 2.5 million ordinary shares at the price of US$100 each.' As of 23 June 2008, the Stock Exchange of Thailand listed MDX's registered capital at US$44 million (1407 million baht) (Stock Exchange of Thailand, 2008).

REFERENCES

ADB (Asian Development Bank) (1989) *Project Completion Report of the Sedawgyi Hydropower Project*, Asian Development Bank, www.adb.org/Documents/PCRs/MYA/36112-MYA-PCR.pdf

ADB (2008a) *Approved Projects, Lao PDR*, www.adb.org/Projects/approvals.asp?ctry=LAO&year=ALL&query=&browse=l&p=ctryLAO, accessed July 23 2008

ADB (2008b) *Asian Development Bank and Myanmar: 2008, A Fact Sheet*, www.adb.org/Documents/Fact_Sheets/MYA.pdf, accessed 15 July 2008

Associated Press (2008) 'Myanmar ethnic leader shot dead', Associated Press, 14 February

Bangkok Post (2007) 'Thai dam project in Burma suspended', *Bangkok Post*, 4 September

Bartle, A. (2005) 'Hydropower plays a leading role in Myanmar's power development plans', *Hydropower and Dams*, no 2, pp119–123

Boonlom, O. (1997) 'Court verdict lends new twist to case on back taxes', *The Nation* (Bangkok), 2 September

Bosshard, P. (2004) *China Exim Bank and China Development Bank Case Studies*, International Rivers Network, www.irn.org/programs/china/Exim_Bank.pdf

Cheung, R. (2004) 'Beijing to approve Yunnan dams plan', *South China Morning Post*, 16 April

Conservation International (2007) *Biodiversity Hotspots: Mountains of Southwest China*, Conservation International, www.biodiversityhotspots.org/xp/Hotspots/china/Pages/default.aspx, accessed 18 June 2008

Council of the European Union (2007) *Council Common Position 2007/750/CFSP of 19 November 2007 amending Common Position 2006/318/CFSP Renewing Restrictive Measures against Burma/Myanmar*, Official Journal of the European Union, 19 November

EarthRights International (2005) *Flooding the Future: Hydropower and Cultural Survival in the Salween River Basin*, 28 April, www.earthrights.org/burma

EGAT (Electricity Generating Authority of Thailand) (1981) *Report No 846-2409*, EGAT, Thailand

EGAT (2003) *The Hydropower Development: A Choice for Electric Power in Thailand*, Hydropower Engineering Division, EGAT, Thailand

EGAT (2005) *Thailand Power Development Plan, PDP 2004*, EGAT, Thailand, www.onep.go.th/eia/SEA/NationalWorkshop/PresentationMeeting_SEAworkshop/Power_Plan.ppt

EGAT (2006) *Annual Report*, EGAT, Thailand

EGAT (2008) *PDP 2007 (Rev 1)*, EGAT, Thailand (in Thai)

EGAT (undated) 'Welcome to Electricity Generating Authority of Thailand', www.egat.co.th

Energy for Environment Foundation (2006) *Thailand's Energy Import from Neighbouring Countries and National Security Seminar*, www.efe.or.th/index.php?option=content&task=view&id=244&Itemid=29

Fan, X. (2005) 'Shuidian da kaifa dui shijie yichan yiji guojia fengjing lüyoudi de yingxiang' ['The influence of great hydropower development on World Heritage and National Scenic areas'], Fengjing Mingsheng Zazhi [National Park of China], www.fjms1984.com.cn/20040101/ca689550.htm, accessed 1 March 2006 (in Chinese)

Foundation for Ecological Recovery (2003) *Briefing Paper:* Salween Hydropower Project, TERRA (Towards Ecological Recovery and Regional Alliance) Project, www.terraper.org/articles/BriefingSalweenThai-BurmaJune03.pdf

Gray, D. (2006) 'Villagers protest joint plans to dam Salween', Associated Press, 30 May

Greacen, C. (2006) *Decentralizing Thai Power: Towards a Sustainable Energy System*, Greenpeace Southeast Asia, November

Greacen, C. and Palettu, A. (2007) 'Electricity sector planning and hydropower in the Mekong region', in Lebel, L., Dore, J., Daniel, R. and Koma, Y. S. (eds) *Democratizing Water Governance in the Mekong Region*, Mekong Press, Chiang Mai

Habito, C. and Antonio, E. (2007) *Sustainable Development Strategies in the Greater Mekong Subregion: Status, Needs and Directions*, United Nations Environment Programme, www.rrcap.unep.org/projects/nsds/meeting/GMS%20NSDS%20Paper%20Final%20Edited%20Complete.pdf

He, Y. and Feng, J. (2004) 'Kexue fazhan yu Nujiang shuidian kaifa' ['Scientific development and Nujiang hydropower exploitation'], *Zhongguo Nengyuan Wang* [*China Energy Network*], www.china5e.com/news/water/200410/200410140036.html, accessed 21 November 2005 (in Chinese)

IED (2007) 'Power supply for sustainable development: The case of rural electrification-decentralized energy option in ASEAN', in *Powering ASEAN: Technology and Policy Options (2007)*, 6–7 September, www.ied-sa.fr/gb/..%5documents%5CRE-DEO_092007.pdf, accessed 20 May 2008

International Water Power & Dam Construction (2007a) 'Works well advanced at Myanmar's Yeywa RCC Dam', *International Water Power & Dam Construction*, 4 December, www.waterpowermagazine.com/story.asp?sc=2048023

International Water Power & Dam Construction (2007b) 'Power machines JV lands Shengsigou, Yinpan contracts', *International Water Power & Dam Construction* 23 November, www.waterpowermagazine.com/story.asp?storyCode=2047961

IUCN (World Conservation Union), International Water Management Institute, Ramsar Convention Bureau and WRI (World Resources Institute) (2003) *Watersheds of the World: CD-Rom*, World Resources Institute, Washington DC

IWHR (China Institute of Water Resources and Hydropower Research) Office (2006) *Yunnan Huadian Nujiang Shuidian Kaifa Youxian Gongsi Lingdao Lai Wo Yuan Fangwen* [*Yunnan Huadian Nujiang Hydropower Development Co Leaders Visit IWHR*], IWHR, www.iwhr.com/News_View.asp?NewsID=3157, accessed 18 April 2006 (in Chinese)

Japan Ministry of Foreign Affairs (2002) *Grant Assistance to Myanmar for the Project for Rehabilitation for Baluchaung No 2 Hydro Power Plant*, www.mofa.go.jp/announce/announce/2002/5/0510.html

KDRG (Karenni Development Research Group) (2006) *Dammed by Burma's Generals: The Karenni Experience with Hydropower Development From Lawpita to the Salween*, KDRG, www.salweenwatch.org/downloads/dammed-eng.pdf

Levett, C. (2007) 'Singapore banks refuse to do business with Burmese airline', *Sydney Morning Herald*, 27 October

Lintner, B. (2007) 'Myanmar's generals hit where it hurts', *Asia Times Online*, 2 November

Lwin, Y. (2006) 'Traders urged to use Euro for transactions', *Myanmar Times*, www.myanmar.com/myanmartimes/MyanmarTimes16-303/n004.htm

Ma, J. (2005) 'Revised plan for dam project', *South China Morning Post*, 6 March

Magee, D. (2006a) *New Energy Geographies: Powershed Politics and Hydropower Decision Making in Yunnan, China*, Geography Department, University of Washington, Seattle, WA

Magee, D. (2006b) 'Powershed Politics: Hydropower and interprovincial relations under Great Western Development', *The China Quarterly*, no 185, pp23–41

McDonald, K. (2007) *Damming China's Grand Canyon: Pluralization without Democratization in the Nu River Valley*, Environmental Science, Policy and Management Program, University of California, Berkeley, CA

MDX PCL (2007a) *Annual Report, 2007*, Stock Exchange of Thailand, Thailand

MDX PCL (2007b) *Disclosure of Investment to the Stock Exchange of Thailand on May 17, 2007, SET Ref No M06/025/07*, Stock Exchange of Thailand, Thailand

Mertha, A. C. and Lowry, W. R. (2006) 'Unbuilt dams: Seminal events and policy change in China, Australia and the United States', *Comparative Politics*, vol 39, no 1, pp1–20

Myanmar Department of Hydropower Planning (2006) *Overview of Water Resources Development in Myanmar*, Ministry of Electric Power No 1, Government of the Union of Myanmar, Myanmar, November

Myanmar Ministry of Foreign Affairs (2003) *Special Loan Agreement from PRC*, 29 August, www.mofa.gov.mm/news/aug29_fri_special.html, accessed 18 February 2008

New Light of Myanmar (2007a) *MOU on Upper Thanlwin Hydropower Project Inked*, www.mission.itu.ch/MISSIONS/Myanmar/07nlm/n070407.htm

New Light of Myanmar (2007b) 'Tar-hsan hydropower project construction begins', 4 April, http://mission.itu.ch/MISSIONS/Myanmar/07nlm/n070405.htm

Niksch, L. A. and Weiss, M. A. (2008) *Burma: Economic Sanctions*, US Congressional Research Service, Library of Congress, http://digital.library.unt.edu/govdocs/crs/permalink/meta-crs-10716:1

Nujiang Lisu Autonomous Prefecture Government (2005) 'Nujiang zhongxiayou shuidan guihua baogao' ['Nujiang Middle and Lower Reach Hydropower Plan Report'], Nujiang Lisu Autonomous Prefecture Government, Nujiang Lisu

Pathan, D. (2005) 'Thailand's war on drugs', in Jelsma, M., Kramer, T. and Vervest, P. (eds) *Trouble in the Triangle: Opium and Conflict in Burma*, Silkworm Books, Chiang Mai, Thailand

Power Machines Company (2006) *Power Machines Company*, http://english.power-m.ru/

SHAN (Shan Herald Agency for News) (2005) 'Wa business firm signs removed', Shan Herald Agency for News, 2 October

SHAN (2006) 'China cases in on Hutgyi', Shan Herald Agency for News, 28 June, www.shanland.org/articles/environment/2006/News170606

Shi, J. (2008) 'Project to dam the Nu River set to move ahead', *South China Morning Post,* 17 February

Siripol, V. (2007) *Power Trade Policy in GMS: An EGAT Perspective*, EGAT, Thailand

SSEO (Shan Sapawa Environmental Organization) (2006) *Warning Signs: An Update on Plans to Dam the Salween in Burma's Shan State*, Chiang Mai, Thailand

Stock Exchange of Thailand (2008) MDX Public Company Limited website, www.set.or.th/set/companyprofile.do?symbol=MDX&language=en&country=US, accessed 23 June 2008

TERRA (2006) *Chronology of Salween Dam Plans*, www.terraper.org/new_detail/Salween%20Chronology% 20Nov06%20final.pdf

The Economist (2008) 'Crony charity', *The Economist*, 19 June

Thu, K. (2006) 'Hutgyi power plant moves downstream to raise output', *Myanmar Times*, www.myanmar.com/myanmartimes/MyanmarTimes17-334/b003.htm

Thu, K. (2007) 'Chinese firm takes 51 percent interest in Tasang hydropower project', *Myanmar Times,* 19 November, www.mmtimes.com/no393/b004.htm

UNESCO (United Nations Educational, Scientific and Cultural Organization) (1992–2008) *World Heritage: Three Parallel Rivers of Yunnan Protected Areas*, UNESCO, http://whc.unesco.org/en/list/1083, accessed 2 May 2008

US Department of the Treasury (2008) *Treasury Sanctions Additional Financial Operatives of the Burmese Regime*, www.treas.gov/press/releases/hp837.htm

US Drug Enforcement Agency (2005) 'Eight high-ranking leaders of Southeast Asia's largest narcotics trafficking organization indicted by a federal grand jury in Brooklyn, New York', News release, 24 January

US Government (2003) *Burmese Freedom and Democracy Act of 2003*, US Department of State, Washington, DC

Vatcharasinthu, C. and Babel, M. S. (1999) *Hydropower Potential and Water Diversion from the Salween Basin, Panya Consultants*, Online Burma/Myanmar Library, www.burmalibrary.org/docs/SW05.htm

Wei, G. (2005) 'Shei shi daba beihou de liyifang' ['Who are the interest groups behind big dams?'], *Zhongguo Qingnian Bao* [*China Youth Daily*], 11 July, http://cydgn.cyol.com/content/2005-09/06/content_1097803.htm (in Chinese)

World Bank (2008) *Myanmar*, http://go.worldbank.org/0M1X5X3A90

Xiang Gang Shangbao (2006) *Yuan Nujiang dianzhan jihua shu lueduoxing kaifa* [*Original Nujiang Hydropower Plan: A Case of 'Predatory Development'*], 25 October (in Chinese)

Xinhua News Agency (2007) 'Myanmar cooperates with Asian neighbours in hydropower development', 6 June, http://english.peopledaily.com.cn/200706/06/eng20070606_381388.html

Xu, J. and Wilkes, A. (2004) 'Biodiversity impact analysis in northwest Yunnan, southwest China', *Biodiversity and Conservation,* vol 13, no 5, pp959–983

Xu, Y. C. (2002) *Powering China: Reforming the Electric Power Industry in China,* Ashgate, Aldershot

Yeh, E. T. and Lewis, J. I. (2004) 'State power and the logic of reform in China's electricity sector', *Pacific Affairs,* vol 77, no 3, pp437–465

Yundian Xinwen [Yunnan Electric Power News] (2008) *Nujiang shuidian tieding kaifa, liang dianzhan lieru '11–5' guihua* [*Nu River Hydropower Development Unalterable, Two Stations Written into 11th Five-Year Plan*], www.gxi.gov.cn/fzgh/fzgh_ghdt/fzgh_ghdt_ws/200803/t20080331_34715.htm, accessed 1 June 2008 (in Chinese)

Zhou, P. (2003) 'Liuku shuidianzhan jiepai' ['Inauguration of Liuku Hydropower Station'], *Yunnan Ribao,* 19 July, www.xb.yn.gov.cn/new/15.htm (in Chinese)

Part II

Livelihoods and Development

6

Irrigation in the Lower Mekong Basin Countries: The Beginning of a New Era?

Chu Thai Hoanh, Thierry Facon, Try Thuon, Ram C. Bastakoti, François Molle and Fongsamuth Phengphaengsy

Introduction

The image of irrigation often conflicts with that of the Mekong River Basin – a monsoonal region with a wet climate and periodic floods. In the countries of the Lower Mekong Basin (LMB), however, the wet season that runs roughly from June to October is, in many places, followed by a dry season for the rest of the year. The rain-fed uplands in Laos and Cambodia receive the most rainfall (3000mm) and the Korat Plateau in northeast Thailand receives the least, between 1000mm and 1600mm (MRC, 2003a). In the LMB countries, irrigation is a key means of securing monsoon crops – shifting from a single crop (mainly rain-fed wet season rice) to multiple cropping systems – and increasing crop yields. It is estimated that water abstraction for agriculture accounts for around 90 per cent of all water diversions (Cambodia: 94 per cent; Laos: 82 per cent; Vietnam: 86 per cent; and Thailand: 91 per cent) in the region (MRC, 2003a). Thailand and Vietnam have extensively developed their irrigation infrastructure; while investments have declined in the last few years, hydropower development is going rapidly ahead (especially in Vietnam and Laos; see Chapter 2 in this volume). Laos, because of its sparse population, and Cambodia, due to the recent history of war and political turmoil, still have a low degree of infrastructure development; but more investments are expected in the coming years.

The Government of Laos (GOL) is intent on developing irrigation in the highland valleys as well as along the Mekong River corridor, although the schemes installed during 1996 to 2000 have shown relatively poor performance. In Cambodia, the agricultural sector contributed 31 per cent of the national gross domestic product (GDP) in 2007 (http://en.wikipedia.org/wiki/Economy_of_Cambodia) and provides employment to nearly 80 per cent of the total labour force in the country (MAFF, 2007). Cambodia has set ambitious targets for irrigation development and institutional reforms, although experience in the past decade has been mixed. Vietnam's agricultural policies must be seen in the context of the country's ongoing transition to a market-based economy. Of particular relevance to the natural resources of the Mekong Basin are Vietnam's plans to expand irrigation, improve existing schemes and expand the delta water management systems to deal with acid-sulphate soils and salt intrusion. Thailand is promoting basin-wide coordination of water resources management, the improvement of irrigation management efficiency, and the strengthening of farmer groups and water user associations (WUAs) (MRC, 2003b). It has also repeatedly proposed massive irrigation and water diversion projects, notably in the northeast region or *Isaan* (see Chapter 10).

Despite these past investments, the expansion of irrigated areas in these countries has slowed down during the last few years. The question is whether this trend heralds the end of large-scale public irrigation or, in fact, whether a new irrigation era is expected? This chapter revisits the development of irrigation in the LMB countries in the past and analyses the possible trends of irrigation in the future by considering several drivers – in particular, the recent increase in food prices around the world that affects food security in many countries in the Mekong region. It addresses both the question of how to improve and reform existing schemes and the rationale and scope for further development.

Revisiting irrigation in the Lower Mekong Basin countries

Agriculture and irrigation in the Lower Mekong Basin countries

Agriculture in the highlands is typically less productive than in the lowlands and the deltas, where the bulk of crop production takes place. Upland crops are usually rain fed, with relatively limited irrigation, and agriculture policies strongly support cash crops such as coffee, tea and rubber. The lowlands and deltas are relatively flat and nutrient rich and are under intensive rice cultivation, with some exceptions in the floodplains due to soils or flooding conditions. Farming systems in these regions include wet season rice, dry season irrigated rice, flood recession rice, floating rice, and multi-crop production systems that have gradually incorporated fish, shrimps, vegetables and fruit trees.

Rice production has increased rapidly during the past decade: by 81 per cent between 1993 and 2000 in Cambodia; by 38 per cent between 1990 and 1999 in Laos; by 33 per cent between 1994 and 2001 in northeast Thailand; and by 27 per cent between 1995 and 1999 in the delta and central highlands of Vietnam. Although high-quality jasmine rice and glutinous rice with low yield is still grown in many places, particularly in northeast Thailand and Laos, this increase was mainly due to the use of high-yielding varieties, increased irrigation and cropping intensity, and expansion of the area under cultivation (MRC, 2003a).

Despite these large investments in irrigation, in 2003 the ratio of irrigated land to total arable land was only 26.8 per cent (see Table 6.1), lower than the 45 per cent of Asia as a whole. And this rate has remained relatively stable during the past two decades.

Table 6.1 *Irrigated area in the Lower Mekong Basin countries until 2003*

Country	Irrigated land: 1000ha				Share in arable land and permanent crops (%)			
	1979–1981	1989–1991	1999–2001	2003	1979–1981	1989–1991	1999–2001	2003
Laos	107	135	174	175	13.3	15.7	18.2	17.0
Cambodia	120	240	270	270	5.8	6.3	7.1	7.1
Thailand	3007	4248	4973	4986	16.4	20.6	25.8	28.2
Vietnam	1685	2867	3000	3000	25.6	44.8	36.4	33.4
All LMB countries	4919	7490	8417	8431	17.7	23.6	26.1	26.8
Annual growth rate*		257	93	5				
(as percentage)		(5.2)	(1.2)	(0.1)		0.6	0.2	0.2

Note: * Compared with record in previous column.
Source: FAO AQUASTAT (2004)

Out of the five categories of the global irrigation system typology (Molden, 2007), four can be found in the LMB countries:

1 large-scale public paddy irrigation systems in humid areas;
2 small- to medium-scale community-managed (and built) systems;
3 commercial privately managed systems producing for local and export markets; and
4 farm-scale individually managed systems producing for local markets around towns or cities.

Large-scale public paddy irrigation systems in humid areas

These irrigation systems, such as the Lam Pao (50,000ha) in Thailand, the Stung Chinit (12,000ha) in Cambodia and the Quan Lo Phung Hiep (250,000ha) in

Vietnam, were developed to produce paddy. In some cases, they have gone through a process of incremental development by gradually increasing water control and cropping intensity; but they face some challenges in terms of economic and financial viability, and technical and managerial upgrading.

Small- to medium-scale community-managed (and built) systems

Many of these systems are found in the highland areas of the LMB countries. They are characterized by their small size, and private or community investment and management. Often these systems divert water from small streams through temporary (or concrete) weirs. These systems form the basis of the economies of their communities and typically show a large variety of cropping patterns. Public-sector involvement includes mostly the rehabilitation or improvement of weirs (often making them out of concrete).

Commercial privately managed systems producing for local and export markets

These systems do not yet represent a large share of irrigated areas in the LMB countries, but they are becoming more important to local economies, especially with non-rice crops such as rubber, palm trees, coffee and upland cash crops.

Farm-scale individually managed systems producing for local markets around towns or cities

These systems develop around towns or cities in the LMB countries to take advantage of local markets for high-value crops such as fruits and vegetables. They are highly dynamic and volatile, often characterized by large short-term returns on investment, and face environmental and health-related problems for both farmers and consumers when they use wastewater.

The size of an irrigation system in the Mekong countries is largely related to its funding sources that are either national or local budgets or international loans or grants. Table 6.2 gives details on the extent of irrigation across the LMB countries. The most basic systems provide only supplementary water during the wet season, while more intensive schemes provide water for two to three seasonal crops per year.

In the LMB region, most of the irrigated areas are concentrated in northeast Thailand and the Mekong Delta in southern Vietnam. Figure 6.1 shows the distribution of irrigation schemes with different sizes in the Lower Mekong Basin.[1]

Table 6.2 *Irrigated areas in the Lower Mekong Basin countries*

Location	Number of schemes	Area of wet season irrigation [d] (ha)	Area of dry season irrigation [d] (ha)	Area of third season irrigation [d] (ha)	Irrigated area[a d] (1000ha)
Laos*	2532	224,232	151,940	0	224,232 [b]
Thailand* (total)	14,494	–	–	–	4,770,018
Royal Irrigation Department (RID) (medium/large)	788	–	–	0	3,781,128
RID (small)	11,567	–	–	–	90,963
Department of Energy Development Promotion (DEDP) (pumping)	2129	–	–	–	606,044
Royally initiated projects	2245	–	–	–	291,883
Cambodia*	1012	248,842	181,506	0	392,117
Vietnam Mekong Delta	120	1,964,223	1,358,669	281,497	1,964,223
Vietnam Mekong Highlands	76	36,008	7290	–	36,008
Total[c]	18,234	2,473,305	1,699,405	281,497	7,386,598

Notes: Dash (–) indicates information not available.

* Except for Vietnam, data given is for the whole country.

[a] Where there is no comprehensive wet or dry season cropping data available, the irrigated area has been taken as the common measure of the irrigation area.

[b] The total irrigated area in Laos has been recorded at 280,000ha, the difference being many small schemes which have not been formally inventoried or mapped.

[c] This is the total of all schemes and areas where data is available (i.e. many small unrecorded schemes are omitted).

[d] The data refer to the potential area under irrigation if the schemes were operating to full capacity, which is not always the case.

Sources: MRC (2003b, 2005); RID (2007) for Thailand

Irrigation development and issues in the Lower Mekong Basin countries

In the LMB countries, irrigation development, especially the traditional irrigation systems in highland areas, began centuries ago. For example, in Cambodia, the history of irrigation development goes back as far as the 3rd century AD, with further development during the Angkorian period between the 10th and 13th centuries (Higham, 2001). In Thailand, traditional farmer-managed irrigation systems, mostly found in the northern part of the country, were established as early as 700 years ago during the period of King Mengrai (Surarerks and Chulasai, 1982).

Large-scale irrigation development was initiated by the French in the Red River and Mekong deltas in Vietnam (see Chapter 8), and by the Thai monarchy in the Chao Phraya Delta at the end of the 19th century. During the Cold War (1946 to 1989), large-scale state-built irrigation was seen as a response to concerns of food security and poverty reduction while serving US geopolitical interests (Barker and

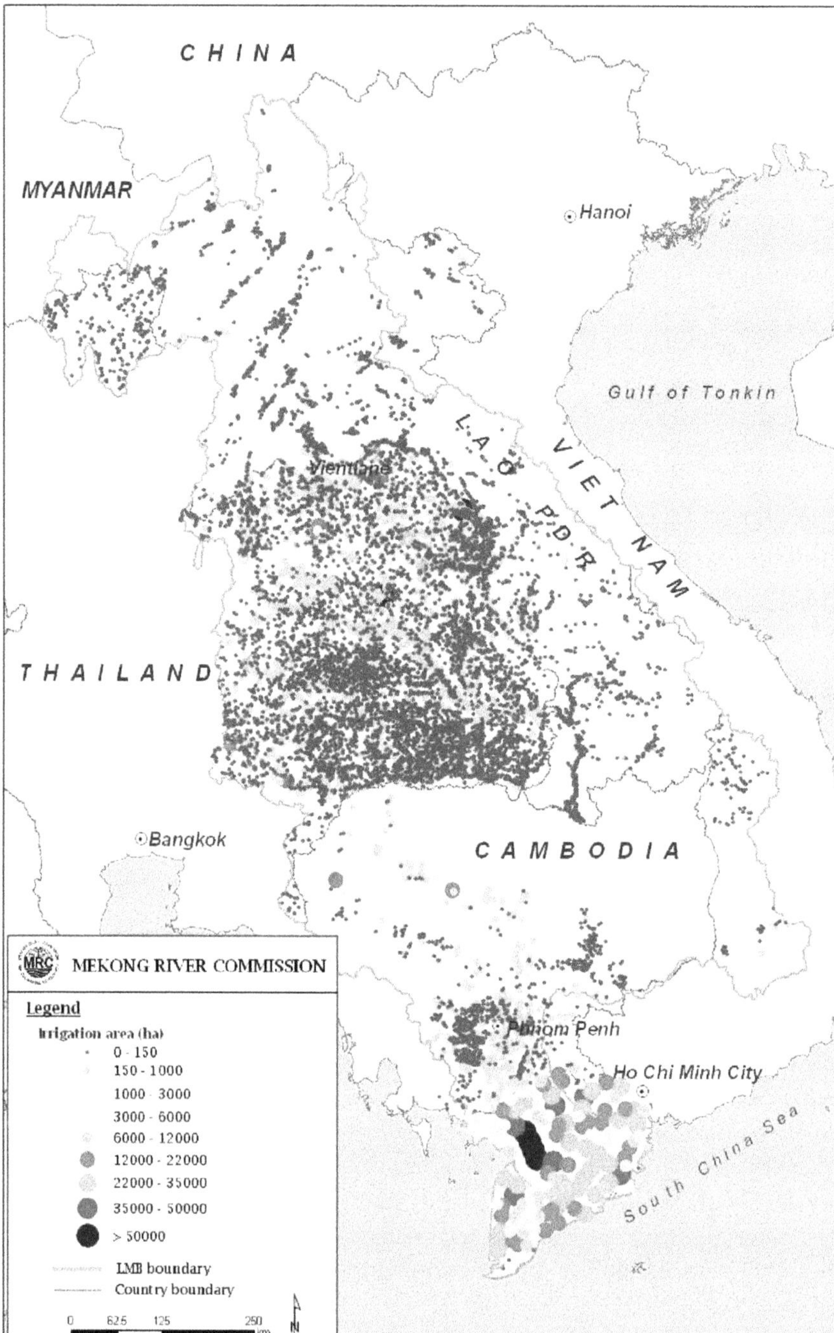

Figure 6.1 *Irrigation projects by area in the Lower Mekong Basin*

Source: MRC (2005)

Molle, 2004). Because of the Indochina wars, Thailand was the only country in the region that could continue to substantially develop its irrigation infrastructures at that time.

Irrigation is generally seen as a core part of rural development, with an important role in poverty reduction in the LMB countries, and food security in Cambodia and Laos. In Thailand and Vietnam, there is more focus on the intensification of irrigated cropping. There is also a move to diversify away from rice and to alternatives with higher financial returns such as aquaculture or cash crops. However, irrigation systems tend to be designed specifically to suit rice production, which makes it difficult for farmers to diversify into non-rice crops (MRC, 2003a).

Despite the great achievements in rice production in the LMB countries, there is a general consensus that irrigation systems have not lived up to expectations because of low performance in terms of control, water productivity, yields and quality of service delivery to farmers. The overall progress of the modernization agenda has remained relatively modest. The concepts of irrigation modernization are not fully understood and properly adopted (FAO, 2007). In some cases, modernization has been used to continue receiving funds for rehabilitation, operation and maintenance, or further capital-intensive interventions. In most countries, participatory irrigation management (PIM) and/or irrigation management transfer (IMT) has made very modest progress in improving system productivity and raising the cost recovery rate. Significant underinvestment in operation and maintenance and poor management continue to be the norm rather than the exception.

The reduction in financial benefits from agriculture between 1980 and 2000 has put governments under pressure to lessen irrigation management costs. Irrigation fees are being introduced with mixed success. Attempts to introduce an irrigation service fee in Thailand resulted in mass protests (MRC, 2003b). The alternative, and in some ways a reaction to excessive state control and bureaucratic government management, is to hand over schemes to farmers. But the effectiveness of this alternative needs to be investigated and questioned; huge investments in smaller schemes are at risk unless farmers have the incentives and capacity to ensure that they are adequately maintained. These institutional reforms do not capture the complexity of the hydrological cycle, the multiple functions of irrigation systems, and the relationships between different levels of management.

The situation among the individual LMB countries shows variations. Laos exhibits a low level of investment/infrastructure that contrasts with the fact that agriculture provides the largest share (40 per cent) of foreign currency income, 52 per cent of the GDP, and 85.5 per cent of employment (Molle, 2007). The Water Vision of the country stresses that 'The national economic development process is to be based on the wealth of natural resources, especially water and water resources', which particularly includes irrigation and hydropower (Nonthaxay et al, 2002). Significant improvements have been achieved in the agriculture sector, with an increase in dry season rice area from 2700ha in 1976 to 110,000ha in 2000, and irrigation has shifted the average rice yield from 3 tonnes per hectare (t/ha) in rain-fed wet season to 4t/ha to 4.5t/ha in the dry season. This increase in area reflects the US$30 million to US$40 million investment in 7000 electric pumps installed along the Mekong and its major tributaries during the late 1990s (CES and AFD, 2007).

Irrigation coverage further increased from 19,170 irrigation schemes with a service area of about 295,000ha in the 1999/2000 wet season to 24,000 schemes serving 310,000ha in 2005 (Pheddara, 2007). This was mainly due to the large investment in the National Pump Installation Management Project (NPIMP), mostly along the Mekong River in the southern part of the country. However, most irrigation schemes are based on traditional weirs and are found in the northern and central regions. Medium-scale public schemes are largely confined to the Nam Ngum Valley near Vientiane. The government *Strategic Vision for the Agriculture Sector* (MAF, 1999) is to invest heavily in irrigation with a focus on small- and medium-scale schemes under farmers' management, both along the Mekong Corridor (pumping) and in the highlands (valley bottoms). In 2001, the government set ambitious targets of increasing irrigation schemes from an actual coverage of 36 per cent of agricultural land to 50 per cent in 2005, and 80 per cent in 2020, with 50 per cent of the area cropped in the dry season (Nonthaxay et al, 2002). But until 2004 the irrigated area reported was only 20 per cent of the national paddy area (Anonymous, 2004) and dropped to 17 per cent in the three following years.

Problems found in the pumping schemes along the Mekong illustrate the variety of difficulties faced by the irrigation sector. The area effectively cultivated in irrigation schemes, in general, totals only 70 per cent of the design command area. Dry season cropping intensity in pumping schemes was found to be around 50 per cent of the wet season cropping area. Problems are technical (no on-farm distribution systems; deteriorated networks; damaged or out-of-order pumps), economic (rice is often less attractive than other on-farm activities in the dry season; the price of fuel became unbearable; operation and maintenance costs are not covered by water fees; schemes deteriorated) and institutional (lack of technical capacity in managing pumping stations; hasty turnover of technical and financial responsibilities to farmer associations; weak water user groups or associations; top-down design and location of schemes), while full-cost repayment policies conflicted with a choice of community based on higher poverty index (CES and AFD, 2007).

New investments in irrigation as well as hydropower (often combined) are contingent upon loans by development banks and private-sector involvement, both explicitly welcomed by the government (Richardson, 2002).

In Cambodia, during the early 1990s the Asian Development Bank (ADB) approved a Special Rehabilitation Assistance Loan (SRAL) for emergency rehabilitation of infrastructure, including some irrigation systems. The need and the scope for further investments in irrigation were emphasized, as agriculture dominates the economy and rural infrastructure was destroyed by the war (ADB, 2000). Irrigation in Cambodia is largely limited to the use of receding flood waters. Recession rice is gradually replacing lower-yielding traditional floating rice. Due to a limited ability in controlling water, the second irrigated rice crop in the dry season is grown only in approximately 10 per cent of the total wet season rice production area (MRC, 2003a). Renovation of irrigation schemes emerged as one of the most urgent rural development interventions for increasing agricultural productivity and for poverty alleviation (Öjendal, 2000).

The Ministry of Water Resources and Meteorology (MOWRAM) *Strategic Development Plan 2006–2010* targeted investments of over US$110 million for the irrigation reform programme. Objectives with the largest estimated costs generally involve significant infrastructural work, while those with lesser estimated costs generally relate to human resources capacity, management systems and the ministry's legal and regulatory basis (MOWRAM, 2007). In practice, the government has allocated only US$10 million per year to this ministry for urgent rehabilitation and construction work, mostly the construction of reservoirs for irrigated areas.

Large irrigation investment and irrigation modernization have been initiated with the support of external donors. The government aims to expand the irrigated area from 407,000ha (the actual area cultivated with rice is 2.5 million hectares) to 1.8 million hectares or more in the future through various interventions (Sinath, 2007), such as rehabilitation, reconstruction and development of gravity irrigation systems of various scales to provide supplementary irrigation in the wet season and full irrigation in the dry season. In addition, interventions include the provision of pumping stations (fixed and mobile), pumping generators and fuel; implementing and strengthening participatory irrigation management and development (PIMD) through farmers' water user communities (FWUCs) in order to ensure physical and financial sustainability, and modernization of the irrigation systems and flood protection facilities. Cambodia's policies on water reforms are strongly influenced by international financial institutions and aim to reduce the financial 'burden' to the state.

The recent master plan published by the Japan International Cooperation Agency (JICA, 2007) shows that the Tonle Sap and its catchments areas are the most important region for poverty alleviation, as well as overall economic development in the country. However, the study based on 320 existing irrigation systems on 4 out of the 12 tributaries flowing into the Tonle Sap reported some major problems, including:

- low ratio of farms under irrigation;
- lack of comprehensive rehabilitation work;
- deterioration of plot bunds;
- low ratio of establishment of FWUCs;
- insufficient canal capacity; and
- lack of irrigation structures.

The management reforms have been shaped by power struggles between agencies and built without due consideration of field experiences or local expertise (Molle, 2005, 2007; Roux, 2005). Water resource management and development come under diverse institutions of MOWRAM and the Ministry of Agriculture, Forestry and Fisheries (MAFF). Poor coordination among water-related institutions, together with unclear definition of their functions and responsibilities have often resulted in overlapping activities and financial costs. Other problems faced in irrigation management include rampant corruption, weak state administration capacity, ailing political legitimacy, under-educated civil servants, political factionalism, local-level conflicts, and a limited national budget (Öjendal, 2000). The problems in organizing local participation and the lack of project coordination at scheme level are still main constraints. This is one of the reasons why certain schemes in Cambodia resulted in cost overrun and why their benefits have been reduced (see Box 6.1).

Box 6.1 Issues in participation and integration: Lessons from the Stung Chinit Irrigation Scheme (SCIP) in Cambodia

The Stung Chinit Irrigation Project (SCIP) provides a striking example of the mismatch between project planning and reality (Try, 2008). The scheme was built during the Pol Pot period in 1977 by using forced labour from Kampong Thom and Kampong Cham provinces. The planned total command area of SCIP was 12,000ha. The project was designed to increase agricultural productivity and farmers' incomes and stimulate the rural economy of Kampong Thom Province by providing irrigation and drainage, agriculture extension, rural roads and markets. But due to some structural shortcomings, the scheme could not be operated up to its design capacity. In 1997, the Asian Development Bank (ADB) financed technical assistance to assess the feasibility of rehabilitating the SCIP. The consultants focused on the full development of the system, including an upstream dam and storage to provide wet and dry season irrigation over 12,000ha. In 2000, the full proposal for the Stung Chinit Irrigation and Rural Infrastructure Project was submitted to the ADB for approval. The revised plan decreased the proposed irrigated area to 7000 ha in the wet season and 2000 ha in dry season. The overall project implementation period was supposed to start in 2001; in practice, construction only started in 2006, and then was extended up to July 2008. However, by that time the scheme could only irrigate 2000ha in the wet season and about 300ha in the dry season.

The delay in construction resulted in higher construction costs and ultimately affected the expected benefits to farmers in the SCIP. The revised design was not fully fitted to local conditions, such as drainage requirements, fish migration, etc. Moreover, the rehabilitation plan could not secure the active involvement of all actors associated with the project; therefore the participation of the water users could not be achieved as expected; for example, some farmers were not willing to join the project because it would result in a loss of grazing and agricultural land. In the dry season, farmers in the project area show limited interest in rice cultivation as it is often affected by insects and they also find alternative sources of livelihood by going upstream for logging or engaging in farming activities that provide more benefit than rice cultivation.

Since 1961, the strategies defined by comprehensive National Economic and Social Development Plans (NESDP) have guided irrigation development in Thailand. In the beginning, emphasis was put on the construction of large- and medium-scale irrigation projects to expand irrigable areas as much as possible in order to prevent drought (Budhaka et al, 2002). During the 1960s and up to the mid 1970s, external development ideas, loans and grants influenced irrigation development. Many of the irrigation systems are underutilized, with low benefits, and continued financial support of government was necessary to maintain the systems. During the late 1970s and the 1980s, the focus of the water policy of successive Thai governments was directed towards the completion and upgrading of distribution systems and on the rapid development of small-scale irrigation infrastructure. Irrigation, however, fell short of its promise regarding cropping intensity and diversification towards cash crops. While contract farming has occasionally been successful (Dolinski, 1995), experience from the Lam Pao schemes (Burt and Styles, 1999) or Nam Oon (World Bank and NESDB, 2005) in northeast Thailand have provided the same lessons about dry season cropping areas remaining at low levels because of lack of tertiary canal service, unreliable water supply, limited market organization and opportunities, labour constraints, and unwillingness of farmers to face the health hazards brought about by pesticide use. During the 1990s a river basin approach was adopted, while irrigation systems had to operate under growing pressure of commercial agriculture (especially dry season farming), as well as competition for water from other non-agricultural sectors (Bastakoti and Shivakoti, 2008).

The direction of irrigation development is reflected in Thailand's National Water Policy and Vision (Budhaka et al, 2002). After the adoption of participatory irrigation management policy, more emphasis has been given to participation of users in system management, especially water allocation, operation and maintenance at tertiary levels (FAO, 2007). A national strategy has been formulated to improve irrigation efficiency and water management in existing systems, while expanding new small and medium systems. The government emphasized the Common Irrigators' Organization framework to integrate local people (beneficiaries) within the irrigation systems (Shivakoti, 2000, 2003). Although there are examples of involvement of local communities in the management of irrigation systems (Molle et al, 2001; Shivakoti and Bastakoti, 2006; Bastakoti et al, 2008), participation has remained limited. In many cases, water user groups were organized at the tertiary level and had no influence on the allocation of water at higher levels or on the quality of the supply of water to their canal; they were thus quickly undermined (Molle et al, 2002).

The government is currently implementing an initiative from the King of Thailand to encourage farmers to diversify out of rice to produce an intensive mix of crops, fish and livestock. No large-scale projects have been developed recently due to lack of economically viable suitable storage sites, as well as environmental concerns. Improving the efficiency of existing water management projects and promoting basin-wide coordination of water resources are key priorities of Thai agricultural policy (RID, 2007).

Although approximately 4 million hectares are irrigated during the wet season, in 2003 the government launched the idea of a national 'water grid' that would triple the area of irrigated land, and in July 2003 announced that it would target 200 billion baht (equivalent to US$5 billion in July 2003) to solve the problem of water scarcity in Thailand, mainly in the poor north east region (see Chapter 10). With the fall of the Thaksin administration, the project was put on the back burner; but, recently, an underground pipeline that would divert water from the Mekong River to *Isaan* (*Bangkok Post*, 2008) was included by the government in the 'top-priority' megaprojects, with a total value of at least 500 billion baht (about US$15 billion). The new project aims to divert water from the Mekong through underground tunnels to Loei and Udon Thani provinces, where reservoirs will act as distribution centres to send the water through small pipelines to farms in other provinces during the dry season. In April 2008, Thailand's cabinet approved a budget of 10 billion baht (about US$334 million) to build irrigation infrastructures within seven years, from 2008 to 2015 (Agroinfo, 2008).

Vietnam, in line with the proverb *Nhat nuoc, nhi phan, tam can, tu giong* ('Water is the first determining factor, then fertilizer, hard work and crop variety'), is still involved in massive investments for rural and water infrastructures. Vietnam started to invest massively in modern irrigation development in 1975, and between 1988 and 1994 the annual expansion rate of irrigation reached 4.6 per cent. Particularly in the Mekong Delta, improved water control permitted a shift from a single crop to two or three crops annually (see Chapter 8). This concerned, in particular, the flood-prone area of the upper delta and the coastal zone where salinity intrusion has been controlled by constructing dikes and by gating streams. Pump irrigation plays an important role in Vietnam and accounts for 26 per cent of its total irrigation area (FAO, 2007). Rapid adoption of small private pumps, particularly in the Mekong Delta, has greatly facilitated crop diversification. But extraction of more fresh river water for irrigation will facilitate further intrusion of salinity in the main branches of the river. Salinity control in coastal areas also leads to other problems such as pollution being retained inland and poor drainage, and conflicts between rice farmers who require fresh water and shrimp farmers who prefer brackish water for their shrimps.

The government, with World Bank support, is focusing on five sub-projects in the Mekong Delta covering 535,000ha, which includes irrigation, salinity control, flood protection and institutional development. Irrigation in the Vietnam Highlands is less developed, with rice and coffee as the main crops. Since 1995, coffee irrigated with groundwater has yielded some of the best financial returns in Vietnam's agriculture.

The national authorities have recognized the challenges faced by the water sector under the '*doi moi*' economic reforms and initiated a series of reforms in the country's water sector. These included the enactment of the Vietnamese Water Law in 1999, the Decision on the Establishment of a National Water Resources Council in June 2000, the establishment of basin-level committees to oversee the management and allocation of water in the Red River Basin, Mekong Delta and Dong Nai Basin (Molle and Hoanh, 2008), and the creation, in November 2002, of the Ministry of Natural Resources and the Environment (MoNRE). In 2006, the total budget of the Ministry of Agriculture and Rural Development (MARD) was about US$200 million, ranked second behind the Ministry of Transport. Planning and management of small- and medium-scale irrigation have been decentralized, with the provincial authorities now autonomous and self-financing many of the irrigation activities.

Based on many surveys of the whole country, Tiep (2002) reported that, on average, existing irrigation systems supply water to only 50 to 60 per cent of the design command area.[2] Less than 5 per cent of expenditure for irrigation and drainage has been devoted to operation and maintenance (O&M). Inadequate cost recovery and deteriorating infrastructure are major concerns, and ways are being sought to involve greater water-user participation in operation and maintenance (Barker et al, 2004). Although water-fee charges are higher than in most other Asian countries, fee collection covers only half of O&M costs. However, in October 2007, the government issued a new policy that exempted farmers of irrigation fees at the national level.

THE BEGINNING OF A NEW IRRIGATION ERA?

The wider economic context of rice production and irrigation

The last 50 years have seen massive investments in large-scale public surface irrigation infrastructure as part of a global effort to increase the production of staple food, ensure food self-sufficiency and avoid famine. Investment in irrigation accelerated rapidly during the 1960s and 1970s, with irrigated areas in developing countries expanding at 2.2 per cent a year and reaching 155 million hectares in 1982 (see Figure 6.2). The private and community-based investments in developing countries grew rapidly since the 1980s, propelled by cheap drilling technology, rural electrification and low-cost small pumps (Molden, 2007). Global irrigated areas increased from 168 million hectares in 1970 to about 300 million hectares at the end of the 1990s. The food price index fell from around 310 in 1974 to around 90 in 2000 to 2002.

As areas best suited to irrigation had already been developed first, the costs of further development increased, while prices of staple cereals declined, in part because of the very success of irrigation (Barker and Molle, 2004). These two factors made irrigated agriculture economically less attractive than in the past. The underperformance of large-scale irrigation also reduced donors' interest. Declining cereal prices at the end of the last century slowed growth in input use, investment in crop research and irrigation infrastructure (Sanmuganathan, 2000). More competition for water from other sectors also reduced the scope for further

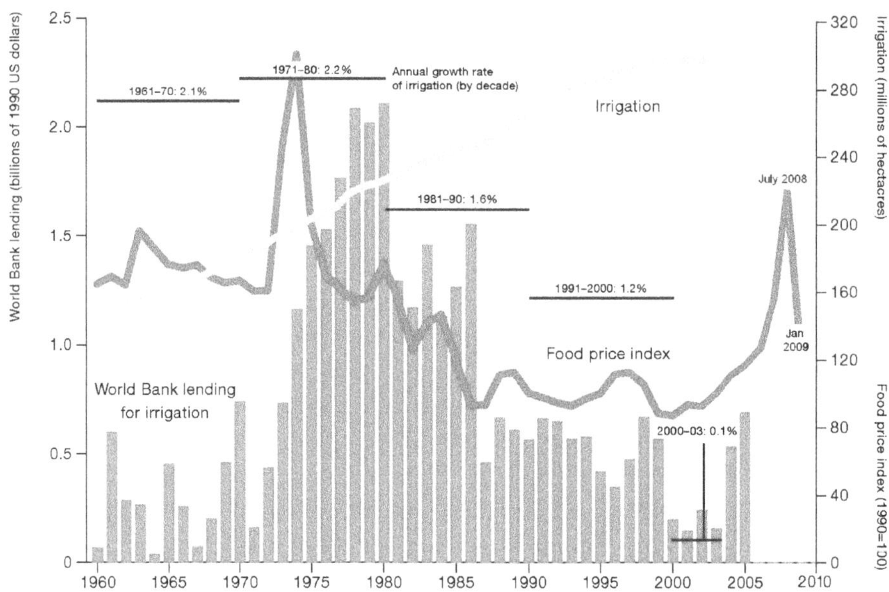

Figure 6.2 *Food price index and irrigation expansion during 1960 to 2009*

Source: Molden (2007) with updated FAO data for 2009

development of irrigation. At the global level, irrigation is predicted to expand at lower rates compared with that from 1960 to 1990; but its contribution to total agricultural production is expected to exceed 45 per cent by 2030 as yields continue to increase and cropping patterns shift to higher-value crops (FAO, 2003). This means more water withdrawals for irrigation.

The situation in the world food market has recently witnessed a radical turn. Rice prices have gradually increased from US$200 per tonne (Thai white rice, 100 per cent B second grade) in 2003 to US$376 per tonne at the end of 2007, and peaked in May 2008 to at US$963 per tonne (FAO, 2008). The food price index soared up to over 210 in 2008, then dropped to 146 in January 2009 (see Figure 6.2). The low annual growth rate of irrigation area during the last ten years (only 0.1 per cent during 2000 to 2003) is not considered to be the main reason of this upsurge, but reflects the lack of attention to, and investment in, the agriculture sector at a global level.

Although both the Organisation for Economic Co-operation and Development (OECD) and the United Nations Food and Agriculture Organization (FAO) (OECD–FAO, 2008) project that high food prices will not last long and will gradually come down because of the transitory nature of some of the factors that are behind the recent hikes, they also conclude that, after falling from their current peaks, prices will remain at higher average levels over the medium term than during the past decade.

With the recent soaring prices of food (particularly rice), more farmers in the LMB countries have reverted to rice because of their advantages – of which irrigation is a major factor – in rice production compared with other countries. The governments of Thailand and Vietnam have encouraged farmers to increase paddy crop production to take advantage of the rising rice prices in the world market in 2008; triple cropping has increased (e.g. in some parts of central Thailand and in Hau Giang and An Giang provinces in the Mekong Delta) and others have switched back to rice farming (like shrimp farmers in the Ca Mau Province, reversing a trend that began over five years ago (*VietnamNet*, 2008). In Cambodia, MAFF and the FAO are launching an emergency project under the FAO-led Initiative on Soaring Food Prices (ISFP) to help farmers boost agricultural production, in particular through the provision of seeds and fertilizers (which are petroleum-based and thus out of reach of poor farmers as oil prices increased) (*China View*, 2008). In Laos, almost 75 per cent of households that have adopted modern rice varieties now have a surplus or are self-sufficient in rice (ADB and OED, 2008). However, even though the government reported that the country produced more than 2.7 million tonnes of rice in 2007, the lack of infrastructure and high fuel price make transportation difficult, and shortages persist in remote mountainous areas (especially those occupied by the Lao-Soung and Lao-Theung tribes) (Vorachak, 2008).

Under the ISFP, many countries have indicated that they would accelerate or strengthen their irrigation rehabilitation, construction and water management programmes, while major donors have emphasized the importance of public investment in irrigation infrastructure and policy reform. The investment flow into the irrigation sector is thus expected to increase significantly in coming years.

Reinventing irrigation in the Lower Mekong Basin countries

The question of whether, where and how to invest in irrigation is much more intricate than it may seem. A broader definition of investment includes public investment in irrigation and drainage development, modernization, institutional reform, improved governance, capacity-building, management improvement, creation of farmer organizations and regulatory oversight, as well as farmers' investment in joint facilities, wells, and on-farm water storage and irrigation equipment (Molden, 2007).

National, local and environmental contexts

The context of irrigation varies from one country to another depending upon factors such as the degree of food security, actual infrastructures and potential, percentage of the population working in the agricultural sector, or the existence of employment opportunities in other sectors. Strategies underpinning the evolution and future development of irrigation in the socio-economic context of the LMB countries are presented in Table 6.3 (FAO, 2007), which distinguishes between different types of irrigation and three main socio-economic contexts. These contexts can be found simultaneously in different areas of the LMB countries, while the general trend is that areas move from A to B and, in some cases, C.[3]

Challenges for the irrigation sector

The irrigation sector faces two main challenges: strengthening both the hardware (hydraulic infrastructures) and software (institutions) of existing schemes, and finding ways of improving decision-making with regard to new investments. These challenges are examined in this section.

Scheme rehabilitation and modernization

As discussed earlier, the debate between rehabilitation of irrigation systems and modernization seems to have been resolved ten years ago in favour of modernization, understood as the transition from supply-driven to demand-driven management, or the adoption of service-oriented management, with supportive re-engineering of management set-up and infrastructure. In practice, very little modernization has happened on the ground, with, on the one hand, a persistence of traditional design standards and processes (both for rehabilitation and new projects), and, on the other hand, rather weak institutional reforms that have not significantly altered the relationship between the irrigation service providers and the farmers, as well as the management and operation practices of irrigation agencies. Irrigation investment costs in Southeast Asia were estimated to be almost the lowest in the world (Inocencio et al, cited in Molden, 2007), reflecting a combination of expected standards and economic performance.

Meanwhile, the concept of modernization has been revisited to take into account a new understanding of the complexity of the rice irrigation systems prevalent in the LMB characterized by multiple roles related to floods and groundwater recharge, a recognition of the importance of fish and aquatic resources for local livelihoods, an acceleration of the transformation of rural economies, and increased attention to supplying water for energy, growing cities and industries. With the soaring food prices, the integration of LMB farmers within the global market will continue. It has been reaffirmed that:

> *... in the present context and under future perspectives, modernization of the irrigation systems and their management to increase their flexibility and insert them in river basin management, taking into account the multiple functions of agricultural water management, is*

Table 6.3 *Irrigation types in the Lower Mekong Basin countries based on water source and their evolution scenarios*

Irrigation water source classes	Main characteristics	National and sub-national stage (see notes)		
		A Developing (focus on domestic agriculture)	B Intermediate (focus on export agriculture)	C Post-agriculture (focus outside agriculture)
1 Reservoir – gravity	Water is stored in large reservoirs, distributed via canal networks to the fields mainly by gravity	Large reservoirs are too expensive for rice but planned as multi-purpose structures	Anticipate on multiple uses, not economically justified by agriculture alone	Optimizing multiple use economically justified, but sites available for new systems are limited
2 Off-river – gravity	Water level in the river is raised by a small weir so that water can be distributed by a canal network	Low costs, hence comparative advantage compared with other options	Improved and (endless) modernization; inherent limitations of supply	Reduced, merged or neglected due to low reliability, or converted to type 3 or 4
3 Off river – pumping	Water is pumped into a canal network for distribution to the fields	Affordable investment, but requires subsidized O&M	Probably reduced due to energy costs, particularly for paddy	Crop diversification because economically unjustified for rice; sites available for new systems are limited
4 Conjunctive groundwater-surface water system	Both gravity-fed surface irrigation and groundwater pumping	Highly flexible, decided by farmers based on market demands	Highly flexible, decided by farmers based on market demands	Highly flexible, decided by farmers based on market demands
5 Integrated management in delta systems	Multi-purpose canal networks in the delta, including dike, dams and sluice gates, for navigation, irrigation, drainage and settlements	Expanded quickly in newly reclaimed lands	Optimized for multiple use, but expensive drainage when facing environmental issues, particularly in peri-urban agriculture	Optimized for multiple use with more crop diversification and urban services
6 Small-scale water storage	Rainwater harvesting structures	Provided water to individual or groups of farms for supplementary irrigation by pumping	Modernized with improved water-use technologies	Optimized multiple water use, with focus on high-value cash crops.

Notes: Economic and agriculture situation in each stage:

Stage A: rely on rice production due to pressing needs for food security. Very little attention paid to ecosystems and water-dependent livelihoods; Stage B: on the way to diversification with a quick demographic transition under improvement of food security and rice export for foreign currency earning; Stage C: highly diversified agriculture under resources competition with high environmental concerns and diet changes; water is linked with multifunctionality of agriculture and classical environmental issues.

Strategy and policy in each stage:

Stage A: water resources development and rice irrigation expansion with strong government financial support and external assistance; Stage B: stabilization and modest development of rice irrigation areas, development of small systems and increase in financial self-sufficiency; Stage C: reduction/decommissioning of rice irrigation areas, specialization for improved water productivity, protection of environment and water quality, and government investment in scheme modernization.

Source: adapted from FAO (2007)

> *more required than ever. A fast pace of change is the one certainty, the other certainty being that unless management adapts, the discrepancy between stated and actual policies will widen.* (FAO, 2007)

In addition to the rising prices of agricultural commodities, which may change the economic and financial equation of investment in modernization, rising energy prices may provide additional incentives to attempt to improve water delivery service by irrigation canal systems and decrease the farmers' needs to pump water from available water bodies or from groundwater.

The question then is: will modernization plans continue to remain mostly on the shelves or will the LMB countries see an effective change of investment and management strategies, away from rehabilitation with classic PIM (with maybe the token addition of IMT), to deliberate modernization? This obviously requires proper policy and suitable institutional arrangements that cannot be determined within the irrigation or agriculture sectors only, but should be considered within the socio-economic conditions of each irrigation system.

The answer then is: cautious optimism. The region should start seeing a broader implementation of modernization concepts and practices, more internalized by irrigation agencies and management than dictated by international agencies and donors. A pragmatic approach to professionalization of management and improvement of service should pave the way in the longer term to more substantial service orientation:

- There is a greater awareness of the present deficiencies of the irrigation systems as knowledge does exist, efforts to develop tools have been substantial and effective, and efforts to develop capacities have been effective where implemented.
- Substantial capacity-building programmes are under way (Vietnam) or are being planned (Thailand) in the context of innovative investment programmes (Vietnam) or agencies' own resources (Thailand), focusing on details of design and operation.
- More importantly, perhaps, assessment and performance evaluation indicators and methodologies are being introduced or revised to be consistent with service orientation and modern management concepts, and their results are being used to review current strategies and to shape investment. They include rapid appraisal of performance, Mapping System and Service for Canal Operation Techniques (MASSCOTE) (Renault et al, 2007), benchmarking, service-oriented irrigation management, balance sheets, and improved data collection and processing. This has led to the recognition of the need to change design standards.

Institutional reforms

To improve irrigation system performance governments and multilateral lending agencies continue to implement reforms in the LMB countries, including water pricing and cost recovery policies, setting up of water user associations, and institutional/legal reforms and policies.

Cost recovery and associated water charges have been the subject of intense debate and controversy (Molle and Berkoff, 2007). A reduction of government funding was expected for irrigation programmes in LMB through the application of cost recovery principles. However, implementation is slow because farmers are not willing to accept cost recovery measures when service delivery is poor and results in an overall reduction in benefits. Vietnam, however, notably in the pump-irrigation schemes of the Red River Delta, where pumping costs are difficult to compress, shows reasonable rates of recovery. Allocating the costs resulting from the maintenance of the Mekong 'delta machine' is also a huge and growing challenge (as shown in Chapter 8). Yet, the decree recently issued by the government that cancelled irrigation fees across the country shows the political nature of taxation/subsidies schemes and also suggests a measure of the social volatility of the countryside and, possibly, a perception of increased independence of the governments in the region on the policy prescriptions associated with funding from international financial institutions.

Under such conditions, a sensible option for reducing public funding in irrigation would be a progressive rise in water charges, corresponding to increased accountability and transparency on the part of service providers and progressive transfer of authority to users, matched by increased profitability of irrigated agriculture (Molden, 2007). However, this option can probably only be implemented widely in the LMB countries for the systems at stage C of the evolution pattern, as presented in Table 6.3. The issue is, therefore, for irrigation agencies that have adopted or declared an intention to move towards service-oriented management (RID of Thailand, MARD of Vietnam) to do so without the incentives and accountability generated when farmers pay irrigation service fees. The demand for improved service will need to come, in this case, initially from the agencies themselves, while in the mid term, improved service delivery to farmers and deliberate efforts to reduce farmers' pumping costs, and the setting up of new institutional arrangements associating farmers to main decisions on management objectives or service agreements, might set the stage for a renewed dialogue on service fees. Meanwhile, costs do need to be recovered from the irrigation agencies themselves, by shifting their budgetary allocations from new system development to better provisioning of O&M for existing projects; from local governments to which responsibility has been shifted through decentralization; from other water users; and from farmers themselves for the levels of system devolved to them.

PIM has also featured prominently in the region, notably in Thailand and Cambodia. A recent survey in Cambodia and an assessment of nine of the FWUCs piloted by MOWRAM as part of the PIMD programme has clearly shown that these organizations are at different stages of development, maturity and overall functionality. These differences are brought about by internal factors, including the degree of organizational management, leadership and decision-making, understanding of financial aspects, resources for system operation and maintenance, and external drivers of functionality, such as physical infrastructure, capacity of agencies to assist in the implementation of PIMD, availability of markets for products, and inputs. While the overall picture is somewhat bleak, there has been positive improvements in the overall operation of the systems, in increased cultivated areas and cropping intensity, and in the number of farmers receiving water (Perera, 2007).

Based on a study by the FAO, Table 6.4 provides a list of conditions for successful institutional reforms, as well as reasons for failure; the latter are currently more prominent in the region than the former.

Table 6.4 *Main conditions for success and reasons for failure of institutional reforms*

Conditions for success	Reasons for failure
Strong political backing A clear role for the different stakeholders	Lack of political support Resistance of public agencies and water users
Support for the empowerment of institutions at all levels (including water user associations and local governments)	Insufficient resources
The autonomy of the water user associations	Poor quality and predictability of water supply undermining local organizations
The legal framework needed to accommodate the proposed changes in authority	Lack of legal support for the proper involvement of water users
Capacity-building of the people governing the transferred system	Lack of coordination among stakeholders involved with project implementation
Functioning infrastructure	Transfer of dilapidated or badly designed infrastructure that is dysfunctional and does not fit within local culture and context
Success in recovering operation and maintenance costs	Cost overrun and reduced benefits

Source: adapted from Molden (2007) based on an FAO study

While some issues can be addressed satisfactorily through successful modernization of the systems, allocation of sufficient resources for support and capacity-building, and legal support, a key question for the future of PIM in the region is whether this would be sufficient and whether the PIM models adopted by the countries should not be revisited altogether. The PIM models of small resource-poor WUAs, progressively federated into resource-poor organizations conceived as instruments for conducting operation and maintenance activities without having a real say in water management introduced to the region, have largely been copied from other countries where, after more than 30 years, they have not proven any more successful than in the LMB countries. It is time for the region to explore new and more diverse options, which may be based on larger organizations with a critical mass, and on the characteristics of traditional water management organizations, or build on the strength of existing institutions (such as irrigation teams in Vietnam) or on processes that allow flexibility in designing farmer institutions to suit particular management objectives and circumstances, rather than follow an overly prescriptive model. They should be able to meet the changing characteristics of farming – more part time, more commercial, more woman oriented and gender sensitive – and to integrate other water users (including fishing and aquatic resources), and roles and functions of the irrigation systems.

The separation of regulatory and management functions to avoid conflicts of interest in the water sector has been a key reform promoted by the ADB in the region. The Ministry of Natural Resources and Environment (MoNRE) has been established in both Vietnam and Thailand, for example (Molle and Hoanh, 2008). The MoNRE is supposed to be the manager of water resources, while the MARD (in Vietnam) and the RID (in Thailand) are supposed to concentrate on building and operating the irrigation systems. Management of water resources is supposed to be done at the level of river basins; but river basin organizations have not been empowered at a level where they could be influential in decision-making, in general, and the definition of basin plans and water allocation, in particular.

They have faced the power of traditional line agencies (MARD, RID) and have lacked legal and political backing (especially in Thailand) to challenge the established positions of these agencies. Environmental regulation remains incipient and the decision-making power with regard to new dams or inter-basin transfer projects remains entrenched in political and line agencies circles. Yet, experiences in some pilot river basins in Thailand suggests that long-term processes have a potential to incrementally shift governance in the direction of a greater participation of concerned stakeholders.

Other 'soft' reforms concern the agronomic and economic environment of irrigation and refer to the necessity of providing adequate agricultural extension services and enhanced access to markets (TWGAW, 2006).

Development of new irrigation areas

At a very general level there are, of course, good reasons to invest in irrigation over the next decades. A growing and more urbanized population means that more food must be produced at a time when competition for water grows and environmental impacts must be reduced. Irrigation can be a path out of poverty for the rural poor and more investments will probably be needed to respond to global warming as more climate variability and sudden and extreme climate events are expected in the future. In the LMB, the projected range of temperature change is 1°C to 2°C. The hot period of the year will extend longer, rainfall intensity will increase (Snidvongs,

2006), and so will the extreme rainfall and winds associated with tropical cyclones in Southeast Asia (Christensen et al, 2007), including the LMB.

However, the rationale underpinning the 're-engagement' in the financing of water projects frequently draws heavily on general arguments of food security, poverty alleviation and economic development. As illustrated by the case of the Thai Water Grid (see Chapter 10), benefits are presented in terms of area served, putative drought-proofing of agriculture, and increased cropping intensity and incomes. The huge costs incurred (with US$10,000 per hectare as a good basis), the necessity to shoulder operation and maintenance costs (including costly pumping operations), and, above all, the question of potentially better alternative investments, both in the water sector and outside of it, are glossed over. Estimates by the Thai Water Resource Department that 19,000 villages are facing a 'chronic shortage of water' or estimates of 'coming water shortages' in Thailand and the alarm sounded by the National Economic and Social Development Board (*Bangkok Post*, 2004) sound like *ad hoc* justifications. The justifications for large-scale investments are usually raised by repeatedly stressing the impact of droughts and floods and by looking at the benefits of the projects alone, disregarding costs. Debates are also frequently 'securitized' by government leaders who conjure up ominous threats to national or food security.

The principal challenge in the region is thus the risk of present circumstances related to food and energy security being exploited to promote projects, irrespective of their intrinsic quality or lack thereof, that are pushed by politicians willing to please their constituencies, and government agencies and private companies in search of business opportunities. This occurs in lieu of giving renewed attention to the sector at the highest levels, as well as mobilizing resources, in order to engage in substantial and meaningful reform and to promote ambitious but rational and balanced approaches, exploring the best options and addressing the poor performance of existing systems and their causes.

In Cambodia, the expansion of irrigation within provinces surrounding the Tonle Sap Lake has been the top priority of existing and new donors. For instance, the joint European Union–Royal Government of Cambodia ECOSORN project (Economic and Social Re-Launch of Northwest Provinces in Cambodia), established in 2006, injected 26 million Euros in integrated rural development, with irrigation renovation as a priority activity. The Northwest Irrigation Sector Project is jointly funded by the ADB and Agence Française de Développement (AFD): more than 50 irrigation schemes in four provinces (Pursat, Battambang, Batteay Meanchey and Siem Reap) are being reviewed for rehabilitation and development (ADB, 2003). The ADB and AFD are still working on selection criteria for irrigation scheme rehabilitation in order to avoid a repetition of the problems faced in Stung Chinit (see Box 6.1).

However, the definition of such criteria is seen as a constraint by the government, which tends to seek new donors or fund new schemes on its own budget. MOWRAM, under joint projects with Qatar and Kuwait, is planning to spend

US$400 million to build two large dams in the Vaico and Stung Sen rivers to irrigate 400,000ha (*The Mekong Times*, 2008). The ministry aims to build four dams in Pursat Province that would supply irrigation to more than 35,000ha of land and generate 300MW of power. Other proposed dam sites include locations in Battambang, Kampong Chnang and Banteay Meanchey provinces, and the ministry is consulting with engineers from China and South Korea. To achieve these giant plans, the ministry will need more than US$4 billion (*The Phnom Penh Post*, 2008).

In sum, the situation in the irrigation sector is evolving towards a scenario that is similar to the hydropower sector:

- Social and environmental standards associated with 'traditional' lending are seen by governments as a burden and a constraint to large-scale development projects.
- This encourages them to deal with funding institutions and construction companies that have not adopted similar safeguards.
- Projects are capital intensive and are promoted by 'iron triangles' and other forms of financial or political interest groups linked to the completion of these projects.
- Governance, as a result, is weak and secretive regarding planning; purported benefits are highlighted, while impacts and costs are downplayed or ignored.

While irrigation projects (e.g. Stung Chinit in Cambodia, the Mekong pumping schemes in Laos, or many schemes in northeast Thailand), even those heavily funded and studied by external financing partners, are showing deficiencies that are difficult to address, new projects to irrigate hundreds of thousands, or even millions (in Thailand) of hectares are being proposed. The involvement of intended 'beneficiaries', as well as consideration of environmental impacts, required labour force and market availability will need to be addressed when these projects face public consultation. Environmental and social impact assessments have now become mandatory – as embodied, for instance, in Thailand's previous constitution. Unsuccessful rehabilitation or construction projects generate huge costs, as well as recurring expenditures that may come to be a disproportionate burden to the benefits generated.

The regional community has recently developed recommendations for the development of new large-scale irrigation projects (FAO, 2007). Before committing to new large-scale irrigation developments, a comprehensive assessment of options for land and water use should be made. If a new large-scale irrigation development is proposed, it should be examined by a wide-ranging feasibility analysis to determine whether it is ecologically, physically, economically, politically, socially and culturally sound. This should take place before progressing into the formal legal, often rigid and relatively narrow, 'impact assessment' process. The Commonwealth Scientific and Industrial Research Organization's (CSIRO's) 5-Way methodology and the World Commission on Dams' (WCD's) guidelines, where relevant, are international references. Irrigation scheme design should also be flexible enough to take account of the inevitability of future demand changes, include credible input from different stakeholder groups, make sure funds to meet full operation and maintenance costs will be available, and include future monitoring of impacts upon ecosystems and livelihoods.

Conclusions

Agriculture provides the livelihood of 75 per cent of the population of the LMB countries and is a key driver of national development in each of the riparian countries. Irrigation has expanded and intensified across the four countries; but irrigation systems have not lived up to their expectations and have faced a number of problems. The differences between stated policies and actual practices are generally large, while policy changes have little impact; institutional reforms do not capture the complexity of basin-wide water management, the multiple functions of irrigation systems, and relationships between different levels of management. PIM/IMT initiatives, furthermore, have made very modest progress, while there is significant underinvestment in operation and maintenance, and poor management remains pervasive.

However, the recent soaring prices of food in the global market has alarmed regional governments and fuelled calls for further reinvestments in irrigation. In a broader sense, these include public investment in new schemes, scheme modernization, institutional reforms, improved governance and the creation of farmer organizations. Changes in governance, however, only emerge slowly and will remain dependent upon the democratization of society and the evolution of the relationships between the state and the various forms of civil society.

Recent announcements of large investments in new irrigation schemes in Cambodia or Thailand seem to have surfaced with no reference to the difficulties and the limitations faced by existing schemes. In Cambodia, the financial capacity to shoulder operation and maintenance costs, the access to markets, the managerial capacity of farmers, and the problematic relationship between the state and villagers preclude the enthusiasm conveyed by big numbers labelled in dollars or area to be equipped. In Thailand, notably its northeast region, the high investment costs per hectare and environmental constraints such as soil salinity, not to mention the actual low interest in dry season cropping, suggest that massive investments will face severe setbacks. In both settings, the existence of more attractive off-farm activities or market constraints remain important limitations, even if increased output prices have made agriculture temporarily financially more attractive. In the Mekong Delta, large increases in the abstraction of fresh water in the dry season combined with rising sea level are likely to worsen salinity intrusion in the main river channels.

The question posed in the title of this chapter thus has to remain for the moment. The coming years will tell us whether the current opportunities to address the real challenges of poverty and food security of the LMR have been used wisely. The risk remains that the large sectoral and private interests that benefit from massive capital investments will prevail over more carefully targeted investments in irrigation or agriculture, more decisive reform and a necessary focus on improving the performance of existing assets.

The large-scale transformation of waterscapes through irrigation comes with risks and costs that are often downplayed, but which must be constantly reassessed and remembered by those with responsibility for decision-making, as well as those directly benefited or affected by these transformations. New systems may still be developed in predominantly agrarian economies, in ecosystems with comparative advantages; but their planning and appraisal process should be reformed in order to include improved water governance.

NOTES

1 The density of spots in northeast Thailand is partly misleading because most of the irrigation shown is local and effected through mere diversion of small streams to bunded plots.
2 But in many cases part of the remaining area is irrigated through local village-based pumping stations.
3 These evolution scenarios were developed at the end of 2005; so some conclusions made at that time may need to be revisited due to, on the one hand, rising food prices and, on the other, rising energy prices.

REFERENCES

ADB (Asian Development Bank) (2000) *Report and Recommendation of the President to the Board of Directors on a Proposed Loan to the Kingdom of Cambodia for the Stung Chinit Irrigation and Rural Infrastructure Project (SCIRIP)*, Asian Development Bank (ADB), Manila

ADB (2003) *Report and Recommendation of the President to the Board of Directors on a Proposed Loan to the Kingdom of Cambodia for the Northwest Irrigation Sector Project*, Asian Development Bank (ADB), Manila

ADB and OED (Operations Evaluation Department) (2008) *Learning from Rice in Lao PDR: Evaluation News*, 23 April 2008, www.adb.org/documents/oed/news/2008/learning-rice-laopdr.pdf

Agroinfo (2008) *Special Report on Rice Market in the First Half of 2008 and Upcoming Prospect*, Agroinfo, Institute of Policies and Strategies, Ministry of Agriculture and Rural Development, Vietnam

Anonymous (2004) *Lao PDR: Water Resources Coordinating Committee Country Paper*, National Water Sector Apex Body, www.adb.org/Water/NWSAB/2004/Lao_PDR_Country_Paper.pdf

Bangkok Post (2004) 'New projects to ease chronic shortages', *Bangkok Post*, 24 March

Bangkok Post (2008) 'Megaprojects "a top priority" for new government: Big plans for diversion of water from Mekong', *Bangkok Post*, 4 February

Barker, R. and Molle, F. (2004) *Evolution of Irrigation in South and Southeast Asia*, Comprehensive Assessment Research Report 5, Comprehensive Assessment Secretariat, Colombo, Sri Lanka

Barker, R., Ringler, C., Tien, N. M. and Rosegrant, M. (2004) *Macro Policies and Investment Priorities for Irrigated Agriculture in Vietnam*, Comprehensive Assessment Research Report 6, Colombo, Comprehensive Assessment Secretariat, Sri Lanka

Bastakoti, R. C. and Shivakoti, G. P. (2008) 'Irrigation systems under market pressure and changing institutional settings: Comparative perspective from Nepal and Thailand', Paper presented to 12th Biennial Conference of International Association for the Study of the Commons (IASC), Cheltenham, England, 14–18 July 2008

Bastakoti, R. C., Shivakoti, G. P. and Lebel, L. (2008) *Community Organizations in Water Resource Governance: Rural–Urban Interface of Irrigation Management in Thailand*, USER Working Paper WP-2008-02, Unit for Social and Environmental Research, Chiang Mai University, Chiang Mai

Budhaka, B., Srikajorn, M. and Boonkird, V. (2002) 'Thailand country report on investment in water', in *Investment in Land and Water: Proceedings of the Regional Consultation at Bangkok, Thailand*, 3–5 October 2001, FAO, Bangkok

Burt, C. M. and Styles, S. W. (1999) *Modern Water Control and Management Practices in Irrigation: Impact on Performance*, Water Report 19, FAO, Rome, www.watercontrol.org/publications

CES (Community and Environment Services) and AFD (Agence Française de Développement) (2007) 'Irrigation diagnostic study. Draft final report, Volume 1', CES, Vientiane

China View (2008) 'Cambodia, UN-FAO launch emergency project for farmers', *China View*, 30 August, http://news.xinhuanet.com/english/2008-08/30/content_9741554.htm

Christensen, J. H., Hewitson, B., Busuioc, A., Chen, A., Gao, X., Held, I., Jones, R., Kolli, R. K., Kwon, W.-T., Laprise, R., Magaña Rueda, V., Mearns, L., Menéndez, C. G., Räisänen, J., Rinke A., Sarr, A. and Whetton, P. (2007) 'Regional climate projections', in Solomon, S., Qin, D., Manning, M., Chen, Z., Marquis, M., Averyt, K. B., Tignor, M. and Miller, H. L. (eds) *Climate Change 2007: The Physical Science Basis*, Contribution of Working Group I to the Fourth Assessment Report of the Intergovernmental Panel on Climate Change, Cambridge University Press, Cambridge, UK, and New York, NY

Dolinsky, D. J. (1995) *Assessment of Contract Farming at Lam Nam Oon, Thailand: A Combined Effort of USAID and the Royal Thai Government*, Working paper no 193, CDIE, Arlington, VA

FAO (United Nations Food and Agriculture Organization) (2003) *World Agriculture towards 2015/2030: An FAO Perspective*, FAO and Earthscan, Rome and London

FAO (2007) *The Future of Large Rice-Based Irrigation Systems in Southeast Asia*, Proceedings of the Regional Workshop on the Future of Large Rice-Based Irrigation Systems in Southeast Asia, Ho Chi Minh City, Vietnam, 26–28 October 2005, FAO RAP Publication 2007/28

FAO (2008) *The FAO Rice Price Update – August 2008*, www.fao.org/es/ESC/en/15/70/highlight_533.html

FAO AQUASTAT (2004) 'FAO AQUASTAT Database', www.fao.org/nr/water/aquastat/dbase/index.stm

Higham, C. (2001) *The Civilization of Angkor*, Phoenix, London, UK

JICA (Japan International Cooperation Agency) (2007) *Basin-Wide Basic Irrigation and Drainage Master Plan Study in the Kingdom of Cambodia, Interim Report*, JICA, Ministry of Water Resources and Meteorology, and Ministry of Agriculture, Forestry and Fisheries (MAFF), Phnom Penh

MAF (1999) *The Government's Strategic Vision for the Agricultural Sector*, Ministry of Agriculture and Forestry, Laos

MAFF (Ministry of Agriculture, Forestry and Fisheries) (2007) *Annual Conference 2007 on Agriculture, Forestry and Fisheries: Taking Stock of 2006–2007 Performance and Setting Direction for 2007–2008*, MAFF, Phnom Penh

The Mekong Times (2008) 'Cambodia to spend US$400 million on irrigation', *The Mekong Times*, 11 August, issues no 130

Molden, D. (ed) (2007) *Water for Food, Water for Life: A Comprehensive Assessment of Water Management in Agriculture*, Earthscan, London, and International Water Management Institute (IWMI), Colombo, Sri Lanka

Molle, F. (2005) *Irrigation and Water Policies in the Mekong Region: Current Discourses and Practices*, Research report 95, International Water Management Institute (IWMI), Colombo, Sri Lanka.

Molle, F. (2007) 'Irrigation and water policies in the Mekong region: Trends and challenges' in Lebel, L., Dore, J., Rajesh, D. and Koma, Y. S. (eds) *Water Governance in the Mekong Region*, Mekong Press-Silkworms, Chiang Mai

Molle, F. and Berkoff, J. (eds) (2007) *Irrigation Water Pricing Policy in Context: Exploring the Gap between Theory and Practice*, CABI Publishing and International Water Management Institute, Colombo

Molle, F. and Hoanh, C. T. (2008) *Implementing Integrated River Basin Management: Lessons from the Red River Basin, Vietnam*, Research report, Mekong Programme on Water, Environment and Resilience (M-POWER), Chiang Mai, Thailand, International Water Management Institute (IWMI), Colombo, Sri Lanka, and Institut de Recherche pour le Développement (IRD), Montpellier, France

Molle, F., Ngernprasertsri, N., Savakon, S. and Chompadist, C. (2001) *Patterns of Social Interaction and Organisation in Irrigated Agriculture: The Case of the Chao Phraya Delta*, DORAS Project, Research report no 9, Submitted to the European Union, Kasetsart University, Bangkok

Molle, F., Ngernprasertsri, N. and Sudsawasd, S. (2002) 'Are water user organisations crucial for water management? A post-mortem analysis of water user groups in Thailand and the prospect for reincarnation', Paper presented to the Sixth Conference on Participatory Irrigation Management, Beijing, 20–26 April 2002

MOWRAM (Ministry of Water Resources and Meteorology) (2007) *Strategic Development Plan 2006–2010*, MOWRAM, Phnom Penh, Cambodia

MRC (Mekong River Commission) (2003a) *State of the Basin Report: 2003, Executive Summary*, Mekong River Commission, Phnom Penh

MRC (2003b) *State of the Basin Report: 2003*, Mekong River Commission, Phnom Penh

MRC (2005) *Irrigation Water Use Assessment: Annex, Programme to Demonstrate the Multi-Functionality of Paddy Fields over the Mekong River Basin*, Mekong River Commission, Vientiane, Laos

Nonthaxay, P., Boulapha, C. and Phanrajsavong, C. (2002) *National Water Vision to Action: A Framework for Integrated Water Resources Management in the Lao People's Democratic Republic*, Water Resources Coordination Committee Secretariat, Lao PDR (second draft)

OECD–FAO (Organisation for Economic Co-operation and Development–United Nations Food and Agriculture Organization) (2008) *Agricultural Outlook 2008–2017 Highlights*, OECD and FAO, Rome

Öjendal, J. (2000) *Sharing the Good: Modes of Managing Water Resources in the Lower Mekong River Basin*, PhD thesis, Göteborges Universitet, Sweden

Perera, L. R. (2007) *Factors Affecting the Formation of FWUCs in Institutional Building for PIMD in Cambodia: Two Case Studies*, Working paper 113, International Water Management Institute (IWMI), Colombo

Pheddara, P. (2007) 'Large rice-based irrigation systems in Lao PDR', in FAO (ed) *The Future of Large Rice-Based Irrigation Systems in Southeast Asia*, Proceedings of the Regional Workshop on the Future of Large Rice-Based Irrigation Systems in Southeast Asia, Ho Chi Minh City, Vietnam, 26–28 October 2005, FAO RAP Publication 2007/28

*The Phnom Penh Pos*t (2008) 'US$4billion dams in planning: Proposed 10 dams will boost irrigation, generate electricity', *The Phnom Penh Post*, 5 September

Renault, D., Facon, T. and Wahai, R. (2007) *Modernizing Irrigation Management – the MASSCOTE Approach: Mapping System and Services for Canal Operation Techniques*, FAO Irrigation and Drainage Paper 63, FAO, Rome

Richardson, M. (2002) 'In its water, Laos sees power to cut poverty', *International Herald Tribune*, 11 March

RID (Royal Irrigation Department) (2007) *Report on Information of Irrigation Projects – 2007*, RID, Bangkok, Thailand

Roux, J. (2005) *Water Governance in Cambodia: Policy in the Making and Links to Implementation*, Imperial College London

Sanmuganathan, K. (2000) *Assessment of Irrigation Options*, WCD Thematic Review Options Assessment IV.2, World Commission on Dams, Cape Town, www.dams.org/docs/kbase/thematic/drafts/tr42_finaldraft.pdf

Shivakoti, G. P. (2000) 'Participatory interventions in farmer-managed irrigation systems in Northern Thailand: Dynamism in resource mobilization', Paper presented at Constituting the Commons: Crafting Sustainable Commons in the New Millennium, the Eighth Conference of the International Association for the Study of Common Property, Bloomington, Indiana, US, 31 May–4 June

Shivakoti, G. P. (2003) 'Facilitating policies for improved governance and management of Asian irrigation systems', Paper presented at a Colloquium at the Workshop in Political Theory and Policy Analysis, Indiana University, 22 September 2003

Shivakoti, G. P. and Bastakoti, R. C. (2006) 'The robustness of Montane irrigation systems of Thailand in a dynamic human-water resources interface', *Journal of Institutional Economics*, vol 2, no 2, pp227–247

Sinath, C. (2007) 'Large rice-based irrigation systems in Cambodia' in FAO (ed) *The Future of Large Rice-Based Irrigation Systems in Southeast Asia*, Proceedings of the Regional Workshop on the Future of Large Rice-Based Irrigation Systems in Southeast Asia, Ho Chi Minh City, Vietnam, 26–28 October 2005, FAO RAP Publication 2007/28

Snidvongs, A. (2006) *Vulnerability to Climate Change Related Water Resource Changes and Extreme Hydrological Events in Southeast Asia*, A final report submitted to Assessments of Impacts and Adaptations to Climate Change (AIACC), Project No AS 07, *Southeast Asia START Regional Centre* (SEA START RC), Bangkok, Thailand, The International START Secretariat, US

Surarerks, V. and Chulasai, L. (1982) *Water Management and Development in Northern Thai Irrigation Systems*, Chiang Mai University, Chiang Mai

Tiep, N. X. (2002) 'Water resources and food security in Vietnam', Paper presented at the national workshop on Water, Food and Environment, 24–25 September 2002, Hanoi, Vietnam

Try, T. (2008) *Participation, Fit and Integration: Early Lesson from Stung Chinit Irrigation Scheme and Rural Infrastructure Project, Cambodia*, Mekong Programme on Water, Environment and Resilience (M-POWER) and Unit for Social and Environmental Research (USER), Chiang Mai, Thailand

TWGAW (Technical Working Group on Agriculture and Water) (2006) *Participatory Irrigation Management and Development (PIMD) in Cambodia – Review of Experiences: Modes of Intervention, Results and Lessons Learned*, Section 1: Main Report, TWGAW, Ministry of Water Resources and Meteorology (MOWRAM) and Ministry of Agriculture, Forestry and Fisheries (MAFF), Phnom Penh, Cambodia

VietnamNet (2008) 'Farmers switch to rice as price triples', *VietnamNet*, 24 April, http://english.vietnamnet.vn/biz/2008/04/780000/

Vorachak, P. (2008) 'Laos – rising world rice price impact', *VOA News*, 5 April, www.voanews.com/lao/archive/2008-04/2008-04-05-voa2.cfm

World Bank (2004) *Vietnam: World Bank Helps Modernize Irrigation Schemes and Improve Dam Safety in Vietnam*, Press release, 9 April, www.worldbank.org.vn/news/press44_01.htm

World Bank and NESDB (National Economic and Social Development Board) (2005) *Thailand Northeast Economic Development Report*, Joint Report of Thailand's National Economic and Social Development Board and the World Bank, Bangkok

7

Landscape Transformations and New Approaches to Wetlands Management in the Nam Songkhram River Basin in Northeast Thailand

David J. H. Blake, Richard Friend and Buapun Promphakping

INTRODUCTION

The Nam Songkhram Basin in northeast Thailand in many ways represents a classic case study in failed developmentalism. A keenly contested ecological and political land and waterscape that is undergoing rapid change, the Nam Songkhram Basin brings out the contradictions implicit in human–nature tensions that have been an influential feature in recent development discourse and practices throughout the wider Mekong region (see Sneddon and Fox, 2008). Similar to other parts of Thailand, state agencies portray the Nam Songkhram Basin as both a flood disaster zone and a drought zone necessitating state-managed intervention and infrastructure development for intensifying agricultural productivity. Not surprisingly, the Nam Songkhram River Basin has been the subject of a number of large-scale water resources management projects over the last half century. Yet only a few of these large schemes have realized full implementation, while many other smaller stand-alone irrigation projects and heavily state-subsidized agribusiness ventures have been subsequently abandoned. Some of these abandoned projects rise out of, or indelibly mark, the Songkhram land–waterscape as vivid reminders of the consequences of narrow sectoral and non-participatory governance.

The Nam Songkhram River has also attained a special status in the eyes of some environmental-based organizations that have portrayed it in terms of its local livelihoods and conservation significance. The basin's most distinctive feature is its biologically rich and ecologically diverse wetlands ecosystem, partly resulting from extensive annual flooding and its interdependence with the mainstream Mekong hydrology. This relatively predictable pattern of flood and recession is the main driver of natural productivity, including the rich capture fisheries and harvest of wetlands products (Blake, 2006). Rather than being victims of, or vulnerable to, natural patterns of flooding, a recent study found local people to be resilient and adaptable to the annual inundation and dry period, traditionally basing their livelihood strategies around its seasonal cycles (Friend et al, 2006)

This chapter considers some of the conflicting representations of the Nam Songkhram Basin and how state policies have, to a great extent, guided development planning that is based on flimsy assumptions and misconceptions about what constitutes 'the problem' (Breukers, 1998). It examines the main actors and stakeholders involved and how the present socio-political and bio-physical landscapes have been shaped by tensions surrounding the dominant state development paradigm that views the floods as an obstacle to agricultural development and a natural disaster in need of an engineering solution, rather than recognizing the wetland's benefits of natural aquatic biomass production, aquifer renewal and soil fertility maintenance for dry season livestock grazing and small-scale crop cultivation.

It also examines recent participatory initiatives that have been field tested in the Nam Songkhram Basin, precipitated by a growing realization concerning the socio-economic and ecological value of riverine ecosystems in the Lower Mekong Basin and a sense of inadequacy about past methodological approaches. In particular, this chapter contrasts the contributions of an environmental flows (*E-Flows*) approach and a participatory local action-research (*Tai Baan Research*) approach to the knowledge base and considers to what extent they have the potential to inform more relevant development practice. The chapter attempts to critically analyse the main lessons drawn from recent novel approaches tested *in situ*, set against five decades of state-led development plans and a background of sustained environmental degradation and social dislocation evident at local and basin-wide levels.

THE NAM SONGKHRAM BASIN

The 13,128km^2 Nam Songkhram Basin in upper northeast Thailand (see Figure 7.1), although encompassing a mere 1.7 per cent of the entire Mekong Basin surface area, is nevertheless regarded as an important Mekong tributary both domestically for Thailand and as an integral ecological component of the wider Lower Mekong Basin (Blake, 2006). The lower part of the Nam Songkhram Basin forms a large

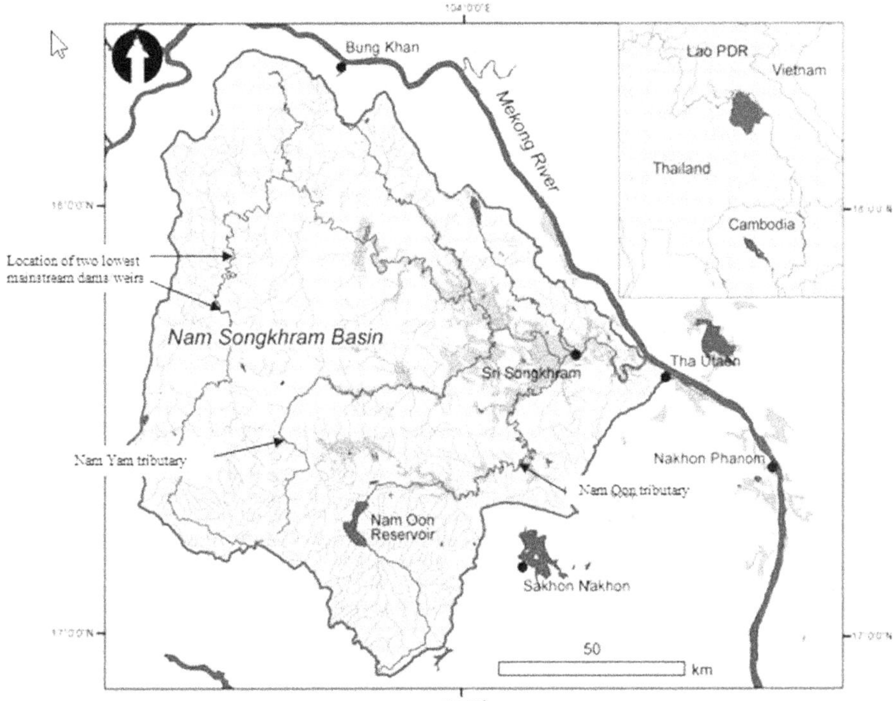

Figure 7.1 *Hydrological map of the Nam Songkhram Basin showing areas of floodplain (darker shading) and relative regional location (inset)*

Source: adapted from Kummu et al (2006)

seasonally inundated riverine floodplain wetland complex that is closely linked to the hydrology and ecology of the mainstream Mekong River (Blake and Pitakthepsombut, 2006a; Kummu et al, 2006). This interconnectivity between the two rivers has ensured that there is high aquatic biodiversity and bio-productivity found in the Nam Songkhram River and associated wetlands, which in turn has allowed the establishment of important fisheries,[1] primarily based on migratory fish species (Blake, 2006; Hortle and Suntornratana, 2008). A distinctive feature of the floodplain strip bordering the Nam Songkhram River are the seasonally flooded forests (*paa boong paa thaam*),[2] formerly extensive but now much reduced in area and quality. Recent estimates by a Khon Kaen University team analysing satellite data suggest that only 73.17km^2 of seasonally flooded forest remained in 2005, an 18.3 per cent decrease from 2001 data (Suwanwerakamtorn et al, 2007).

The Lower Nam Songkhram Basin wetlands form a distinctive geographical area extending to over 3000km^2, of which approximately 960km^2 may be temporarily flooded in the average wet season and as much 1850km^2 in a 1-in-

50-year flood (Khon Kaen University, 1999). River levels rise and fall by up to 12m inter-seasonally, reflecting the region's annual extremes in rainfall and hydrology. This annual 'flood pulse' phenomenon[3] both underpins and is considered the 'driver' of the river floodplain ecosystem's productivity and biodiversity (Blake, 2006; Lamberts, 2008). Some observers have likened the seasonal flood event and associated backwater effect with occasional backflow influence from the Mekong River in the Lower Nam Songkhram Basin wetlands to a lesser version of the well-known Tonle Sap Lake flood pulse phenomenon (Blake and Pitakthepsombut, 2006b; Kummu et al, 2006). Empirical evidence and modelling has shown that the level of the Mekong mainstream is pivotal in determining the level of the lower Nam Songkhram River and, thus, the height, duration and extent of inundation on the floodplain (Sarkkula et al, 2006; Blake, 2009).

The richness and diversity of the aquatic resource base is reflected in Nam Songkhram fishers' catches and capture gear, which attest to a high level of specialization and technological ingenuity by local people adapting to the constantly changing surrounding environment (Kohanantakul, 2004). While almost 190 species of fish have been identified by fishery researchers in the Nam Songkhram Basin through scientific approaches, local people have recorded 79 separate fishing gears used to harvest them, although not all are in use today (Blake and Pitakthepsombut, 2006b). A Mekong River Commission (MRC)-funded study found that between 80 and 93 per cent of all households fish, with the average household catch estimated at 207kg per year (Hortle and Suntornratana, 2008). The same study found that despite extensive habitat modification and the spread of markets for other food sources, the wild capture fishery continues to contribute most to household intake of animal protein. The importance of fisheries and wetlands resources to local livelihoods is also evident in the local language, culture and belief systems (Tai Baan Research Network of Lower Songkhram Basin, 2005a, 2005b).

Population estimates of the Nam Songkhram Basin vary between about 1.45 million (Blake, 2006) and 1.94 million people (Hortle and Suntornratana, 2008). It is a complex river basin not only ecologically, but also politically and administratively, spanning parts of four provinces with numerous state agencies involved in different aspects of natural resources management, often with contradictory priorities, objectives and goals. Some of these multifarious agencies' conflicting roles and activities are examined in Breukers (1998) and Blake and Pitakthepsombut, (2006a). At the same time, there are diverse non-state actors drawn from civil society and local people's movements concerned with development and resources management issues in the Nam Songkhram Basin and regionally. This plurality of stakeholders has influenced the way in which the river basin as a concept is regarded and multiple worldviews are constructed by different actors (Lohmann, 1998).

In addition to fish, the wetlands[4] are important sites for harvesting many other categories of natural resources, both aquatic and terrestrial, which have traditionally formed the basis of local livelihoods prior to the advent of the modern 'development' era (approximately 1961 onwards) and are still important for tens of thousands of households. These non-fish natural wetlands products include numerous species of edible plants, fungi, insects, birds, mammals, amphibians and reptiles. Additionally, the wetlands are a source of medicinal plants, honey, earthworms (for export to Taiwan), leeches, fuelwood, fencing, reeds for mats and handicrafts, tool- and implement-making materials, plant-based dyes, construction timber, spiritual and ceremonial-associated plants, etc. Apart from direct-use value wild products such as these, there are agricultural crops and products that are wholly or partly dependent upon a wetlands ecosystem, including paddy rice and a variety of livestock (mostly buffalo and cattle). Local communities have been able to benefit and, in many cases, prosper from the aquatic resource abundance that has defined the Lower Nam Songkhram Basin and been described by one Thai anthropologist and historian as forming a distinctive local culture.[5]

Despite the presence of this rich biodiversity, which has long been both intensively and extensively utilized and traded by local communities, its value and role in the local economy has not been recognized by external actors until relatively recently (Choowaew et al, 1994). Wetlands biodiversity forms an important part of what has been termed the 'hidden harvest' (IIED, 1997) that has consistently been ignored and undervalued during conventional cost-benefit analyses of water management infrastructure projects in the Mekong Basin. Even then, this recognition has been confined to a limited narrow constituency at a time when the natural resource base has been moderately to severely degraded and further rapid change is occurring as a result of policy and planning decisions set in motion decades ago (Blake and Pitakthepsombut, 2006a). Outside of the state development discourse, the Nam Songkhram River has frequently been portrayed as a naturally flowing river, free of dams and other large water management infrastructure ('The last undammed Mekong tributary in Thailand', e.g. Saskaki et al, 2007). While this image is no longer valid since several dams and weirs have been built across the upper and middle river in the past decade (Blake and Pitakthepsombut, 2006a), the Nam Songkhram Basin may contain relatively less flow control infrastructure than other north-eastern river basins and is without a mainstream dam structure on the lower 300km or so (see Figure 7.1 for dam locations). However, this is more likely to be a result of its inherent unsuitability for damming along most of its course, rather than a shortage of plans by various state agencies to dam it. Many have never come to fruition; but Nong Gaa 'weir' and its upstream neighbour Ban Muang 'weir'[6] were planned as irrigation dams by the now-defunct Accelerated Rural Development Office and executed as combined irrigation and flood control structures by the Department of Water Resources (DWR) under the Ministry of Natural Resources and the Environment (MoNRE) on the middle reaches of the river between Udon Thani and Sakhon Nakhon provinces. Built between 2000 and 2004, neither dam required an environmental impact assessment (EIA) and neither has had post-completion studies conducted to ascertain its impacts or efficiency.

On completion, the DWR handed over the dams to 'the people' (*prachachon*), although no one locally claims any ownership and the dams have effectively been abandoned by the state, serving neither an irrigation nor flood control role (nor any other apparent useful purpose), but checking the flow of the river especially during the dry season, while causing multiple environmental impacts, including increased erosion and elevated salinity levels (first author's observations in 2005, 2006 and 2007). It should be noted that above Nong Gaa 'weir', the lowest mainstream dam at present, the river is engineered and modified over considerable lengths right up to the headwaters below Phu Phan hills (Blake, 2007).

OF GRAND VISIONS AND FAILED EXPERIMENTS

Central to an understanding of the Nam Songkhram Basin's current situation is an appreciation of past state-led visions, plans and projects that have helped to transform the physical, political and social landscape of the region (e.g. Bello et al, 1998; Sneddon, 2003; Pye, 2005). In this respect, the Nam Songkhram Basin is little different from other parts of northeast Thailand, which have been subject to grand water resources visions and master plans for over five decades, some of which have never progressed beyond the drawing board, while others have reached advanced stages of implementation before partial or total abandonment by users (Sneddon, 2003; Floch et al, 2007; Molle and Floch, 2007). Molle and Floch (2008), with reference to Thai bureaucrats' and politicians' visions of transforming the northeast into an irrigated and agriculturally fertile region ('*Green Isaan*'), have termed the phenomenon as the 'Desert Bloom Syndrome' and compared it to similar schemes on the African and North American continents. Indeed, despite its somewhat distinct climate (significantly higher precipitation) and hydrology (greater relative flows and notable influence of Mekong hydrology) compared to other basins in the northeast (Blake, 2006), the Nam Songkhram Basin has long been incorporated as an integral target of the large trans-basin regional irrigation megaprojects foisted on *Isaan* by the state (Breukers, 1998; Floch et al, 2007), which inherently assumes the entire region as an area of water shortage. While plans for diverting Mekong waters into the Nam Songkhram Basin stayed on the drawing board, the Royal Irrigation Department (RID) with US Agency for International Development (USAID) financial support (from 1977 to 1991) constructed a large storage reservoir and irrigation project on the Nam Oon tributary (see Figure 7.1) during the late 1970s (USAID, 1995). This major project required the relocation of 1639 households (Floch et al, 2007) to create a reservoir with a storage capacity of 520 million cubic metres.

The developmental discourse until relatively recently has been dominated by modernization models and thinking, loosely based on notions of prosperity through market liberalization, adoption of science, technology and progress (Bello et al, 1998). These development aims of a modern nation state partly arose out of the 1950s Cold War and ideologically led American presence, providing significant funding for the United Nations Economic Commission for Asia and the Far East (ECAFE) and the subsequent Mekong Committee, forerunner to the MRC (Floch et al, 2007). However, the Lower Songkhram River Basin (LSRB) presented an additional 'problem' for the state planners and hired domestic and international consultants charged with 'developing the arid northeast' – namely, prolonged and reliable floods that annually inundated vast tracts of low-lying land on either side of the river channel (i.e. the floodplain). Rather than recognize the benefits and opportunities that this phenomenon presents (e.g. in terms of natural aquatic biomass production, aquifer renewal and soil fertility maintenance for dry season livestock grazing, and small-scale crop cultivation during flood recession), the floods were viewed by the state as an obstacle to agricultural development and a 'natural disaster' ('*utokapai*') in need of an engineering solution (Friend et al, 2006). This dominant paradigm still holds sway in key state water management agencies and was repeated by officials during the Mekong floods of August 2008 (*The Nation*, 2008).

Thus, naturally occurring dry season conditions and the wet season floodplain inundation in the Nam Songkhram Basin became recurring 'natural disasters'. This view entered into the standard lexicon of state planners and their contracted experts and technocrats as the principle problems begging solutions to free the region of poverty and deliver 'development' to the 'backward region' (Breukers, 1998; Blake, 2006). Essentially, this dogma has remained unchallenged until the present time in the unending so-called 'drama of development' (Lohmann, 1998). This over-simplistic, but ubiquitous, development tenet was articulated in a summary commissioned by the Interim Committee for Coordination of Investigations of the Lower Mekong Basin[7] (1983) to justify a massive project entitled The Nam Songkhram Basin Irrigation and Flood Control Development as: 'The Nam Songkhram Basin … includes large areas of fertile land that are not being used due to flooding, while other areas lie fallow during the dry season due to lack of water.' The empirical observation that local residents might be utilizing the diverse terrestrial and aquatic resources year round in non-agricultural uses (both intensively and extensively) apparently did not occur to the authors of the report, or if it did, was dismissed as being counter to predominant notions of central Thai agrarian society and the basis of 'development' itself.

Instead, this study called for the construction of a large dam (curiously termed as a 'regulator') at the mouth of the Nam Songkhram and extensive poldering using raised embankments and pumps (along Dutch lines) of vast areas of floodplain to create flood-free areas for intensive crop cultivation. Tributaries and the upper Nam Songkhram River were to be dammed to create several large shallow reservoirs with surrounding (higher elevation) areas being served by pump irrigation systems that could 'ultimately irrigate all irrigable land in the basin' (NEDECO/TEAM, 1983). This project's lofty aims became a standard point of reference for all subsequent state plans to tame the 'unruly' Nam Songkhram River.[8] Although progress in turning the vision to reality over the next decade was largely governed by frequent changes in political fortune, traditional bureaucratic lethargy and interdepartmental competition, by the early 1990s a new impetus was growing to construct a large irrigation project in the Nam Songkhram Basin (see Box 7.1) as part of the Khong-Chi-Mun Project (see Chapter 10). At the same time, rapid economic growth based on industrial exports and easy credit for newly created agribusiness ventures had allowed a politically well-connected company based in Bangkok (SunTech Group Ltd and associated businesses)[9] to acquire significant tracts of floodplain land in the Lower Nam Songkhram Basin, much of it former public land (*tee satarana prayote*) accessed by village communities (Blake and Pitakthepsombut, 2006a) that had been utilized by several surrounding communities under a common property regime (Blake, 2008). One source estimated that this company and its subsidiaries had gained possession of approximately 60,000 rai (9600ha) of low-lying flood-prone land that was progressively converted from primarily flooded forest ecosystem to intensive agricultural crop cultivation (*Watershed*, 1999).

Box 7.1 The Nam Songkhram Project

In 1983, the Nam Songkhram Project, building on earlier proposals made by the Interim Mekong Committee, proposed constructing a low dam with liftable water gates (termed a 'regulator') near the mouth of the Nam Songkhram River that would create an upstream reservoir extending to 255km^2. This new water source would then be used to irrigate a planned 90,400ha of farmland using a series of electric pumping stations around the perimeter for growing 'high-value crops'. The project was to be implemented by the Department of Energy Development and Promotion (DEDP), under the Ministry of Science, Technology and Environment that was also responsible for the controversial Khong-Chi-Mun Project. Numerous studies were conducted during the late 1980s and 1990s in order to justify the Nam Songkhram Project, some of which were never fully made public. The estimated cost of the project was at least US$400 million in 1995. The DEDP started purchasing land for the headworks at the dam site before cabinet permission to proceed with the project had been granted. The project ignited strong local and national resistance, with opponents fearful of the damage a dam might do to the river's ecology, productive fisheries and local livelihoods. It precipitated a villager-led and non-governmental organization (NGO)-supported opposition group to be formed in 1996 called the Nam Songkhram Conservation and Rehabilitation Club (partly inspired by the Pak Mun Dam resistance movement), which also campaigned for the land rights of communities affected by agribusiness company activities, especially loss of land to eucalyptus plantations. Despite the strong opposition to the dam and apparent outright rejection by the Thai Council of Ministers in 2002, it has remained on the agenda of the Royal Irrigation Department's plans for water development in the region up to the present day.

Source: Breukers (1998); Blake and Pitakthepsombut (2006a, 2006b)

The promotion and subsidization of export-focused agribusiness was an integral component of the state vision for north-eastern regional development, both as a part of the modernization thrust and as a way of utilizing the rich natural and human resources close to their source, encompassed in successive five-year national development plans through the 1970s and 1980s. This vision was encouraged by significant sums of foreign aid being channelled into promoting agribusiness and contract farming models by donors such as USAID at Lam Nam Oon Irrigation Project (USAID, 1995). While large sums were made available to agribusiness companies that benefited from direct financial subsidies and tax breaks for industrial-type agriculture projects, relatively little state funds were made available for promoting more bottom-up, participatory or sustainable models of rural development (Bello et al, 1998). At the same time, there was a fundamental shift in the labour market as rural people left the land in ever greater numbers; according to Baker and Phongpaichit (2005), by the mid 1990s almost four-fifths of total cash income of rural households was earned off-farm, including 43 per cent from wage work. Despite physically living in the city, these migrants were still technically considered as rural 'farmers' by the state, virtually located by house registration documents in their home villages (Rigg, 2003). Through exploiting a power nexus that included military, political and bureaucratic connections and weaknesses in Thai land laws and local institutions, the SunTech Group was able to systematically purchase large amounts of common property floodplain land at very low prices (Blake, 2008). In line with state policy for the region to promote agribusiness, SunTech also constructed a state-subsidized modern vegetable canning factory (see Figure 7.3) in 1988 with an adjacent dam on a wetland site in Sri Songkhram District near its own tomato and sweet corn plantations, which at the time were reportedly the largest of their kind in Thailand (Blake and Pitakthepsombut, 2006a). Previously, the same company grouping had also been involved in setting up an intensively reared dairy herd and milk-processing plant and cultivating fast-growing pulpwood plantations[10] on floodplain common lands that were extensively utilized by local people for fishing, livestock grazing and harvesting non-timber forest products (NTFPs) (Guayjaroen, 2001; Blake, 2008).

Significantly, despite the state assistance, tax breaks and low capital cost of (supposedly fertile) land, all of these agribusiness projects have subsequently failed and are now idle assets dominating the landscape (see Figures 7.2 and 7.3). One argument offered as to why SunTech's management failed to achieve success from its agribusiness model on the Nam Songkhram floodplain is that it failed to appreciate the challenging biophysical conditions that accompany an environment naturally under water for two to four months a year and the need to adapt the agronomic approach to the environment, not vice versa (see *Watershed*, 1996). However, this reality has not stopped a state-business elite and certain politicians from continuing to push for long-dormant irrigation projects to be resurrected, perhaps as an effective way of recapitalizing their idle land assets and recovering

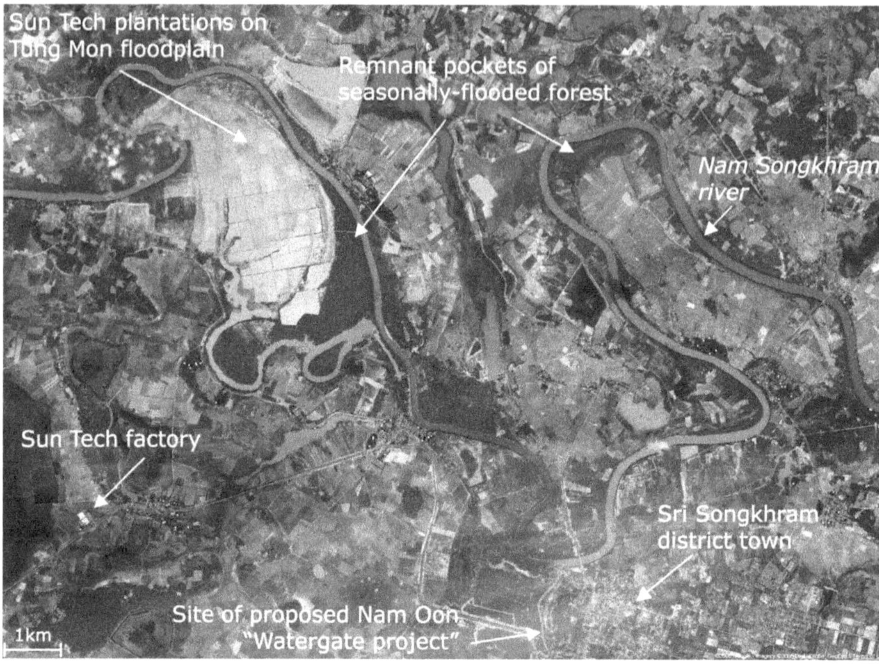

Figure 7.2 *Dry season 2005 satellite image of Tung Mon and the Lower Songkhram Basin floodplain area west of Sri Songkhram district town, Nakhon Phanom province*

Source: David J. H. Blake

bad debts through the provision of compensation for land flooded, or of providing an outlet for the RID to justify and spend its vast annual budget, which in 2001 totalled 33.5 billion baht, or about US$1 billion.[11]

ALTERNATIVE VISIONS, NEW APPROACHES

By the early 1990s, large-scale infrastructure development projects in Thailand were in the early stages of being obliged to consider rudimentary environmental issues following the enactment of the 1992 Environment Law, which required such projects to undertake an environmental impact assessment (EIA) before permission to proceed could be granted. In 1994 the National Environmental Board (NEB) rejected the EIA for the proposed Nam Songkhram Project (which had basically been fashioned from the earlier large-scale Dutch-inspired irrigation plans for the basin) after it was found that they were carbon copies of EIAs conducted for the Khong-Chi-Mun Project (Breukers, 1998). New studies were called for to

Figure 7.3 *SunTech Group vegetable processing factory complex in Sri Songkhram district*[12]

Source: David J. H. Blake

satisfy the NEB requirements, which were subsequently conducted by a Khon Kaen University team of academics and, in due course, were also rejected as being inadequate because of persistent doubts about project benefits (Blake and Pitakthepsombut, 2006a).

As the state-led model of 'development' encroached ever further into sensitive ecosystems and north-eastern rural dwellers' everyday lives, so a counter-discourse emerged (Breukers, 1998; Lohmann, 1998). This alternative discourse matched a growing national environmental NGO movement with local communities and activists witnessing the destructive tendencies of large infrastructure projects on local livelihoods. Initially only a handful of concerned Thai academics and domestic NGOs were involved in opposition to the Nam Songkhram Project, such as the Project for Ecological Recovery (PER) and its sister organization, Towards Ecological Recovery and Regional Alliance (TERRA). These organizations were instrumental in spearheading primary information-gathering from local people, setting up community resistance networks and wider dissemination of an alternative viewpoint that ran counter to the official techno-centric purported benefits from the project. Nationally, in the previous decade there had been several high-profile cases of combined villager and NGO resistance to large infrastructure projects, the most well known being the infamous Pak Mun Dam case (see Chapter 3). Nam Songkhram riparian communities living in the reservoir flood zone, concerned about the possible loss of homes, forests, rice fields and capture fisheries in the name of 'development', started to ask fundamental questions from state officials and to organize intercommunity flooded forest conservation networks. As word spread of the Nam Songkhram Basin's value as a wetlands ecosystem, so outside organizations (significantly, both state and non-state) began to take a greater interest in the Nam Songkhram Basin. Following the 1995 Agreement on the Cooperation for the Sustainable Development of the Mekong River Basin signed by Laos, Cambodia, Thailand and Vietnam, more decisive natural resources management projects emerged regionally, including the four-nation Mekong Wetlands Biodiversity Conservation and Sustainable Use Programme (MWBP), which established a so-called 'demonstration site' on the Lower Songkhram River Basin (LSRB) wetlands in 2003.[13]

After the recognition of the LSRB as a 'wetlands of international significance'[14] by the Office of Environmental Policy and Planning (1999), further studies on fisheries (e.g. Boonyaratpalin et al, 2002; Suntornratana et al, 2002), socio-economics and culture (e.g. Brenner, 2003) emerged. The later studies by Thai and foreign researchers articulated a more nuanced and rounded view of the complexity of the wetlands and multi-component livelihoods of the local people (e.g. Blake and Pitakthepsombut, 2006a, 2006b) than seen in the earlier, rather sectoralized, EIA and environmental mitigation and management reports produced (e.g. NEDECO/TEAM, 1983; Khon Kaen University, 1999). However, the studies were still mostly being conducted by outsiders for outsiders according to individual and institutional agendas and worldviews, with the information obtained being extracted and not shared or validated back at the community level where it was first generated.

It should be recognized that through certain important statutes enshrined in the 1997 Thai Constitution that encouraged greater local participation in natural resource management (sections 46, 56 and 79), decentralization initiatives, new national environmental legislation and ratification of various international conventions (e.g. the Ramsar Convention[15] in 1998), an environmental conservation discourse began to develop amongst various sectors of Thai society. Of particular relevance was the reorganization of local government to allow elected representatives to work at sub-district level (*tambon*) as a move towards bureaucratic devolution. While greater social and political pluralism were increasingly evident, the state-led approach still emphasized control over people, spaces and resources, where local people were considered 'threats' to conservation goals, and main strategies proposed were to raise environmental awareness, education to overcome local ignorance, and ring-fence 'protected areas' against encroachment. With broad, seasonal floodplain wetland sites, such as the LSRB, and their complex local livelihood mix, cultural heterogeneity, power relations, ill-defined boundaries, etc., uni-sectoral state agencies were presented particular challenges that the state-centric model found hard to appreciate or accommodate. Even within a supposedly cross-disciplinary agency like the Office of Natural Resources and Environmental Policy and Planning (ONEP), there was a definite tendency to push MWBP towards compiling a Lower Songkhram River Basin management 'master plan' from an early stage of the project before empirical fieldwork had been allowed to articulate the complexity of the situation or explore local visions and realities.

The generation and ownership of knowledge has long been a contested issue in the traditionally hierarchical Thai society with its strong nation-state ethos (Rigg, 2003; Baker and Phongpaichit, 2005). Thus, local ownership and input into research processes initiated by external agencies (including universities) was consistently low or non-existent. This was an implicit barrier for meaningful engagement with the local society to generate a deeper understanding of the opportunities and constraints that communities faced. This reality was recognized by the MWBP Lower Songkhram Basin Demonstration Site staff and partners during the set-up phase, allowing initiation of activities to specifically address the deficiency. While the overarching aim of all MWBP activities was capacity-building for better wetlands management at all levels, the demonstration site was also keen to field test and adapt promising new approaches to research that seemed locally appropriate for gaining knowledge of the wetland resources and engendering co-management practices.

FIELD TESTING OF NEW PARTICIPATORY APPROACHES

Taking the lead from the apparently successful implementation of a so-called '*Tai Baan Research*' approach in communities situated along the dam-impacted Lower Mun River in southern *Isaan*, MWBP facilitated the same Thai NGO that had developed the concept to adapt the methodology to the Nam Songkhram River. Four communities in Sri Songkhram District were selected to take part in the initial Tai Baan Research effort, with about 240 villagers volunteering to participate as co-researchers alongside a handful of young project-employed facilitators. In close cooperation, the facilitators and villagers planned and carried out extensive field-based empirical research on local culture and national resource utilization under six sub-theme headings – namely, fish; fishing gear; flooded forest vegetation; floodplain agricultural systems; livestock-raising; and local ecosystems.

The Tai Baan Research methodology, as applied in the LSRB, can essentially be distilled down to 13 practical steps, according to Blake and Pitakthepsombut (2006b). However, they stress that the list of steps is not prescriptive but should be adapted according to specific local situations and varying socio-political contexts. The authors also point out that the approach requires adequate time, patience, perseverance and adaptability on the part of the researchers (resource users) and external facilitators to succeed. The research methodology required the facilitators to virtually live in the villages, thus getting to know the communities intimately and to gain the trust of the people involved. At the same time, village participants were able to conduct research in their chosen sub-theme as they went about their normal livelihood activities; over time, a detailed picture was constructed linking the various components of their livelihood systems.

Periodically, the villages and sub-theme groups held inter- and intra-village meetings to exchange and validate their findings and observations, making the research similar to conventional scientific knowledge methodologies, but with the added benefits of context and holism. For example, with regard to the floodplain ecosystem, the Tai Baan researchers were able to categorize 28 distinct habitats or 'sub-ecosystems', which they recognized as being important components of the whole floodplain system. This is a more detailed and complex ecosystem classification system than any existing official wetlands classification has achieved. After over two years of field research, results analysis and peer review, the findings were published in Thai language and disseminated, including distributing copies to all of the local researchers (Tai Baan Research Network of Lower Songkhram Basin, 2005a, 2005b). The recognized success of the Tai Baan Research process in Thailand was a significant factor in its later adaptation to Mekong communities in the Stung Treng area of northern Cambodia (also a MWBP demonstration site), and there was an active exchange of villagers and facilitators from the Nam Songkhram to Stung Treng during 2005 to 2006 as a Cambodian equivalent – *Sala Phoum Research* – was developed in line with the local context (see Figure 7.4). A crucial part of the Tai Baan Research process was the close involvement of related government staff at sub-district, district, provincial and, in some cases, national levels, which not only helped to inform external stakeholders about the research findings, but also helped to raise the legitimacy of the process to a level not previously recognized at other sites of Tai Baan Research (e.g. Mun Basin, Salween and Upper Mekong in Chiang Rai Province), where the process was perceived as being more antagonistic to state development goals (Scurrah, in preparation). As a result, several senior ranking government officials in Sri Songkhram District and Nakhon Phanom Province became vocal proponents of Tai Baan Research in government circles, while Tai Baan Network leaders were invited to sit on state-led committees and attended wider forums previously inaccessible to local resource users. Thus, local people engaged in the Tai Baan process were able to articulate a different local reality and alternative aspirations through new channels and media,[16] often for the first time.

Like any research approach that is relatively new and untested across a range of situations, there were certain inherent weaknesses and question marks that could be identified about the methodology employed and results obtained. First, because the research explored locally situated knowledge, the findings from the four communities involved cannot necessarily be extrapolated across the lower basin, although the basic methodology is readily transferable elsewhere. There are also issues of linguistic, ethnic and cultural diversity to consider and adequately reflect in the process and research outputs, which can pose challenges in culturally hegemonic societies. In the Lower Nam Songkhram Basin, for example, there are at least six ethnic minorities represented locally (Lao, Phu Thai, Tai-So, Nyaw, Galeung and Chinese), as well as the dominant central Thai influence, each of whom have their own cultural and linguistic identities. When documents are published or audio-visual material prepared, a balance has to be struck between a full and accurate representation of this local diversity and a compromise for the intended audience, which could be regarded as a dilution of complex local reality and a potential source of contention. There is also the issue of sustainability to be considered (i.e. to what degree is the process dependent upon external sources of funding for its momentum and what measures are available to ensure self-perpetuation following initial assistance?).

Figure 7.4 *Plant specimens from beside the Mekong in Stung Treng Province, Cambodia, during an exchange visit for members of the Tai Baan Research Network in Nam Songkhram with the Sala Phoum researchers in Cambodia*

Source: David J. H. Blake

Influenced by the Water and Nature Initiative's (WANI's, under IUCN) stated goal of contributing to 'improved river basin management by mainstreaming a livelihoods-ecosystem approach to water resources management' and a first Mekong region application in the Huong River Basin of central Vietnam (IUCN, 2005), it was decided to trial an Environmental Flows, or E-Flows, approach (refer to Box 7.2) in the Nam Songkhram Basin during 2006 to 2007. This decision was the result of an extensive dialogue amongst various basin stakeholders from state and non-state institutions, including representatives from the Thai National Mekong Committee (TNMC) and the Mekong River Commission (MRC), who recognized the need for improved tools for water management decision-making in northeast Thailand. It was also seen as being complementary to the ongoing much larger, financially costly and spatially disparate Mekong Integrated Basin Flow Management (IBFM) Project,[17] (and involved some of the same partners), and Thai National Water Policy objectives of encouraging greater decentralization, local participation in planning and application of integrated water resources management (IWRM) principles. Thus, the Department of Water Resources (DWR) Region 3 Office in Udon Thani responsible for water resources planning in the Nam Songkhram Basin took an active interest in the E-Flows study and dispatched a team of hydrologists to assist with flow measurements during fieldwork (see Figure 7.5)

A key point regarding the Nam Songkhram Basin is that it encompasses parts of four provinces, (each with its own natural resources development plans and priorities), and has a plethora of government departments and agencies working on water resources-related issues (frequently with conflicting goals and priorities). With bureaucratic reforms and new discourses raising the importance of civil society, decentralization and local participation in governance seemingly

Box 7.2 Environmental flows

Environmental Flows (E-Flows) is a rapidly growing area of interest amongst certain sectors of environmental and water resources management institutions and is still a young discipline. There are various conceptions about what E-Flows is and what are the most appropriate methodologies to employ from amongst the over 200 identified (e.g. Tharme, 2003), which becomes evident in the discourse between actors and locations. One concept of E-Flows contends that it has strong ecosystems and livelihoods dimensions; thus, any modification of river flows is explicitly considered in terms of ecological, economic, social and cultural issues. Hence, E-Flows requires the integration of a range of worldviews and disciplines, making it inherently holistic in approach but liable to misunderstanding due to complexity. A useful guidebook for understanding E-Flows has been produced by the IUCN (*Flow: The Essentials of Environmental Flows*) and has recently been translated into the riparian languages of the Mekong Basin (see Dyson et al, 2003).

Figure 7.5 *Members of the multidisciplinary E-Flows study team prepare to take to the water in the Lower Songkhram River Basin with flow measuring equipment during the August 2006 flood*

Source: David J. H. Blake

spelled an era of more representative decision-making over natural resources during the early years of the 21st century (Baker and Phongpaichit, 2005). The reality on the ground, however, has proven to be somewhat more complex. Nominally, water resources planning is managed by the DWR under a nascent 'River Basin Organization' system, which in the case of the Nam Songkhram Basin is split into six arbitrary sub-basin committees (which do not meet together as a matter of course), rather than as one complete and distinct basin-wide entity (Blake and Pitakthepsombut, 2006a). This situation, unsurprisingly, has led to much agency overlap, competition and confusion over areas of responsibility, supposed beneficiaries, articulating a coordinated future development model, or visualizing a coherent basin overview. The process of compiling a river basin development plan was tendered to five consultancy companies, some with names similar to transnational companies (DWR, 2004), thus raising suspicions about their impartiality. It is not uncommon to find DWR plans for irrigation projects overlapping with Royal Irrigation Department (RID) plans for irrigation projects, which in turn may eclipse or overlap with Tambon Administration Organization

(TAO) plans for more localized irrigation projects as they vie for different budget sources. The two common factors to all these projects is a simple belief that the primary problems of the area are drought and floods (see Breukers, 1998) and all offer a prescriptive solution involving river engineering, landscape change and radical flow alteration.

As a means of addressing the institutional malaise and confusion described above, an E-Flows (see Box 7.2) study was proposed to offer a potential alternative approach that fitted in with a growing IWRM rhetoric at national level. It was anticipated that some of the complexity of the basin that had eluded earlier grand project plans could be captured and communicated through a multidisciplinary approach, using river flows as the basis of a common understanding between the relevant sectors and disciplines. After reviewing existing literature and knowledge on the subject, the E-Flows study team prioritized the following main objectives for the study:

- Improve understanding of E-Flows concepts and the importance of managing flows to ensure downstream ecological, economic and socio-cultural benefits.
- Build local technical and institutional capacity to apply concepts and integrate E-Flows principles into basin management plans.
- Initiate stakeholder dialogue across the four provinces of the Nam Songkhram Basin to consider implications of possible future development scenarios for the basin (and beyond) on the environment and society.
- Understand ecosystem roles in people's livelihoods, especially seasonally flooded forests, by establishing the present-day relationships between flow regime, ecology and human-level dependencies.
- Develop an appropriate set of tools and methodologies that can potentially be applied to other Thai river basins for those interested in applying E-Flows approaches at a later date.

The fieldwork was carefully planned to coincide with the peak flooding period (early September) and period of minimum flows (March), so the river system could be assessed at its two hydrological extremes and related to the wider ecosystem and floodplain resource usage by local people. Local and international specialists in the fields of freshwater wetlands ecology, fisheries, socio-economics, hydrology/geomorphology, land use/agriculture and wetlands vegetation were gathered together to create a cross-disciplinary synergy and learning environment that would not have been possible in a traditional type of EIA or mono-disciplinary expert input study. Following the field surveys, the specialists met on several occasions to discuss their observations and provide an 'expert judgement' (King et al, 2000) or opinion on the present-day ecological health and degree of human-induced degradation of the three representative sites studied, validated internally within the group initially and then disseminated to a range of basin stakeholders at a seminar. In this manner, the Intermediate E-Flows Assessment Process (Tharme,

2003) was able to provide an approximate 'health check' of the riverine ecosystem in 2006/2007 and make predictive statements based on reasoned arguments about potential future condition under three development scenarios. The first scenario assumed the construction of the Nam Songkhram Dam and smaller Nam Oon 'Watergate Project' (see Figure 7.2 for project location) inundating much of the floodplain; the second scenario assumed the implementation of the pan-regional Water Grid irrigation project with inter-basin water transfer and intensive agriculture dominating; while the third scenario assumed a continuation of the existing 'business-as-usual' approach, with many more small irrigation and 'flood protection' projects constructed alongside promotion of agricultural intensification, but no major irrigation infrastructure built. These scenarios were considered and assessed by the E-Flow team at a scenarios workshop held in May 2007, prior to the multi-stakeholder meeting a few weeks later (Blake, 2009).

An Intermediate E-Flows Assessment, by definition, allows only a relatively narrow timeframe for field surveys and direct empirical research, which necessarily limits some of the more detailed observations that might be made from a temporally and spatially more extensive study. A considerable portion of research time was devoted to a literature review and the gathering of secondary data sources. While these were useful for providing background knowledge and context for both individuals and the group, as a whole, the gaps in understanding and knowledge of the Nam Songkhram Basin became more apparent to the team in each discipline, although some (e.g. fisheries) can draw on a wider research literature than others (e.g. floodplain wetland botany). Integrating local knowledge terms and concepts, such as that derived from the Tai Baan Research approach, within the more techno-scientific approach and language of the E-Flows Assessment scenario workshop presented a number of challenges to the team. The resolution of this challenge needs to come from an open dialogue about best methods and approaches to employ at each step of the E-Flows process, where no one discipline should dominate over the rest, and taking into account methodological and conceptual diversity. Thus, it must be implicitly recognized that the interpretation and application of E-Flows methods will, to a large extent, depend upon the agenda and goals of the institution, or even the individuals applying them, the resources available, as well as the circumstances of the river's flow regime.

Old plans, new disguise?

It is frequently observed in Thailand that large-scale irrigation projects proposed by the state are rarely abandoned entirely, but tend to periodically reappear, sometimes dressed in new guises. A good case in point is the resurrection (ironically, on World Environment Day) of the controversial Kaeng Sua Ten Dam Project in Phrae Province (*Bangkok Post*, 2008). The Nam Songkhram Project has proven no exception. Despite a decision in March 2002 by the Thai Council of Ministers to respect the findings of the NEB and to reject the DEDP plan to develop the Nam Songkhram Project, the project has subsequently reappeared after a period of several years' slumber, pushed by influential actors. While the DEDP was formally dissolved in 2002, the unfulfilled plans for the Nam Songkhram Project were passed on to the RID in the Ministry of Agriculture and Cooperatives. Meanwhile, the newly created Department of Water Resources (DWR) was given the mandate to plan, manage and conserve water resources nationwide under the provisions of the National Water Policy of October 2000 (see Blake and Pitakthepsombut, 2006a). The RID retained the option to develop the Nam Songkhram Project and included it in long-term basin plans of the Basin Development Committee, while senior politicians and even the then Prime Minister Thaksin Shinawatra, making a flying visit to Sri Songkhram District in August 2005 during a period of natural flooding, called for its rapid construction in order to combat 'disastrous' flooding and to provide dry season irrigation (Blake and Pitakthepsombut, 2006b). Smaller versions of the Nam Songkhram Project, designed to provide dry season irrigation by permanently inundating significant areas of floodplain land for storage reservoirs, were also inserted into provincial and basin development plans by the RID, including a 300 million baht (approximately US$8.5 million) plan to impound the largest Songkhram tributary, the Nam Oon, at the point just above its confluence with the Nam Songkhram (see Figure 7.2). The reservoir area also happened to coincide with some of the richest remaining areas of *paa boong paa thaam* and abandoned plantation land belonging to SunTech Group Ltd, now largely colonized by the invasive noxious weed *Mimosa pigra*.

Not to be outdone, the DWR has formulated ambitious water infrastructure plans of its own to rival the Nam Songkhram Project, also drawing on the legacy remaining from DEDP's Khong-Chi-Mun Project to create a national Water Grid Project (Molle and Floch, 2007; see Chapter 10). While the supposed problems and solutions essentially remained identical to those of the Khong-Chi-Mun Project (with the exception of transferring 'surplus' water transboundary from rivers in Laos, as opposed to pumping it from the Mekong mainstream), the ambitious scope and budget required to implement the project had considerably increased from earlier estimates to US$5 billion (Molle, 2007).

Sneddon (2003) refers to the 'manipulation of water' and 'reconfiguration' of river basins that these grandiose projects inevitably entail and how they 'constantly rework the politics of scale through the extension and contraction of networks of humans and non-humans'. Under the proposed Water Grid Project, although details are still sketchy, a favoured model calls for the promotion of intensive cultivation of cash crops (including for biofuels and plantation forestry) by private farmers and agribusiness interests (Anonymous, 2005), apparently in a manner not dissimilar to failed past attempts such as that attempted on Tung Mon and other floodplain wetlands of the Nam Songkhram Basin over the past three decades. It is instructive that even the abundant free water provided by the RID's Nam Oon Project and agribusiness model courtesy of USAID does not seem to have convinced local people to remain on the land and farm, with much land left vacant during the 2006 to 2007 dry season (first author's personal observations). Despite political turbulence and changes in government over the last two years, there have been repeated attempts to kick-start the vote-catching Water Grid Project by some leading politicians and senior bureaucrats, expressed most recently by Prime Minister Samak Sundaravej, describing it as 'a high-priority project', an announcement that attracted opposition from a range of civil society organizations (*The Nation*, 2008).

The spectre of renewed attempts to implement large-scale irrigation projects in the Nam Songkhram Basin has taken on added resonance since the Office of Natural Resources and Environmental Policy and Planning announced in 2006 that it would like to consider submitting the Lower Songkhram River Basin as a possible candidate site for the Ramsar List of Wetlands of International Importance (Blake and Pitakthepsombut, 2006b), and as the significance of the LSRB wetlands becomes progressively better understood by wider external actors. This potential extra dimension as an internationally recognized wetland with high biodiversity conservation significance, weighed against the alternative state-sponsored vision as a site of intensive irrigated agriculture using ex-basin water resources and extreme landscape transformation (as seen at heavily degraded sites such as the Tung Mon floodplain shown in Figure 7.2), raises fundamental questions about the nature of these competing visions. For example, can wetland biodiversity and current livelihood values be maintained if a massive inter-basin irrigation project, accompanied by inevitable native forest destruction and conversion to agricultural land, is allowed to run its *course to completion*?

As a site of competing stakeholders, discourses and visions of 'development', especially with regard to water management, it is a difficult (perhaps impossible) task to succinctly encompass the entire spectrum of actors and voices that exist and vie for attention on the multiple scalar (spatial and temporal) levels of the Nam Songkhram Basin. To learn more about the roles, authority and responsibilities of the various state agencies and other non-state actors involved in the basin, the reader is encouraged to refer to Blake and Pitakthepsombut (2006a, Chapter 6, p70) for a fairly comprehensive listing. While the Nam Songkhram Basin has largely avoided the outright conflict and violence seen in the Mun Basin, largely concerning competing visions over state infrastructure development projects, nevertheless, fault lines and conflicts exist between and amongst actors at every scalar level, from national state agencies and ministries, right down to the local level. Tensions at the national level between competing agencies are apparent in tussles over control of generous national budgets allotted to water management and political manoeuvrings to promote particular megaproject visions. Also apparent and relatively well documented are struggles between state and non-state actors (e.g. Breukers, 1998; Lohmann, 1998) over the Nam Songkhram Project, land disputes and issues of participation in decision-making. Less apparent are conflicts at sub-district and community level, which may be equally divisive, hard to solve and may detract from goals of integrated natural resources management. At the heart of these disputes are usually issues of unequal power distribution between actors, sometimes coupled with unsatisfactory methods of conflict resolution. One such protracted dispute over land use and water management split the community of Ban Pak Yam, which had formerly been a site of Tai Baan Research and nearly led to violence between the parties on several occasions.

At the same time as Thailand grapples with the complexity of multi-stakeholder management opportunities, new external threats start to loom over the contested waterscape of the LSRB. It is not only upstream hydropower developments on the Lancang-Mekong in China's Yunnan that have the potential to significantly alter the hydrological characteristics of the Nam Songkhram's flood pulse; hydropower developments closer to hand, such as the series of dams being built or planned for the Nam Ngum and Nam Theun-Kading river basins in Laos, could seriously deplete capture fisheries production for communities dependent upon the Nam Songkhram floodplain for their livelihood. The proposed Nam Theun 1 Hydropower Project alone is predicted to cause an 8 per cent reduction in Mekong wet season flows downstream of the confluence of the Nam Theun-Kading river (Norplan and EcoLao, 2007), which will inevitably reduce the extent and magnitude of the critical flooding on the LSRB floodplain wetlands and diminish living aquatic resources production. Cumulatively, the scale of hydropower development planned for Mekong tributaries in Laos could be devastating for fishery production for Mekong-linked floodplain fisheries in Thailand, as well as in Laos, Cambodia and Vietnam. Yet, ironically, Thailand is a major investor in the same projects that could further impoverish vulnerable parts of its rural population and degrade critical wetlands habitats. Any attempts to objectively define and measure losses and gains of various sectors of the economy through the alluring terminology of 'trade-offs' are fraught with difficulty and risks of major oversight (Friend and Blake, 2007).

Conclusions

The Nam Songkhram Basin has been an arena of externally imposed 'developmentalism' and associated contestation amongst and between actors, accompanied by sustained natural resource commercialization and degradation, for the past five decades. The LSRB's extensive floodplain viewed as a relatively remote and ecologically resilient wetland, subject to climatological and hydrological extremes, has helped to maintain it as an agricultural frontier for longer than other nominally protected forest zones in upland areas of northeast Thailand. This apparent irony, the inherently rich natural resource base, and cheap and plentiful land available on the floodplain helped attract a steady stream of migrants to the area from other northeastern provinces up until a few years ago (Blake and Pitakthepsombut, 2006a). By the same token, a lack of legally secure land tenure rights and policy opportunities also attracted the attention of national politicians and business entrepreneurs, who were able to systematically turn vast common property regimes into private land assets and new monocultural agricultural regimes. However, despite strong state support and large investments of technology and capital, these ventures still managed to fail, leaving behind a socio-ecologically transformed landscape with an uncertain future. Similarly, grand visions and plans put forward by politicians, bureaucrats and technocrats to turn the basin into one large irrigation scheme have so far come to little, despite decades of external 'expert' studies, hundreds of closed meetings and strong central government encouragement at various times. This chapter has argued that it is a past failure of government to engage with the local, the grassroots and the complex ecosystem–livelihoods relationships that has led to the present confused situation, where state agencies appear to be working towards incompatible development outcomes. At the same time, it is recognized that villagers themselves are not homogeneous actors and also display complex patterns of contestation at the local level.

Over the past decade or so, new and alternative representations of the values and potential of the Nam Songkhram Basin have emerged, partly as a result of local resistance struggles over key resource management issues, particularly land, forests and fisheries. External institutions have engaged with diverse local actors to pilot novel participatory approaches to wetlands and river basin management and ecosystem understanding. Tai Baan Research and an Intermediate E-Flows Assessment have both thrown considerable light on the complex human–nature and ecosystem–hydrology–livelihoods relationships that exist in a tropical riverine floodplain wetlands system. They approach the 'problem' from different angles than conventional research methodologies and, as such, are complementary rather than contradictory.

This increased awareness of complexity through participation in the process could reduce the tendency to make simplistic conclusions about the best ways in which to solve assumed problems and impose exogenous 'development' solutions. Tai Baan Research and E-Flows both provide practical and logical alternatives to the state-favoured 'master plan' approach and traditional view that naturally high and low seasonal river flows are problems seeking a solution. On the socio-economic side, traditional stereotyped images of rural livelihoods were challenged in both approaches – for instance, the dominant notion that villagers should be regarded as primarily 'rice farmers who fish', as opposed to, say, 'fishers who farm rice' or 'wetlands product harvesters who occasionally practice agriculture', or other more nuanced overviews of multi-component livelihoods, deriving income from on- and off-farm sources.

Essentially, both Tai Baan Research and E-Flows approaches are holistic approaches to building knowledge of complex wetlands systems that allow more enlightened planning and management of these valuable resources in the Lower Mekong Basin. As such they could be regarded as potentially useful tools that invite use and adaptation to individual cases by river basin or wetland managers, whether they are local community networks, state agencies or NGOs seeking to better understand complex environmental and social linkages. Neither approach should be regarded as a panacea for solving rising environmental-based conflicts in the Mekong region, but used in conjunction with other multi-stakeholder approaches can help to reduce the risk of these conflicts escalating at the local and wider basin levels. It remains to be seen, however, if the incredibly productive Nam Songkhram floodplain wetlands ecosystem will be able to withstand the growing external influences on land and water resources brought about by regional development projects, especially hydropower, agribusiness and the growing pulp and paper industry.

NOTES

1 'Fisheries' in this chapter is used to denote all living aquatic resources that are commonly harvested in wetlands, including fish, plus species of amphibians, reptiles, insect larvae, molluscs, crustaceans, birds and aquatic plants (macrophytes and algae).

2 *Paa boong paa thaam* is the local Lao term for the distinctive, but poorly recognized, seasonally flooded freshwater swamp forest that was until relatively recently a common natural habitat found along Mekong tributary rivers in northeast Thailand and parts of Laos. It is known to harbour a wide plant biodiversity, with at least 191 species being utilized by local people, according to the findings from Tai Baan Research (Tai Baan Research Network of the Lower Songkhram Basin, 2005a)

3 The 'flood pulse' concept first coined by Junk et al (1989) is now a widely accepted scientific term and ecological concept understood to occur in lowland river-floodplain systems in tropical regions with highly seasonal rainfall patterns, helping to create massive bursts of aquatic and terrestrial bio-productivity following episodic flooding events.

4 This chapter refers regularly to the term 'wetlands' as a commonly used term amongst development professionals, especially those from an environmental background, but recognizes that it is a relatively unfamiliar term in the context of the Mekong Basin and does not translate easily into local languages. Therefore, it has been the subject of some debate and confusion as to the term's exact definition and correct application across the basin.

5 Srisakara Vallibhotama, a prominent Thai historian, has described the local culture found in riparian communities such as Ban Pak Yam as being '*wattanatam pla daek*', translated literally as 'fermented fish culture'. This distinction arises from the observation that this humble processed product, somewhat disdained in central Thailand as being an 'uncivilized' element of Lao culture, is at the heart of the local socio-economy and has been used as a staple trading item for centuries (Petchkam, 1997).

6 Both of these structures, although described as 'weirs' by the Department of Water Resources (DWR), are, in fact, 'dams' by any normal definition as they provide a complete barrier to flow for several months of the year. They incorporate liftable metal water gates that were not functioning within a year of completion, having been apparently vandalized by local people for scrap metal. Upstream, the river has been

dredged, widened and earth embankments built, supposedly to protect the floodplain against flooding, although, in practice, this function is not fulfilled.

7 This institution was the temporary name for what was later to become the Mekong River Commission, following the 1995 Mekong Agreement between the governments of Cambodia, Laos, Thailand and Vietnam.

8 Interestingly, the NEDECO/TEAM (1983) study concluded that neither upstream reservoir nor the 'regulator' structure at the mouth of the Nam Songkhram would be economically feasible in controlling floods.

9 SunTech Group Company Ltd is part of a conglomerate of interconnected agribusiness companies that have been operating in the floodplain of the Nam Songkhram Basin since 1978 and later expanded their business operations to Laos and Vietnam under the name Asia Tech Group Ltd. A recent report suggests that the company has now changed its name to Apex Development Public Company Limited (http://wrightreports.ecnext.com/coms2/reportdesc_COMPANY_C764H6820).

10 During the early 1990s, SunTech and its allied companies had a plan to construct a large pulp and paper mill in the LSRB to utilize the eucalyptus and acacia plantations that it had established and promoted with local communities as a way of overcoming the flooding problem with a flood-tolerant tree species and for creating local prosperity (*Watershed*, 1996).

11 RID, which reportedly accounts for around 50 per cent of the Ministry of Agriculture and Cooperatives annual budget (Budhaka et al, 2002), spends the majority of its income on infrastructure development and is committed to expanding the irrigation area throughout the Kingdom of Thailand by ensuring greater water storage capacity. In lowland flood-prone areas, such as the LSRB, this aim is challenging without either sacrificing large areas of land to reservoirs or using small-scale pumped irrigation projects from permanent water bodies.

12 The factory was closed and remaining staff were laid off in 2007.

13 MWBP was a joint programme of the four riparian governments of the Lower Mekong Basin – Cambodia, Laos, Thailand and Vietnam – implemented by the World Conservation Union (IUCN), the United Nations Development Programme (UNDP) and the Mekong River Commission (MRC), in collaboration with other key national and international stakeholders. With core funding provided by the Global Environment Facility (GEF), the programme aimed to address the most critical issues for the conservation and sustainable management of wetlands natural resources throughout the Lower Mekong Basin during two phases. The programme was closed down by donors at the end of phase A in December 2006 despite considerable progress towards its aims over the first two and a half years.

14 It is noteworthy that despite recognizing the international importance of the Nam Songkhram wetlands, apart from one discrete freshwater lake wetland called Bung Khong Long non-hunting area in the north-eastern sector of the Nam Songkhram Basin being declared a Ramsar site in June 2001, no part of the wetland is included in any formal or state protected area, with much of the most valuable habitat officially classed as degraded forest or a wasteland awaiting conversion to agricultural production land.

15 The Convention on Wetlands, signed in Ramsar, Iran, in 1971, is an intergovernmental treaty that provides the framework for national action and international cooperation

for the conservation and wise use of wetlands and their resources. There are currently 158 contracting parties to the convention, with 1758 wetland sites, totalling 161 million hectares, designated for inclusion in the Ramsar List of Wetlands of International Importance.

16 In addition to the research findings being published in an illustrated book format, they were also widely disseminated through other media such as posters, postcards, bookmarks, local radio, video and displays at various events.

17 This project apparently aims 'to provide information and knowledge to decision-makers on the predicted costs and benefits of water resources development in the Mekong Basin in relation to changes in river flow regimes' (MRC, 2006).

References

Anonymous (2005) 'Water Grid: Why and how?', Powerpoint presentation given at Research and Development Institute, Khon Kaen University, 23 February

Baker, C. and Phongpaichit, P. (2005) *A History of Thailand*, Cambridge University Press, Cambridge, UK

Bangkok Post (2008) 'Samak to revive old dam project', *Bangkok Post*, 6 June, Thailand

Bello, W., Cunningham S. and Kheng Poh, L. (1998) *A Siamese Tragedy: Development and Disintegration in Modern Thailand*, Zed Books Ltd, London

Blake, D. J. H. (2006) 'The Songkhram River Wetlands: A critical floodplain ecosystem of the Lower Mekong Basin', Paper presented at the International River Symposium 2006, Brisbane, Australia, 4–7 September 2006, www.riversymposium.com/index.php?element=06BLAKEDavid

Blake, D. J. H. (2007) '14 days on the river', *Bangkok Post*, Outlook Section, 17 November, www.bangkokpost.com/171107_Outlook/17Nov2007_out001.php

Blake, D. J. H. (2008) 'The multi-dimensional commons of the lower Nam Songkhram River wetlands, Thailand', Paper submitted to the 12th Biennial Conference of the International Association of Study of the Commons, 14–18 July 2008, University of Gloucester, Cheltenham, UK (available from author)

Blake, D. J. H. (ed) (2009) *Nam Songkhram River Basin E-Flows*, Forthcoming report for IUCN, Asia Regional Office, Bangkok

Blake, D. and Pitakthepsombut, R. (2006a) *Situation Analysis: Lower Songkhram River Basin, Thailand*, Mekong Wetlands Biodiversity Conservation and Sustainable Use Programme, Bangkok, Thailand

Blake, D. J. H. and Pitakthepsombut, R. (2006b) 'The Songkhram River wetlands of Northeast Thailand and participatory resource research initiatives', in Li, K. Y. (ed) *Knowledge, Culture and Livelihood: The Struggle for Water and People's Everyday Resistances – Asian Exchange*, vol 21, no 2/vol 22, no 1, Asian Regional Exchange for New Alternatives, Hong Kong/Seoul

Boonyaratpalin, M., Kohanantakul, K., Sricharoendham, B., Chittapalong, T., Termvitchagorn, A., Thongpun, W. and Kakkaew, M. (2002) 'Ecology, fish biology and fisheries in the Lower Nam Songkhram River Basin', *Journal of Fisheries*, vol 55, no 6, pp491–514 (in Thai)

Brenner, V. (2003) *Utilization of Floodplain Vegetation in Northeastern Thailand: Compilation of Survey Results from Ban Pak Yam, a Village in the Songkhram River Basin*, SEFUT Working Paper no 8, August, Albert-Ludwigs-Universitat Freiburg, www.sefut.uni-freiburg.de/

Breukers, S. (1998) 'Who defines the "problem", who defines "development"? The case of the Songkhram Irrigation Project', *Watershed*, vol 4, no 2, pp25–32, TERRA, Bangkok

Budhaka, B., Srikajorn, M. and Boonkird, V. (2002) 'Thailand country report on investment in water', in *Proceedings of the Regional Consultation, Bangkok, Thailand*, 3–5 October 2001, FAO Document repository, RAP Publication 2002/09, Originated by Regional Office for Asia and Pacific, Bangkok, Thailand, pp325–338, www.fao.org/docrep/005/ac623e/ac623e0m.htm

Choowaew, S., Chandrachai, W. and Petersen, Jr. R. C. (1994) 'The socio-economic conditions in the vicinity of Huai Nam Un Wetland, Songkhram Basin, Lower Mekong Basin, Thailand', *Mitt. Internat. Verein. Limnol*, Stuttgart, vol 24, pp41–46

DWR (Department of Water Resources) (2004) Handout document to second Sub-Group Meeting, State of the Lower Songkhram Basin, Project for Participatory Water Resources Management in the Mekong Basin of Isaan, Proposed by Sanyu Consultants (Thailand) Ltd, Macro Consultants Ltd, Tesco Ltd, Thai DCI Ltd and Southeast Asia Technology Ltd, Department of Water Resources, Ministry of Natural Resources and Environment (in Thai)

Dyson, M., Bergkamp, G. and Scanlon, J. (eds) (2003) *Flow: The Essentials of Environmental Flows*, IUCN, Gland, Switzerland, and Cambridge, UK

Floch, P., Molle, F. and Losikandl W. (2007) *Marshalling Water Resources: A Chronology of Irrigation Development in the Chi-Mun River Basin, Northeast Thailand*, Working Paper MP-2007-02, Unit of Social and Environmental Research, Chiang Mai University, Thailand, www.mpowernet.org/download_pubdoc.php?doc=3629

Friend, R. M. and Blake, D. J. H. (2007) 'Negotiating trade-offs in water resources development in the Mekong Basin – implications for fisheries and fishery-based livelihoods', Draft paper submitted to the CGIAR Challenge Programme on Water and Food, International Forum on Water and Food, Vientiane, Laos, 12–17 November 2006

Friend, R. M., Janprasart, S., Petchkam, S. and Blake, D. J. H. (2006) 'Climate change vulnerability and adaptation in the Songkhram River Basin, NE Thailand: A preliminary assessment', Final draft (December 2006), Paper presented to International Resources Group Ltd under the Vulnerability and Adaptation to Climate Change Impacts Tasks of the TAS Task order under the EPIQ II IQC

Guayjaroen, P. (2001) 'The return of the flooded forest of Tung Phan Khan: A social process with biodiversity recovery', in *Forest Restoration: How Long Till People Can Be Trusted?*, Seminar at Sima Thani Hotel, Nakhon Ratchasima, 23–25 June, 2000, RECOFTC, Bangkok (in Thai)

Hortle, K. G. and Suntornratana, U. (2008) *Socio-Economics of the Fisheries of the Lower Songkhram River Basin, Northeast Thailand*, MRC Technical Paper no 17, Mekong River Commission, Vientiane

IIED (International Institute for Environment and Development) (1997) *Valuing the Hidden Harvest: Methodological Approaches for Local-Level Economic Analysis of Wild*

Resources, Sustainable Agriculture Programme Research Series, vol 3, no 4, Sustainable Agriculture Programme, International Institute for Environment and Development, London, UK

IUCN (World Conservation Union) (2005) *Environmental Flows Project, Lower Mekong Component, Project 80405-210*, Undertaken in 2003 to 2004 with financial support from DFID, Final Project Report April 2005, www.mekongwetlands.org/Common/download/SESSION%20REPORT_complete_101005.pdf

Junk, W. J., Bayley, P. B. and Sparks, R. E. (1989) 'The flood pulse concept in river-floodplain systems rivers', *Canadian Journal of Fisheries and Aquatic Sciences*, vol 106, pp110–127

Khon Kaen University (1999) *Study and Preparation of Environmental Mitigation Plan for Water Storage in the Basin of the Nam Songkhram Project*, Final Report, Department of Energy Promotion and Development, Ministry of Science, Technology and Environment, Bangkok (in Thai)

King, J. M., Tharme, R. E. and De Villiers, M. S. (eds) (2000) *Environmental Flow Assessments for Rivers: Manual for the Building Block Methodology*, WRC Report no TT 131/00, July 2000, Freshwater Research Unit, University of Cape Town, Cape Town

Kohanantakul, K. (2004) *Kreuang meu pramong nai lum nam Songkhram* [*Fishing Gear of the Nam Songkhram Basin*], Department of Fisheries, Thailand and Mekong River Commission, Vientiane, Laos (in Thai)

Kummu, M., Koponen, J., Veijalainen, N., Sarkulla, J. and Keskinen, M. (2006) *Modelling Interconnected Mekong River and Nam Songkhram Watershed System, Thailand*, Draft manuscript to Third APHW Conference in Bangkok, Thailand

Lamberts, D. (2008) 'Little impact, much damage: The consequences of Mekong river flow alterations for the Tonle Sap ecosystem', in Kummu, M., Keskinen, M., Varis, O. and Suojanen, I. (eds) *Modern Myths of the Mekong*, Water and Development Publications, Helsinki University of Technology TKK-WD-002 (EN), www.water.tkk.fi/global/publications

Lohmann, L. (1998) *Mekong Dams in the Drama of Development*, Cornerhouse, Dorset, UK, www.thecornerhouse.org.uk/item.shtml?x=52021

Molle, F. (2007) 'Irrigation and water policies: Trends and challenges', in Lebel, L., Daniel, R. and Koma, Y. S. (eds) *Democratizing Water Governance in the Mekong Region*, Unit for Social and Environmental Research, Chiang Mai, pp9–36

Molle, F. and Floch, P. (2007) *Water Poverty and the Governance of Megaprojects: The Thai 'Water Grid'*, Working Paper, Mekong Programme on Water, Environment and Resilience, Thailand

Molle, F. and Floch, P. (2008) *The 'Desert Bloom' Syndrome: Irrigation Development, Politics and Ideology in the Northeast of Thailand*, Working Paper, Mekong Programme on Water, Environment and Resilience, http://mpowernet.org/download_pubdoc.php?doc=470

MRC (Mekong River Commission) (2006) 'Integrated basin flow management (IBFM)', Powerpoint presentation at the Mekong Waters Dialogue, Vientiane, 6–7 July 2006

The Nation (2008) 'Dams in China not to blame for flooding', *The Nation*, 16 August, www.nationmultimedia.com/2008/08/16/national/national_30080709.php

NEDECO/TEAM (1983) *Pre-feasibility Study of the Nam Songkhram Basin Irrigation and Flood Control Development*, vol 1, Main report March 1983, Code 4.56.056, Prepared by NEDECO, Arnhem, The Netherlands, in association with TEAM, Bangkok, Thailand, Interim Committee for Coordination of Investigations of the Lower Mekong Basin, Bangkok

Norplan and EcoLao (2007) *Nam Theun 1 Hydropower Project*, Final Report, Cumulative Impact Assessment September 2007, Unpublished document prepared for Gamuda Bhd and EGCO, Vientiane

Office of Environmental Policy and Planning (1999) *An Inventory of Wetlands of International and National Importance in Thailand*, Ministry of Science, Technology and Environment, Bangkok, (in Thai)

Petchkam, S. (1997) *The Economic and Social Changes in Ban Pakyam: A Fishing Village in the Nam Songkhram Basin*, MSc thesis, Thai Studies Department, Faculty of Social Science, Mahasarakham University, Thailand (in Thai)

Pye, O. (2005) *Khor Jor Khor: Forest Politics in Thailand*, Studies in Contemporary Thailand no 14, White Lotus Press, Bangkok, Thailand

Rigg, J. (2003) *Southeast Asia: The Human Landscape of Modernization and Development*, second edition, Routledge, London and New York

Sarkkula, J., Koponen, J., Keskinen, M., Kummu, M., Lauri, H., Rasphone, A., Souksakoun, K., and Veijaleinen, N. (2006) *Hydrological, Environmental and Socio-Economical Modelling Tools for the Lower Mekong Basin Impact Assessment*, Working Paper no 5 – Nam Songkhram Model Applicationsm, WUP-FIN Phase II – Mekong River Commission and Finnish Environment Institute Consultancy Consortium, Vientiane, Laos, www.eia.fi/wup-fin/wup-fin2/publications.htm

Saskaki, T., Worrapornpan, S. and Seesang, S. (2007) 'Seasonally flooded community forest on the banks of the Songkhram River: A research framework', in Sawada, H., Araki, M., Chappell, N. A., LaFrankie, S. V. and Shimizu, A. (eds) *Forest Environments in the Mekong River Basin,* Springer Books, Japan

Scurrah, N. (in preparation) *The Politics of Local Knowledge and Participation: Implications for Expanding Tai Baan Research in the Mekong Region*, Forthcoming (2009) Draft Paper for M-POWER, Unit of Social and Environmental Research, Chiang Mai University, Thailand

Sneddon, C. (2003) 'Reconfiguring scale and power: The Khong-Chi-Mun Project in northeast Thailand', *Environment and Planning A*, vol 35, pp2229–2250

Sneddon, C. and Fox, C. (2008) 'River-basin politics and the rise of ecological and transnational democracy in Southeast Asia and Southern Africa', *Water Alternatives*, vol 1, no 1, pp66–88, www.water-alternatives.org

Suntornratana, U., Poulsen, A., Visser, T., Nakkaew, S. and Talerkkeatleela, T. (2002) *Migration onto the Floodplain of the Songkhram River Basin*, Proceedings of the Fourth Technical Symposium on Mekong Fisheries 10–11 December 2001, MRC Conference Series no 2, Mekong River Commission, Phnom Penh, pp270–282

Suwanwerakamtorn, R., Ratcharoen, A. and Salakkham, E. (2007) 'The seasonally-flooded forest change in the Lower Songkhram River Basin', *Journal of Remote Sensing and GIS Association of Thailand*, vol 8, no 2, May–August, pp36–47

Tai Baan Research Network of Lower Songkhram Basin (2005a) *Tai Baan Research on the Ecology and History of the Seasonally Flooded Forest of the Lower Songkhram River Basin*, Sretthachau, C., Pitakthepsombut, R. and Niyomchat, W. (eds), MWBP, Vientiane, Laos, www.mekongwetlands.org/common/download/IUCNT_ThaiBaanResearch.pdf (in Thai with English Executive Summary)

Tai Baan Research Network of Lower Songkhram Basin (2005b) *Fish Species in the Paa Thaam: Local Knowledge of Fishers of the Lower Songkhram Basin*, Sretthachau, C., Pitakthepsombut, R. and Niyomchat, W. (eds), MWBP, Vientiane, Laos

Tharme, R. (2003) 'A global perspective on environmental flow assessment: Emerging trends in the development and application of environmental flow methodologies for rivers', *River Research and Applications*, vol 19, pp1–45

USAID (US Agency for International Development) (1995) *Agribusiness Program in Thailand: Contract Farming at Lam Nam Oon*, PN-ABS-501, USAID Evaluation Highlight no 38, March 1995, http://pdf.usaid.gov/pdf_docs/PNABS501.pdf

Watershed (1996) 'Asia Tech: Working for the betterment of local people', Interview with Asia Tech President, *Watershed*, vol 2, no 1, pp13–14, TERRA, Bangkok

Watershed (1999) 'Interview with Sinsamout Pakprom', Forum section on Fish and Fisheries of the Mekong River Basin: Obstacles and Opportunities, *Watershed*, vol 4, no 3, pp11–13, TERRA, Bangkok

8

The Delta Machine: Water Management in the Vietnamese Mekong Delta in Historical and Contemporary Perspectives

David Biggs, Fiona Miller, Chu Thai Hoanh and François Molle

Introduction

Since ancient times, a combination of natural and social forces has produced the Mekong Delta. The delta's unique waterscape – with its dense maze of canals, extensive horizons of rice fields, village orchards and aquaculture farms – is the result of natural forces such as rain, floods, sedimentation and tides, and of human constructions such as canals and dikes. This *made* landscape, defined by ongoing canal-building enterprises and other works associated with a rapidly urbanizing human landscape, remains at constant risk of being *unmade* by the destructive and sediment-spreading natural effects of seasonal floods, erosion from daily tidal fluxes, storms and also the man-made effects from poorly placed dikes and other works. Enormous investments are required to keep the waterways free of sediment for irrigation, flood control and transportation. Yet, the same sediment, associated nutrients contained in it, and water flow are crucially important to agricultural productivity, ecological biodiversity and efforts to avoid coastal erosion.

Many present-day challenges facing society in the delta are partly the result of past actions that have tended towards more mechanistic approaches to the water environment premised on ideologies of centralized state control rather than support for local adaptation to change and variability. The clearly delineated physical geography of the delta bounded by coastlines and waterways has inspired grand plans of ambitious engineering. The predictability of this natural and social environment, assumed by such master planning, has been contradicted by periods of intense social conflict and continuing occurrences of catastrophic floods and variations in freshwater availability. The threats posed by these social upheavals and natural hazards have both interrupted and justified certain water resource development programmes. Contemporary conflicts over resources are an expression of agricultural intensification, urbanization, and corresponding demands for irrigation and flood protection that tend to foster dependency upon older technological approaches. In recent times, interest in more adaptive and decentralized approaches to water management has returned, although large-scale 'command-and-control' approaches continue to dominate; yet, modern era institutional, political and technological legacies prevent the easy adoption of new policy alternatives.

This chapter traces the historical and contemporary tensions between adaptive and control-oriented approaches to water in the Vietnamese delta region, considering the causes behind the historical shift from traditional approaches oriented to flexible adaptation, towards modernist policies of centralized governmental control. As with many regions of the world, this shift occurred in the late 19th and early 20th centuries. To what extent are future development choices constrained by the weight of a history of past choices supporting the state's technological domination over water environments? In an era where uncertainty and risk is of growing concern to policy-makers and inhabitants, a critical historical perspective on water resources management may reveal more clearly how past decisions have closed off present-day opportunities to pursue new approaches or to identify where opportunities for alternatives to contested policies might still exist.

Using historical examples that illustrate connections between contemporary problems and past decision-making in water control, this chapter focuses on three issues that are central to water resources development in the delta: *total management schemes, mechanical approaches to water management, and trends in adaptation and disaster response.* After giving a brief overview of past and contemporary issues in the delta waterscape, this chapter considers how an ideology of modern technocratic control came to dominate decision-making processes. This ideology has changed over the decades from its often violent implementation in the colonial era to the imported approaches favoured by state engineers under the advice of foreign consultants with experience in other river basins and deltas. Given the role that war and natural catastrophe played in limiting water resource development in these decades, the third section considers how such social and natural disasters not only disrupted attempts for centralized water management, but also presented some interesting alternatives in methods of individual adaptation and disaster response. By focusing on these three issues, the chapter examines the evolution of a unique

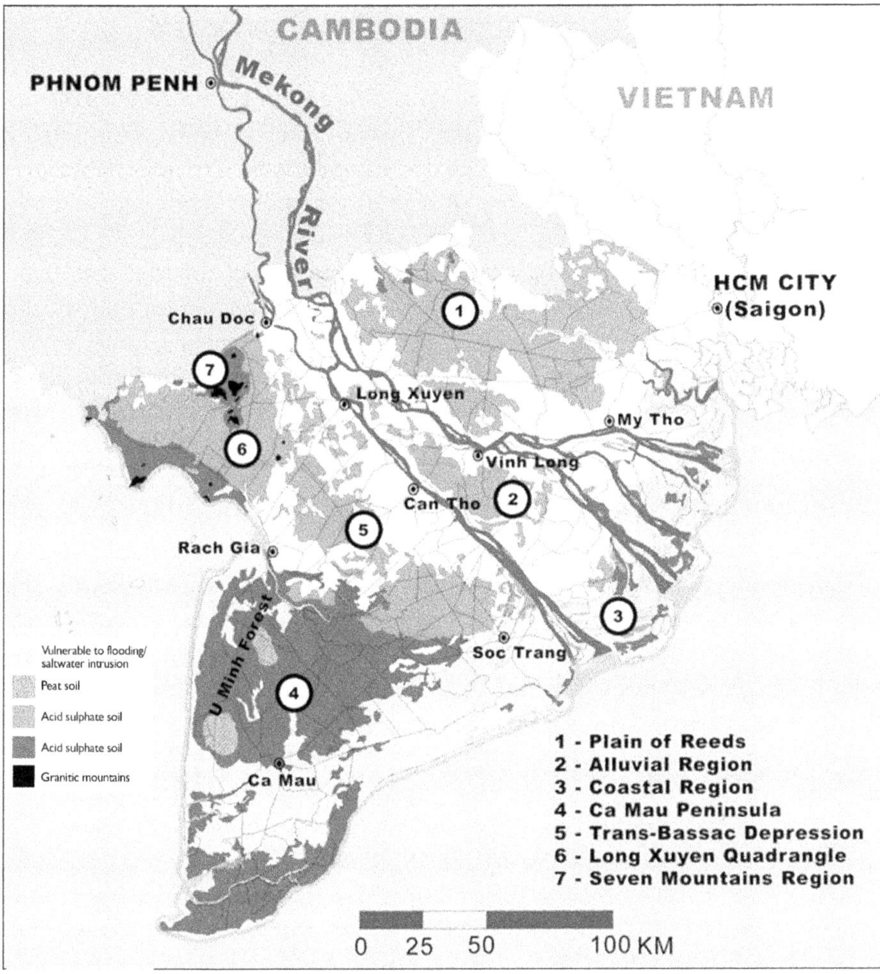

Figure 8.1 *The Vietnamese Mekong Delta*

Source: David Biggs

waterscape that has resulted in the creation of one of the most intensively populated and cultivated regions in the world, one that is now strongly embedded in national and global food economies. The chapter concludes by returning to its main premise that a historically informed analysis of present-day challenges may be employed not only to critique the assumptions of large-scale water management schemes, but to suggest more effective alternatives.

HISTORICAL AND GEOGRAPHICAL OVERVIEW

The Mekong Delta covers an area of roughly 5.9 million hectares and spans the southernmost border between present-day Vietnam and Cambodia, with roughly 4 million hectares in Vietnam. As indicated in Figure 8.1, the Vietnamese Mekong Delta can be divided into a number of regions defined by hydrology and soil chemistry. The most fertile and oldest cultivated regions are the naturally drained freshwater alluvial regions bordering the major channels of the Mekong River. Stretching across the river's two major branches, this area includes most of the delta's largest cities, including My Tho, Can Tho, Vinh Long and Long Xuyen. East of this region is an area subject to year-round or seasonal saltwater inundation referred to as the coastal zone. Especially since 1986, this area has been targeted both for construction of sea dikes and expansion of shrimp aquaculture. There are four major basins subjected to annual monsoon floods from August to December: the Plain of Reeds, the Long Xuyen Quadrangle, the Trans-Bassac Depression and the Ca Mau Peninsula. All of these areas are characterized by acid sulphate soils and are extremely vulnerable to both flooding and, in many areas, saltwater intrusion. Given the environmental hazards in this region, these areas have generally been more politically and economically vulnerable than the alluvial and coastal regions. Two smaller geologic features in the delta include areas of peat soil and a group of granitic mountains that spans across the modern border with Cambodia.

Although the region's population today exceeds 18 million, for most of the delta's history it was relatively sparsely inhabited, estimated to be close to 1 million at the beginning of the 20th century. Archaeological excavations undertaken with the aid of aerial photography during the 1930s revealed canals and settlements in the Long Xuyen Quadrangle, in the upper part of the delta, that were built sometime between 300 BCE and 700 CE.[1] The fact that this society disappeared rather suddenly after 700 CE suggests the precarious nature of this building process, where early society in the delta was vulnerable to flood damage, silted waterways and bays, diseases, piracy and competition from other trading ports in the region. For the next 1000 years, the delta was a sparsely settled coastal frontier of the Khmer Empire with its capitals upstream at Angkor Wat and later Phnom Penh. Early modern water management accompanied the expansion of Vietnamese and ethnic-Chinese groups into the delta region, with a consequent contraction of Khmer interests as the region by 1800 had fallen under Vietnamese political authority. Especially after 1800, waves of settlers and trading interests helped to bring about a 'water frontier' in the Lower Mekong Basin where Thai, Khmer, Viet, Lao, Chinese, Malay and European groups intermixed (Cooke and Li, 2004). In this pre-colonial period, major canal projects such as the Vinh Te Canal (1820 to 1825) both expanded the reach of the Vietnamese state and further separated historic Khmer settlements downstream from a weakened kingdom at Phnom Penh. After several decades of continuing unrest into the 1850s, the French navy then commenced a campaign to conquer the delta and achieved control over the Vietnamese portion of the delta in 1867.

Under French colonial rule and with the introduction of steam-powered dredging in the 1880s, the delta's population quickly rose from some 500,000 in 1860 to over 4 million in 1930. From 1890 to 1930, more than 165 million cubic metres of earth were dredged and the total area put under cultivation rose fourfold to over 2 million hectares (Inspection des Travaux Publics, 1930, p20). From 1930, combined economic, political and environmental troubles stalled further colonial reclamation projects; the escalation of military conflicts in the region after 1945 continued off and on for three decades, so for over 40 years very few new canal projects were undertaken. Nevertheless, as we will see, the war period (1945 to 1975) was important for incubating new strategies of water use that often involved reverting to relying on local resource management, especially in resistance zones controlled by Vietnamese revolutionaries. This period also saw the emergence of internationally supported delta-wide master-planning by the US (Development and Resources Corporation and Republic of Vietnam, 1969) and The Netherlands under the auspices of the Mekong Committee (The Netherlands Delta Development Team, 1974).

Following the end of the Second Indochina War in 1975, the reunified Vietnamese government immediately embarked upon a number of new reclamation projects in war-torn areas, but with little overall effect on raising productivity. It was not until the relaxation of collectivization strategies and the privatization of agriculture in 1986 that production levels and industrial intensification of agriculture began to increase rapidly. Since then, the Mekong Delta has become one of the most productive zones for rice and aquaculture in the world, supplying more than 70 and 50 per cent, respectively, of Vietnam's foreign export amounts of these staples. However, this increased productivity has come at great environmental and social costs as water resources are often degraded and many farmers cannot keep up with the costs of living. Since 1986, the Vietnamese government has moved to shift the cost of maintaining canals and other infrastructure away from the central government to provincial governments and private landowners as part of a broader strategy of (fiscal) decentralization. Such decisions have challenged sub-national and local authorities to find new cooperation models that can sustain and improve their systems. Second, in an attempt to widen the consideration of water resources management from the historical focus on irrigation development, control over the development of waterways and irrigation has shifted from the single domain of a Ministry of Agriculture and Rural Development (MARD) to shared responsibility with the Ministry of Natural Resources and the Environment (MoNRE) (Molle and Hoanh, 2008).

Total management schemes

The first issue that has played a major role in the development of water-use policy in the delta is a long history of delta-wide and basin-wide 'total' water management plans. While increasing communication between the riparian states in such organizations as the Mekong River Commission is generally viewed as a positive step towards encouraging international cooperation towards sustainable water use, the ways in which delta master plans and basin management schemes have been established in the past have produced trends towards technocratic management and solutions that favour major modifications to the river's hydrology. It is these modifications that continue to remain problematic. The Mekong Committee, formed in 1957, initially focused on building a cascade of mainstream dams on the Mekong River south of the Chinese border. The conceptual division of the basin into upper/Chinese and lower/Southeast Asian regions in the 1950s especially reflected US concerns in containing Chinese economic and political influence by focusing development initiatives solely on the lower part of the Mekong Basin located outside Chinese territory. The changing relationship of the US with China after 1949 played a major role in the type of support given to Mekong projects and the Mekong Committee. Before the Communist Chinese military victory over the Nationalists in 1949, US and Chinese engineers worked extensively on projects such as a proposed dam at Three Gorges on the Yangtze (Biggs, 2006). Only after the establishment of the People's Republic of China (PRC) did US agencies pay closer attention to the Lower Mekong Basin. While no mainstream dams have been built in the lower valley,[2] this period of intense international involvement in the development of water resources in the Mekong has had an enduring influence on development planning, as the surveys and feasibility studies produced continue to attract the interest of individual states and private firms lured by the promise of electricity generation and water control. The idea of such developments continues to influence development discourse and negotiations between riparian countries and underpins the recent groundswell of water projects in the basin (see Chapters 1 and 2).

In the Mekong Delta, historically, the pivotal form of water control and management has been the 'Dutch dike' strategy, which involves construction of encircling dikes for settlement, flood control or prevention of seawater intrusion to provide the favourable freshwater conditions for agriculture. The first projects to build such dikes and saltwater dams began under colonial rule in the 1930s as hydraulic engineers and agricultural development cadres sought to clear new lands to raise the colony's production of rice, as well as to defuse mounting social tensions by resettling many thousands of poor tenants to the new lands. French colonial officials, influenced largely by the dike-enclosed landscapes encountered in the Red River Delta – a landscape of distinctly different historical, geophysical, climatic and demographic features – drew up massive plans to relocate farmers from the

Red River Delta into the broad depressions such as the Long Xuyen Quadrangle and Plain of Reeds, as well as the coastal region (see Figure 8.1).

One of several tests to the soundness of the colonial hydraulic infrastructure came in 1937 when higher than normal floods destroyed much of the rice planted in the flood depressions. In an aftermath punctuated by frequent acts of peasant violence aimed at plantation owners, colonial engineers and planners debated development strategies. Rather than reconsider plans to build in flood zones, they instead decided upon greater flood control structures and, simultaneously, the elimination of areas that had grown a flood-tolerant variety of 'floating rice' in favour of faster-growing short-stem varieties. In 1943, during the Japanese military occupation, Vichy Governor-General Decoux created a new budget category entitled 'Aid to rice farmers' that called for 5 million piaster to send entire villages of peasants down the recently completed Trans-Indochinese Railway to populate the first enclosed settlements called '*casiers*' and effectively ended the hydraulic conditions necessary to grow 'floating rice' (Decoux to the Governor of Cochinchina, 1942; Service du Génie Rural, 1943); but these works did not concern the deeply flooded areas of the northern part of the delta and were soon discontinued by warfare.

After the 1954 Geneva Accords brought an end to the First Indochina War, the Republic of Vietnam with US technical and financial support continued this settlement and 'Dutch dike' strategy, especially in politically contested areas such as Long Xuyen and the Plain of Reeds. From 1968 to the war's end in 1975, Vietnamese and many foreign advisory teams continued conducting feasibility studies, developing regional management schemes, and publishing many reports for large-scale settlement and agricultural development initiatives in the delta. As part of President Lyndon B. Johnson's initiative to 'win hearts and minds', David Lilienthal, known during the 1960s as the architect of the Tennessee Valley Authority and in the press as 'Mr TVA', accepted a contract in 1966 to organize with Vietnamese officials a Mekong Delta Development Programme (Development and Resources Corporation and Republic of Vietnam, 1969), as part of the larger Lower Mekong Scheme (Jenkins, 1968). As a 'true believer' in the promise of high technology and regional planning and development to empower grassroots participation, Lilienthal quickly grew sceptical upon seeing the apparent disconnects between the US 'pacification' mission, the violence of the counter-insurgency experiments, and the military conduct of the war. Upon seeing Vietnamese farmers passing his boat travelling on canoes with a modified 'long-tailed' outboard engine, he reflects in his journals that 'even on many technical matters it is *we* who have a lot to learn' (Lilienthal, 1976). With the increasing levels of military violence, however, this proliferation of international consulting firms and management planning initiatives did not coalesce into major effective projects.

Two US moves, the creation of the Asian Development Bank (ADB) and President Nixon's policy of 'Asian regionalism' that involved contracting with local Asian companies instead of US ones to carry out development projects, had a lasting effect on the international nature of water and land-use planning in the Mekong Delta. One Asian firm, in particular, Nippon Koei, exemplified the return of Japanese technicians and investors after the Japanese military had evacuated from the region in 1945. Formed in 1946 during the US occupation of Tokyo, Nippon Koei took some of the more dangerous projects in the Mekong Delta that typically required field surveys and the placement of construction teams in non-secure areas.

Beginning with surveys conducted in 1957, the company sent its engineers to a series of salt intrusion barriers abandoned by the French in 1946. Intermittently into the 1970s, Nippon Koei then fulfilled contracts to redevelop the works, and in 1972 actually completed construction of anti-salinity dikes and barriers through the Tiep Nhut Project, southeast of Soc Trang Province. Funded by the World Bank, the project aimed to protect some 50,000ha of farmland from salt intrusion in the dry season in order to allow the double-cropping of rice. The Go Cong 'pioneer agricultural project', similarly focused on salinity intrusion control, was funded by the ADB. Both projects faced numerous technical, environmental and socio-institutional challenges. For example, Nippon Koei's engineers continued visits to the Tiep Nhut site as late as 1974, when they noticed almost immediately that the new project had stopped saltwater from intruding but had created other problems due to stagnating water inside the dikes in parts of the project.

Not only were such 'Dutch dike' schemes (and the advice of Dutch engineers) employed by the colonial and Saigon governments, but after 1975, the reunified Vietnamese government commenced a 'rice everywhere' campaign due to severe food shortages in the country, especially in the north (The Netherlands Delta Development Team, 1974). Saline water was, in the state's point of view, a constraint to agriculture rather than a resource for aquaculture, as farmers view it today, and flood a threat and constraint to intensification. This 'all rice strategy' was intensified following the severe flood of 1978, when more than 700 people lost their lives in the delta and floating rice crops were devastated, which served to justify investment in flood-protection dikes, canals and pumping stations. This strategy was further strengthened in the 1980s, as the country continued to experience food shortages (Hoanh et al, 2003b; Tuong et al, 2003). Such schemes to manage water across vast territories were first realized slowly, with mainly earthworks and small sluices dug by hand, and it was only after the *doi moi* (renovation) period with market liberalization starting in 1986 that the government was financially able to invest in large-scale plans again. Local authorities determined that people would need to adopt dry season double-cropping across the region. While the relatively easy 'closing off' of the coastal areas continued, huge investments in the diking of polders in the traditional floating rice area of the Long Xuyen Quadrangle were initiated as a means of providing homesteads for the growing population and water control schemes that would allow a shift to high-yielding varieties and multiple cropping.

During the first few years of market-oriented policy after 1986, rice farmers in the Mekong Delta were at first happy with surplus rice production that improved their livelihoods compared with farmers elsewhere. However, since the mid 1990s, the limited income of rice farmers, especially small landholders, has not helped them to keep up with the high speed of economic growth; hence, they have become one of the poorest groups in their communities. Even with the soaring rice price, increasing by 76 per cent between December 2007 and April 2008, several analyses indicate that the high price may not necessarily improve the income of rice farmers – only that of rice-related trading companies.

With *doi moi* policies that stopped the use of quotas in rice production and allowed diversification, some farmers have switched to more high-income crops such as fruit trees and aquaculture. This has only been possible, however, in areas where agro-hydrological conditions allow diversification away from rice and where people have access to necessary credit, knowledge, expertise and markets for non-rice products. Except in the intermediate zone under tidal management, where the expansion of orchards on raised beds has been spectacular, the current water management system through dikes was primarily designed for rice irrigation. Diversification to non-rice crops requires significant modifications – full protection for trees and pumping to highlands for vegetables or fruit. The state faces the increasingly difficult challenge of continuing another cycle of investment into new infrastructure at the same time that it attempts to maintain older works.

While these works were carried out quite independently from events unfolding at the level of the Mekong Basin, this wider scale has recently regained relevance as ideas of harnessing the river's water resources at a large scale have resurfaced (see Chapter 2). The high demand for energy for economic development in the Mekong countries, together with increased concern over climate change and rising fuel costs have fuelled a renewed emphasis on infrastructural solutions that aim to 'climate proof' local economies by ensuring the security of water and energy supply. Mainstream and tributary dams in the lower basin are now back on the planning board after a period when increased awareness of the environmental and social implications of such dams led to limited international multilateral investment in such projects.

China, in particular, is steadily and determinedly pursuing the construction of eight dams on the upper length of the Mekong (Lancang) River (two completed, two under construction), while Laos has 77 dam projects in its pipeline. From the formation of the Mekong Committee until the 1990s, Vietnam was an advocate of Mekong mainstream dams, seeing the potential flood control and dry season flow augmentation function of dams as beneficial to agricultural production in the delta. Environmental impacts such as changes in water flow and quality were of less concern than possible benefits from flood mitigation and regulated supply of water during the dry season. During the mid 1990s, Vietnam turned to opposing Mekong mainstream dams, fearing the impact of projects such as the Thai Khong-Chi-Mun on dry season salinity intrusion in the delta (Hori, 2000). This attitude has, in recent years, shifted again, as the delta's coastal areas are increasingly protected and energy generation has become a priority over food production.

With estimates that as much as 50 per cent of the Mekong River's sediment originates in the upper basin, however, the impending decline in sediment loads in the lower basin is likely to have severe implications for bank erosion, and stream and floodplain morphology throughout the basin, notably in the delta where the river deposits much of its sediments (Chapter 9). Many farmers in the Mekong Delta are dependent upon the sediment and nutrients brought by the seasonal flood waters to maintain soil fertility and, thus, crop productivity.

Even more important is the role of sedimentation in protecting the delta from coastal erosion, as can be seen by retreating deltas, from the Nile to the Yellow Sea. On top of that, the spectre of climate change is likely to put ecological and social systems under increased stress: the predicted impacts on the delta include a rise in sea level that will compound problems of coastal erosion, worsening salinity intrusion in the river's main arms, as well as increases in the incidence of severe floods, droughts, storms, tropical cyclones and heat waves, including unknown ecological changes (Hoanh et al, 2003a; Wassmann et al, 2004). Experts estimate that with a sea-level rise of 1m, Vietnam will suffer a loss of 12.3 per cent of its cultivated land, including 170,000ha of coastal land in the Mekong Delta region (*Vietnam News Briefs*, 2008). They conclude that the country must 'upgrade its sea dyke system, which is deteriorating and unable to combat the sea level rise', with US$606 million needed from now until 2020. One expert stated that building sea dikes was part of an economic and 'national defence security strategy', and that the minimum width of each dike should be 5m to 6m to cope with the Force 9 to 10 storms.

At a conference on 24 March 2008, MARD revealed its continued preference for 'engineered solutions' with a proposal for new plans to raise around 10.7 trillion Vietnamese dong (US$676 million) to further extend and upgrade dikes in 15 vulnerable provinces along the Vietnamese coast, including seven in the Mekong Delta. Moreover, after Cyclone Nargis devastated the Irrawaddy Delta in Myanmar/Burma during early May 2008, some Vietnamese officials proposed to strengthen the infrastructure in the Mekong Delta, concerned about the possible damages of a similar cyclone, although cyclones of such intensity are very rare in the Mekong Delta. These new developments represent the latest step in a long history of state efforts to further human control over the natural flow of water between the delta and the sea. The costs of all these infrastructural solutions, however, may also become unbearable.

The delta as machine: A work without end

From very early in the colonial period, hydrographers and engineers were confronted by the physical and ecological complexity of the water environment in the Mekong

Delta. The extreme flatness of the delta, combined with high sediment content in the rivers, produced hog's-back ridges (*lung tom*, *dos d'âne*) in newly constructed canals that soon interrupted most water traffic when the tide was low. French observers generally saw such areas as 'dead zones' because here the water stilled, and most deeper-hulled boats were forced to wait for the high tide to pass (Direction Générale des Travaux Publics, 1911, p34). In local society, however, such places were traditionally known as 'meeting points' (*giap nuoc*, or water interface). Inhabitants frequently built markets at these intersections of opposing currents because of low flow velocity.[3] The village of Thu Thua, located on a cut between the two branches of the Vam Co River, or the Phung Hiep floating market at a junction of seven canals were such places where people travelling from different places with different goods met to trade (Nguyen Hien Le, 1989, pp23–28).

This tidal ebb and flow of water in the delta was also an important source of clean water and fertile sedimentation. Especially in the four depressions shown in Figure 8.1, if water in the fields was not routinely exchanged, high levels of acid sulphate would soon dissolve aluminium and iron ions and stunt plant growth. This condition was especially severe in newly established fields and along newly dredged canals, only relenting after three to five years of irrigation (Phong et al, 2007). To prevent the build-up of 'alum', farmers frequently exchanged water by draining water from the paddy, opening their bunds and letting water escape while the tide was out. When the tide rose again, fresh water from the river filled the ditches and returned clean water to the paddy. Nature provided a system of irrigation that required almost no extra labour and little organized cooperation between individuals and communities on shared waterways. Engineers, standing in one of the flattest deltas in the world, were sometimes blind to these local functions of terrain and micro-topography and confused by a tropical hydrology that was, to them, exotic; and those who did pay attention to the natural regulation of the river system were often forced to implement water control projects under political pressures (Nguyen Huu Chiem, 1994; Vo Tong Xuan and Matsui, 1998).

Why have local and state governments over the years continued to favour mechanical approaches to water management over projects that work off the natural ebb and flow of the rivers (seasonally) and tides (daily)? In part, the answer stems from the modern global experience of a population explosion and agricultural intensification; however, a significant factor in the ways in which new technology became embedded in the Mekong Delta was the region's unique political ecology. The introduction of steam-powered dredging machines to the Mekong Delta in the late 1880s fitted the political and technical needs of the colonial state by replacing thousands of labourers needed for traditional canal or dike projects. Through the monopoly enterprise that operated these machines, the colonial state also ensured that most of the money spent on this work benefited French interests rather than local ones.

This colonial pattern of funding and organization of infrastructure projects continued into the 1960s as the US Agency for International Development

(USAID) replaced colonial agencies and continued support for big-ticket purchases of US-made equipment, such as a fleet of new diesel-powered dredgers eventually operated by US construction firms undertaking no-bid development contracts. During the late 1950s, the US Operations Mission in Saigon even hired the old French dredging enterprise to operate its dredgers and train new crews. Amidst frequent scepticism both in Saigon and Washington that such aid was merely enriching US and French interests at the expense of the waterway system, the US advisory mission in Saigon continued to call for more expensive equipment as the war intensified (Biggs, 2008). After 1960, US construction firms such as DMJM (now AECOM) and RMK-BRJ (now part of Halliburton) entered Vietnam to fulfil these development contracts to build highways, dredge canals and (after 1965) military bases. For reasons of security and politics in Washington, these firms typically worked with heavy diesel-powered equipment and they tended to propose projects that made use of these machines (Department of the Army, 1972, p133).

Again in the present era of rapid economic growth, concerns about the powerful influence of politically connected contractors over that of local water users have resurfaced, although the politics and technologies of construction have changed considerably. Their interest in capital-intensive methods is shared by hydraulic bureaucracies that seek to expand their budget and power and fulfil professional inclinations towards infrastructures. Infrastructure development plans in ecologically sensitive areas such as the Ca Mau Peninsula (Tuong et al, 2003) are often driven by planning and engineering departments in Hanoi and Ho Chi Minh City, with the sole goal of increasing export rice production; however, construction at different levels is often shared between national, provincial and local firms. State companies typically build principle and primary canals, while secondary and tertiary canals are allocated to provincial firms, and on-farm systems to local firms and farmers themselves.

The following case study (Miller, forthcoming) on dredging politics in Luong Hoa Commune, Tra Vinh, illustrates that operation and maintenance of this 'mechanical' hydrologic system remains a deeply political issue involving complex negotiations between local government, outside construction firms and local water users. Irrigation infrastructure in the commune was initially developed under the Tam Phuong Project, with the assistance of Australian aid in 1985. Yet, from the late 1980s onwards (as the aid dried up), secondary and tertiary canals gradually deteriorated as a result of neglect of essential operation and maintenance (O&M) by the district irrigation enterprise and local farmers, respectively. Farmers were annoyed that the government did not properly fulfil its responsibility in maintaining the secondary canals, so they neglected the maintenance of tertiary canals. This made water access difficult for poor farmers who tended to have land far from canals or on higher land. Recently, 'public service labour' contributions (*lao dong cong ich*) have been phased out in favour of charging a fee and replacing manual dredging with the hire of mechanical dredgers to keep secondary canals clear. Concerns continue with regard to the quality of work undertaken by different dredging companies, as well as (uncompensated) loss of individual farmers' land due to the placement of the dredged material on paddy fields. While most people preferred mechanical dredging to those dug by hand, many poor farmers could not afford the cash fee for canal dredging and wished to contribute labour instead.

This cash fee is just one of the many rising costs associated with water, which has increased considerably with the rise in oil prices due to people's reliance on diesel-powered pumps for irrigation. While these tensions were temporarily resolved by a recent large cash injection into development of irrigation by a Japanese aid project, they indicate continuing challenges between state and local authorities to find a long-term solution to high water costs and maintenance of irrigation infrastructure that ensures fair and efficient water access for small rice farmers reliant on timely water availability (Miller, forthcoming).

Looking at the flows of funding – especially the unusual infusion of cash from Japan in this case – and the actors involved, one issue not yet resolved is what safeguards exist to ensure that contractors respond to the needs of water users. Again, examining the past, historical records reveal that even 100 years ago, determining state and contractor liability to water users was a complicated subject for courts and administrative bodies. During construction of one of the first major colonial projects, Xa No Canal, a group of native landowners in 1901 brought a law suit against the government seeking indemnities for damages to land after a dredger cut the village off from the existing waterways. The matter, pitting a group of relatively wealthy and legally recognized Vietnamese landowners against the colonial Department of Public Works, eventually reached the desk of the Governor General in Hanoi in 1901, who decided the value gained in having property bordering the new canal outweighed the damages done through the destruction of existing irrigation structures (Nguyen Ngoc Chan, 1901). Since then, especially with Vietnamese independence and reunification, government accountability and response to local complaints have improved considerably; however, administrative mechanisms for resolving local grievances continue to lag behind local expectations.

One crucial reason why both resource managers and water users continue resorting to mechanical rather than adaptive 'fixes' to their problems is the endless need to maintain them once established, after settlements and activities have attuned to the changes induced by past constructions. A French inspector studying the colony's early plan to build new infrastructure in the delta in 1881 called such projects '*oeuvres de Penelope*' (works without end). Visiting the colony on a fact-finding mission in 1880 to 1881, Charles Combier assessed the colony's proposal to build canals and elevated railways throughout the delta and criticized the plans on numerous grounds, first on the ethical basis that funding for the projects was to come partly from state-controlled sales of opium and then on technical grounds as engineers had not yet solved the problem of hog's-back ridges – dredgers would thus be constantly returning to clean silt out of the new canals (Combier, 1881). When economic and political events after 1930 prevented the dredgers from keeping up with their routine schedule of clearing channels, large portions of the irrigation infrastructure became degraded and many people abandoned such newly opened areas as the Ca Mau Peninsula, the Plain of Reeds and the Long Xuyen Quadrangle. Considering the story in Tra Vinh described above, replacing public service labour with mechanical dredging will probably require mechanical dredging in the future to maintain deeper channels; thus, by shifting the method of maintaining the waterways to construction equipment, the state and local authorities increase their dependence upon such firms in the future, making them vulnerable to the availability of heavy equipment, spare parts and changing fuel prices.

More crucially, the constant dredging of canals and drains, construction and maintenance of dikes and sluice/control structures, and consolidation or raising of embankments to face higher levels of risks or coastal erosion translate into ever-increasing and non-ending financial costs. The nature of waterscape transformations is such that the state eventually has to cope with the maintenance of this hydro-agricultural 'machine' as people withdraw from earlier works and do their best to adapt to the new conditions that have been created. Thus, what has continued in the Mekong Delta is something French engineers such as Combier (see above) over a century ago worried would become an '*oeuvre de Penelope*': a work without end.

Efforts have been made to decentralize financial responsibility for O&M of irrigation systems; but many provincial and district agencies are barely able to cover the most basic maintenance works from local fees and taxes. Underpinning the entire irrigation system, its construction *and* maintenance is also a larger dependence upon aid from international agencies: aid which entails the imposition of agencies' own contingencies and requirements (such as the purchase of donor country equipment and software). As access to soft aid monies becomes increasingly difficult for Vietnam, funds for maintaining system efficiency are likely to be sourced increasingly from international loans, which transfers the debt burden to future generations. A crucial implication of past choices is the recurring costs for maintaining an ever-more complicated array of hydraulic works: from colonial power to (partly) people, then the state, then, more recently, to provinces, foreign aid and future generations. The challenge with increasing costs is to allocate this cost and the 'political game' is to shift it to other parties.

LOCAL ADAPTIVITY AND RESPONSES TO DISASTERS

While most of this chapter has considered state responses to issues of development and disaster response, this final section addresses the very important ways in which individuals have responded to adversity – social and natural – and to the constraints/opportunities brought about by large-scale state interventions, as well as the problems that this poses in a place governed by top-down policies.

Especially in the current era of market liberalization, historical disconnects between individual action and state projects continue to inform present-day conflicts. Those disgruntled by the government's inability to provide suitable water infrastructure can refer back to a long history of negative experiences with state authorities that did little to support farmers' needs and, instead, encouraged settlers to clear forests and drain swamps. French historian Pierre Brocheux describes how colonial land policies and the all-consuming demand for wood to supply the need for steam engines produced what he terms a colonial frontier society in the delta. Individuals followed in the wake of the steam dredgers, their sampans (*ghe tam ban*) loaded with tools, supplies and basic building materials. They built huts, burned down sections of the forest beyond, and began the back-breaking work of clearing stumps and forming fields. Once land was cleared and agriculturally productive, however, they often found that landlords had already claimed rights to the land. They then either left to clear new unclaimed lands or else worked out some tenancy arrangement (Brocheux, 1985, p123). As economic and social conditions for farmers worsened with the Great Depression after 1930, many living in marginally productive, flood-prone areas joined the nationalist campaigns of the Indochinese Communist Party, forming protests that called for lower interest rates, food for the starving and land to the tiller.

Dissatisfaction with state action continued after independence, with many of the policies followed during 1975 to 1986, notably attempts to collectivize production and redistribute land, causing considerable hardship on the population. State authorities confiscated much of the privately held machinery for ownership by collectives and sent many thousands of people to dig new canals and build dikes by hand. Both collectivization and required labour were met with widespread resistance as many farmers refused to put the requisite care into water management and growing crops. Since the *doi moi* reforms in 1986, much of this resistance has subsided as the state government now tends to side more with large private development interests. The relationships between the state and the citizenry have therefore had a significant impact upon the way in which large-scale transformations of the landscape and local adaptation by rural populations have been interrelated and mediated. As this section shows, farmers have responded to economic pressure and waterscape transformations by counter-strategies, coping behaviours, opportunistic adjustments and innovations.

Past responses to social disasters, such as the destruction caused by the Indochina Wars, and natural disasters such as floods have had far-reaching effects on water management. Life in marginally productive acid-sulphate soil and flood-prone areas such as the Ca Mau Peninsula and the Plain of Reeds, held as 'cradles of the revolution', illustrate people's capacity to adapt to extreme conditions. U Minh's forests, for example, served for decades as an important base area for revolutionaries and guerrillas, which resisted repeated efforts by French, US and South Vietnamese military forces to drain and penetrate the swampy terrain using armoured dredgers, napalm, Agent Orange (a powerful herbicide and defoliant used by the US military in its Herbicidal Warfare Programme during the war), and B52 aerial bombing strikes. They built submerged barriers in the canals and creeks, and waited at key bottlenecks to ambush French vessels. French forces, in turn, used flame-throwers and incendiary bombs to burn away the forest and French aircraft repeatedly bombed the earthen dams that the Viet Minh militias built in efforts to maintain the swamp forests. In his history of one base in U Minh Forest, Bui Van Thanh (1997) recognizes the role that such barriers played in protecting the base; the larger dams required hundreds of labourers working in often dangerous conditions to repair them.

Throughout most of the Second Indochina War (1959 to 1975), the floodplains and wetlands remained under the control of the National Liberation Front. Farmers living in these 'free fire zones' were routinely subjected to aerial bombardment and strafing; but they managed to develop a kind of extreme survivor mentality that required almost constant adaptation to changing environmental and social conditions. In 1971, after three years of intense US and Vietnamese bombing campaigns and operations in the base areas, one US survey estimated that roughly 63,000 people living in the U Minh Forest had fled their homes to request government assistance and relocation. Travelling in the Ca Mau Peninsula, the American provincial adviser to the survey described the waterways as 'wall-to-wall boats' where families had brought on their sampans stores of food, house frames and all personal belongings to re-establish homesteads elsewhere. Perhaps most interesting in the report was the general observation that the overwhelming majority reported that living conditions under the National Liberation Front (NLF) in the flooded lands were 'reasonably good'. It was only the intensified combat and bombing that forced them to move (Pacification Studies Group, 1971).

The same individuals who, in their support for the Vietnamese Revolution, performed heroic measures to maintain flooded wetlands and swamps could not, however, as civilians or leaders in the post-war era control the actions of park managers or thousands of settlers hungry for land (Biggs, 2005). People in such severely flooded areas as Tam Nong District in the Plain of Reeds had traditionally practised flood season cropping of floating or deep-water rice in combination with wild-capture fisheries. While such systems accommodated often volatile environmental fluctuations in flood cycles, they became increasingly unable to support the growing population densities in the post-war period as tens of thousands of new migrants settled into these areas despite their susceptibility to flooding and marginal productivity. In the spring dry season of 2002, over 8000ha (roughly half) of one of the last remnants of cajuput mangrove and peat swamp in the U Minh Forest area burned out of control (Sanders, 2002, p113). This points to the intense pressures of economic growth on the whole country where people had to find every way possible of increasing their immediate income to survive lest they become the 'poor' in society.

Although invasion and reclamation of marginal land is a response to economic needs, innovation has also provided a way out through intensification. Perhaps one of the most important technological changes to occur in this period with long-standing effects on water management was the introduction of portable motors used for water pumping after 1960. Robert L. Sansom in *The Economics of Insurgency* (Sansom, 1970) describes how Pham Van Thanh, a former employee of the French dredging company in Saigon, accompanied a Republic of Vietnam (RVN) military engineer to watch a dredging project in progress nearby. Studying the old German and Japanese diesel engines powering the French dredgers and the centrifugal pumps powering the newer US equipment, Thanh began experimenting with German-made impellers (reversed propellers to create suction, not propulsion) attached to the shaft of US-built engines sold as a 'shrimp-tail' boat motor. After developing a successful water pump, he sold, on average, 600 motors a month through Sansom's period of research in 1967. Sansom's interviews with farmers suggested a rapid farmer-motivated diffusion of the labour-saving devices even against the wishes of US, RVN and NLF authorities (Sansom, 1970).

Other innovations have included the spread of mobile pumping operators (large pump sets on boats) offering collective pumping services to groups of farmers in the upper part of the delta (Lienhard et al, 2001), the acclimation of shrimps to fresh/low salinity water, and the development of raised-bed techniques to cultivate vegetable or fruit trees on lands with clay soils and poor drainage. This technique is best developed in the intermediate zone that takes advantage of tidal management. It has allowed diversification out of rice and much higher land productivity in a context of declining per capita endowments. More generally, Figure 8.2 shows how these areas, close to the city of Can Tho, have historically shifted from one flood-adapted traditional varieties to triple rice-cropping and orchards (for further insight on historical changes, see Nguyen Huu Chiem, 1994; Tanaka, 1995; Vo Tong Xuan and Matsui, 1998; Le Coq, 2001). This evolution is the combined fruit of landscape transformations, external innovations (e.g. high-yield rice varieties), and farmers' innovation. Similar changes unfolded in other agro-ecological parts of the delta, including the flood-prone areas of An Giang Province, where floating rice gradually disappeared in the 1990s due to works in dredging, diking, excavation of secondary canals and land levelling (Lienhard et al, 2001).

In other cases, farmers have worked to revert negative effects from state projects. Coastal polders built by Nippon Koei Co (1966) during the 1960s resulted in severe drainage problems that were mitigated by farmers inside the area using an estimated 1000 portable water pumps to individually move water in or out of their fields. More recently, larger-scale water control schemes funded by national funds and World Bank loans (in O Mon-Xa No, Quan Lo Phung Hiep and South Mang Thit) built during 1994 to 2001 for similar water control objectives have faced identical problems of stagnant water, insufficient through-flow of fresh water into the system during the dry season, and inflexibility to multiple water uses, underlining the continuity of historical problems in contemporary settings (World Bank, 1999).

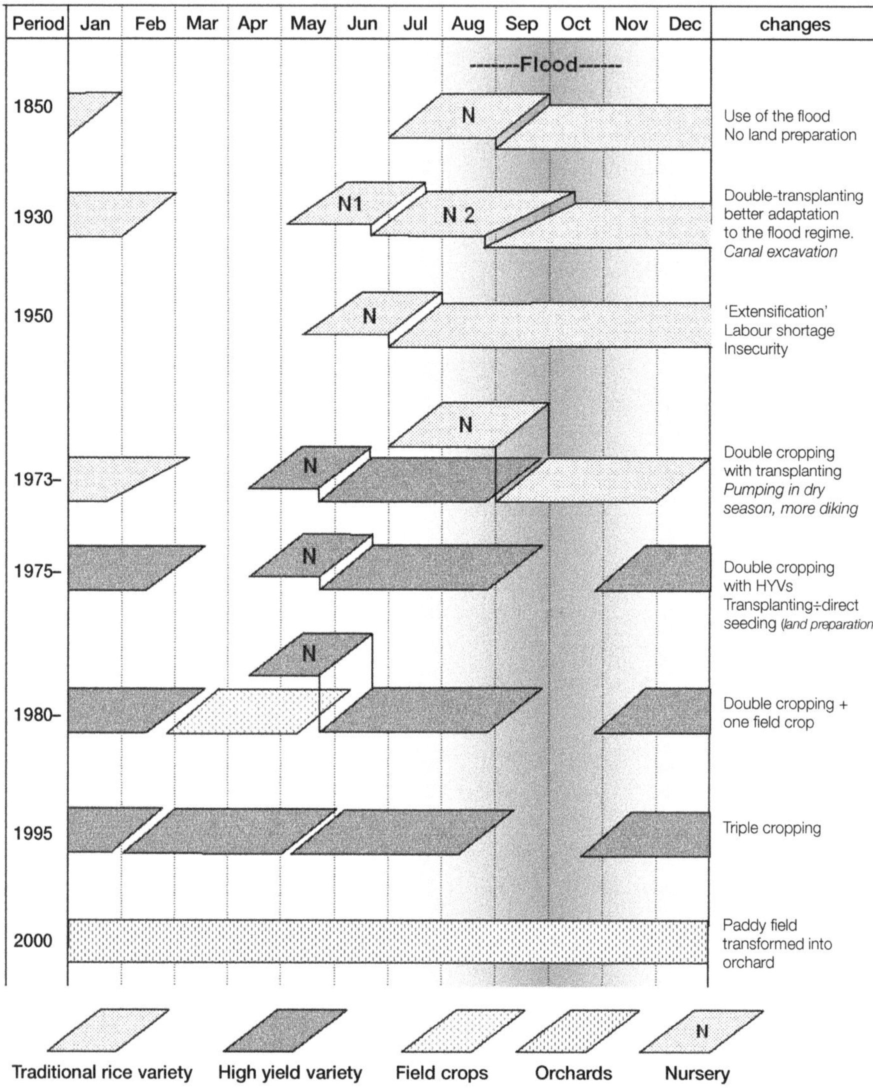

Figure 8.2 *Changes in cropping patterns in the tidal management zone of the delta*

Source: Adapted from Le Coq (2001)

In still other cases, farmers have fiercely resisted environmental changes resulting from state projects. For example, in February 2001, farmers in Bac Lieu Province broke the new Lang Tram salinity-control sluice dam that was planned to close off an area for rice production, while many farmers had already opted to take advantage of salinity and shifted to raising more profitable shrimp instead (Hoanh et al, 2003b). The state was not sufficiently flexible to respond to changing market demands and local desires, thus illustrating the main conflict between adapting to natural resources versus resisting and controlling natural forces to achieve ends conceived by state officials.

What these historical events suggest is, first and foremost, the intense determination of farmers to adapt to local environmental adversities independent of local or state authorities. The reasons for this common resistance to state-initiated projects, such as coastal dikes in the present, are complex; but in many cases they appear to be informed by a deep historical distrust between farmers and state authorities over the best ways in which to 'improve' land, and high ingenuity and capacity of responding quickly to adversity. While such individualist approaches were key to survival during the wars and in times of natural disaster, they nevertheless pose new problems in post-war water management strategies. What appears to be lacking most in the present are the means for farmers and others directly involved in managing the delta's water resources to be involved in contemporary decision-making processes typically dominated by the state. So long as large projects such as coastal dikes or enclosed irrigation districts are conceived without this participation, it is likely that farmers will continue to act independently to realize higher levels of productivity regardless of the environmental consequences or implications for other resource users.

CONCLUSIONS

Returning to this chapter's objective to establish the usefulness of employing a critical historical perspective to better understand challenges to contemporary water management issues, there are two main ways in which historical events continue to influence contemporary decision-making processes in the Mekong Delta. First, there is a phenomenon of what might be called *institutional inertia*, where past institutional arrangements such as the reliance on private contractors to carry out public works since the 1880s have continued to shape the form of state decisions ever since. This first happened immediately following the formal end of colonial rule in 1954 when old French enterprises continued to carry out public works construction to 1960. Besides the political motivations for continuing this mode of public works with the lobbying interests of entrepreneurs and the large sums of money involved in securing foreign development loans, the bureaucracy formed around hydraulic works also became an entrenched power vying for its reproduction. Second, there is also a corresponding *physical inertia* in terms of the historical built landscape and aging technology that works against propositions to make major changes in water resource strategies. As the colonial inspector noted in 1880, past works such as canals and coastal dikes have become 'works without end' that require continuing attention to maintain them against rising sea levels, changing river conditions and fluctuating fuel prices. Entire communities have become dependent upon artificially maintained water levels, and it would be technically and politically difficult, if not unfeasible, for state leaders to require people living in such areas to become 'friends with the flood',[4] particularly now that with more valuable assets they are physically much more vulnerable to their impacts. In sum, institutional and infrastructural path-dependency makes it close to impossible to remedy the past transformations of the waterscape.

However, in the more environmentally hazardous zones such as the Plain of Reeds and the Ca Mau Peninsula, where communities have always lived more on the edge separating prosperity from any number of natural and man-made disasters, perhaps there are more opportunities for state authorities to experiment with alternative small-scale and adaptive strategies for coping with fluctuating environmental conditions and, at the same time, maintaining more stable economic and social conditions. The willingness of many to switch from rice cultivation to aquaculture and the historical ingenuity of local people to evolve suitable and affordable technologies suggest that there may, in the future, be new economic and environmental opportunities to be gained from promoting rather than resisting such actions. Especially in these areas, there are numerous opportunities for developing new models of co-management. Sansom's (1970) story above of one inventor highlights both the rapidity with which delta farmers adopt a technology once proven and the enormous sums of money to be made, with no government subsidies involved, through the creation of new small-scale technologies. As the constant buzzing of gasoline- and diesel-powered boat engines and the crowded rivers of road traffic today attest, the Mekong Delta today is already a richly productive, vibrant zone increasingly shaped by local entrepreneurs and increasing access to foreign capital.

Although living conditions have improved dramatically in the Mekong Delta as in other river deltas, the growing threats of climate change, closing agricultural frontiers and urbanization require new efforts to maintain standards of living and to avoid future catastrophes. Yet, the financial implications of the need to maintain and protect the 'delta machine' are awesome and the distribution of attendant costs has become a central issue of current politics, in general, and of the decentralization process, in particular. Given the physical and institutional inertia of the past, it may not be possible to completely escape the problems associated with aging infrastructure and old ways of doing things. New methods for democratizing water resources policy-making through state–local co-management are needed. These contemporary issues are not unique to Vietnam and the Mekong Delta, but may be found in other river deltas around the world. In the Mississippi Delta, for example, the US government is faced with similar problems in its efforts to rebuild the protective levee infrastructure around the city of New Orleans and to prevent further subsidence of the delta into the sea. The devastation wreaked by Cyclone Nargis in May 2008 in the Irrawaddy Delta in Myanmar is also a further reminder of the deeply political and social nature of disasters and their aftermath. By employing a more critical historical perspective on such issues, it may be possible to gain a clearer sense of both the institutional cum physical inertia that informs such decisions and possibilities that may exist for tapping into the incredible resourcefulness and ingenuity of people who have for decades had to adapt with little government support. By finding ways to incorporate local water users into water resource management and, at the same time, to respond to changing environmental conditions, it may be possible for Vietnam – which relies so heavily on the Mekong Delta for food and commerce – to achieve more stable economic and environmental security.

Notes

1 One of the most comprehensive descriptions of the ancient material culture in the delta is provided by Malleret (1959, vol 1, pp27–33) who gives an excellent discussion of ancient hydraulic infrastructure and pre-Angkor settlements near present-day Chau Doc and the Vinh Te Canal.

2 Efforts have, instead, focused on dam construction and water diversions on major tributaries in southern Laos, northern and north-eastern Thailand, central Vietnam and, more recently, Cambodia.

3 Since the conversion of domestic boat traffic from sail- and oar-powered to motorized vessels in the mid 1900s, the largest markets today are now located at the junctions of large canals and major rivers.

4 During the past years the government has adopted a strategy called *Living with Floods* (Thanh Lam and Tran Dinh, 2008). However it is apparent that the strategy merely emphasizes better preparedness and early warning, not a change in philosophy.

References

Biggs, D. (2005) 'Managing a rebel landscape: Conservation, pioneers and the revolutionary past in the U Minh Forest, Vietnam', *Environmental History*, July, pp448–476

Biggs, D. (2006) 'Reclamation nations: The US Bureau of Reclamation's role in water management and nation building in the Mekong Valley, 1945–1975', *Comparative Technology Transfer and Society*, vol 4, no 3, pp225–246

Biggs, D. (2008) 'Breaking from the colonial mold: Water engineering and the failure of nation-building in the Plain of Reeds, Vietnam,' *Technology and Culture*, vol 49, no 3, pp599–623

Brocheux, P. (1985) *The Mekong Delta: Ecology, Economy and Revolution, 1860–1960,* Center for Southeast Asian Studies, Madison, WI

Bui Van Thanh (1997) 'Bao cao Khai quat Lich su va Phuong huong Bao ton, Phat huy Di tich Lich su Can cu dia Cach mang U Minh Thuong, Kien Giang', *Ky yeu Hoi thao Khoa hoc: Di tich Lich su Can cu dia Cach mang U Minh Thuong*, So Van hoa Thong tin The thao, Rach Gia, Vietnam

Combier, C. (1881) *Rapports présentés à S.E. Ministre de la Marine et des Colonies sur les Grands Travaux Projetés en Cochinchine*, no 4/904(3), Social Sciences Information Centre, Ha Noi, Vietnam

Cooke, N. and Li, T. (eds) (2004) *Water Frontier: Commerce and the Chinese in the Lower Mekong Region, 1750–1880*, Rowman and Littlefield, New York, NY

Decoux to the Governor of Cochinchina (1942) File H62/10, TDBCPNV Record Group, VNA2, 30 September

Department of the Army (1972) *Vietnam Studies: Base Development: 1965–1970*, GPO, Washington DC

Development and Resources Corporation and Republic of Vietnam (May 1969) *Mekong Delta Development Program*, Development and Resources Corporation, New York, NY

Direction Générale des Travaux Publics, Gouvernement Générale de l'Indochine (1911) *Voies d'Eau de la Cochinchine*, Imprimerie Nouvelle, Saigon, Vietnam

Hoanh, C. T., Guttman, H., Droogers, P. and Aerts, J. et al (2003a) *Water, Climate, Food, and Environment in the Mekong Basin in Southeast Asia*, International Water Management Institute, Mekong River Commission Secretariat, and Institute of Environmental Studies, Vietnam

Hoanh, C. T., Tuong, T. P., Gallop, K. M., Gowing, J. W., Kam, S. P., Khiem, N. T. and Phong, N. D. (2003b) 'Livelihood impacts of water policy changes: Evidence from a coastal area of the Mekong river delta', *Water Policy*, vol 5, pp475–488

Hori, H. (2000) *The Mekong: Environment and Development*, United Nations University Press, Tokyo, New York, Paris

Inspection des Travaux Publics, Gouvernement Générale de l'Indochine (1930) *Dragages de Cochinchine: Canal Rachgia-Hatien,* Saigon, Vietnam

Jenkins, D. (1968) 'The Lower Mekong scheme,' *Asian Survey*, vol 8, no 6, pp456–464

Le Coq, J. F. (2001) *Libéralisation économique au Vietnam, intensification et diversification*, PhD thesis, INA-PG, Paris

Lienhard, P., Molle, F. and Hieu Trung, N. (2001) 'Local dynamics in land and water use: A case study in An Giang Province', in Can Tho University and IRD (eds) *Change in Land and Water Use in the Mekong River Delta: Micro and Macro Perspectives*, Can Tho University Press, Can Tho, Vietnam

Lilienthal, D. E. (1976) *The Journals of David E. Lilienthal: Volume VI – Creativity and Conflict 1964–1967*, Harper & Row, New York, NY

Malleret, L. (1959) *L'Archéologie du Delta du Mékong*, École Française d'Extrême-Orient, Paris

Miller, F. (forthcoming) *Negotiating Individual and Collective Interests in Water: Contradictions and Tensions Arising from Transition in Vietnam*

Molle, F. and Hoanh, C. T. (2008) *Implementing Integrated River Basin Management: Lessons from the Red River Basin, Vietnam*, Mekong Programme on Water Environment and Resilience, Vietnam

Nguyen Hien Le (1989) *Bay Ngay Trong Dong Thap: Du Ky va Bien Khao*, NXB Long An, Long An, Vietnam

Nguyen Huu Chiem (1994) 'Former and present cropping patterns in the Mekong Delta', *Southeast Asian Studies*, vol 31, no 4, pp345–384

Nguyen Ngoc Chan to Chief of Dinh Bao District (1901) TTLTQG2, Fonds Goucoch IA 13/232(1), 12 June

Nippon Koei Company Ltd (1966) *Design Report on Ngan Ro Barrage Project*, October 1966, TTLTQG2, NL402, Vietnam

Pacification Studies Group (1971) *The People of the U-Minh Forest Region: A Survey of Their Attitudes in the Context of Past and Present VC/GVN Activities*, RG 472, Office of Civil Operations for Rural Development Support/Plans, Policy and Programs Directorate/CORDS Historical Working Group Files, Box 32, Naraii

Phong, N. D., Hoanh, C. T., Tuong, T. P., Malano, H. and Weaver, T. (2007) 'Modelling acidity propagation in a coastal zone with acid sulphate soils', in Oxley, L. and Kulasiri, D. (eds) *MODSIM 2007 International Congress on Modelling and Simulation*, Modelling and Simulation Society of Australia and New Zealand, December 2007, pp74–80

Sanders, B. M. (2002) *Fire Incident Assessment, U Minh Ha and U Minh Thuong National Park, Ca Mau and Kien Giang Provinces, Vietnam*, J. G. Goldammer/Global Fire Monitoring Centre, Hanoi, Vietnam

Sansom, R. L. (1970) *The Economics of Insurgency in the Mekong Delta of Vietnam,* MIT Press, Cambridge, MA

Service du Génie Rural (1943) *Compte Rendu d'Enquête dans la Province du Chau Doc*, File H62/9, Goucoch Divers, TTLTQG2

Tanaka, K. (1995) 'Transformation of rice-based cropping patterns in the Mekong Delta: From intensification to diversification', *Southeast Asian Studies* vol 33, no3, pp363–378

Thanh Lam and Tran Dinh (2008) 'Rising Mekong tests *Living With Floods* strategy', Inter Press Service, www.ipsnews.net/news.asp?idnews=43862

The Netherlands Delta Development Team and the Committee for the Coordination of the Investigations of the Lower Mekong Basin (1974) *Recommendations Concerning Agricultural Development with Improved Water Control in the Mekong Delta – Main Report*, Committee for the Coordination of Investigations of the Lower Mekong Basin, Bangkok, Thailand

Tuong, T. P., Kam, S. P. et al (2003) 'Impact of seawater intrusion control on the environment, land use and household incomes in a coastal area', *Paddy and Water Environment,* vol 1, no 2, pp65–73

Vietnam News Briefs (2008) *UN: Vietnam is Hardest-Affected Country by Sea Level Rise*, 4 July, http://mailman.anu.edu.au/pipermail/enviro-vlc/2008-July/000382.html

Vo Tong Xuan and Shigeo Matsui (1998) *Development of Farming Systems in the Mekong Delta of Vietnam*, Ho Chi Minh City Publishing House, Ho Chi Minh City, Vietnam

Wassmann, R., Hien, N. X. et al (2004) 'Sea level rise affecting the Vietnamese Mekong Delta: Water elevation in the flood season and implications for rice production', *Climatic Change*, vol 66, nos 1 and 2, pp89–107

World Bank (1999) *Project Appraisal Document on a Proposed Credit of SDR 72.8 Million to the Socialist Republic of Vietnam for the Mekong Delta Water Resources Project*, 15 March 1999, World Bank, Washington DC

9

Hydropower in the Mekong Region: What Are the Likely Impacts upon Fisheries?

Juha Sarkkula, Marko Keskinen, Jorma Koponen, Matti Kummu, Jeff E. Richey and Olli Varis

INTRODUCTION

The human impact upon water resources has increased dramatically all over the world during the last several decades (Vörösmarty, 2000). The Mekong River is one of the few large river basins yet to be irreversibly modified by large-scale infrastructure. While the first dams in a planned cascade have been built in the upper-middle reaches in China, flow regimes in the lower reaches of the mainstream are still, essentially, natural (MRC, 2005). These conditions may not last much longer. The Mekong River Basin is facing the prospects of a major growth in infrastructure projects as surrounding economies continue to expand. Huge hydropower dams as well as diversions for irrigation are planned throughout the basin, some on tributaries and others on the mainstream (e.g. King et al, 2007).

A body of recent research concludes that development in the river and the basin will alter flows and floods in the basin (Adamson, 2001; ADB, 2004; World Bank, 2004). The alterations to flow may have significant impacts upon the river ecosystem, wetlands, floodplains and well-being of people, especially those dependent upon natural resources in the Lower Mekong Basin (LMB).

Potential impacts upon fish form a particularly important issue: fish is a central social, economic and cultural resource in the basin, and forms, together with rice, the foundation of food security in practically all riparian countries. Mekong fisheries are also globally exceptional for their diversity and size, and the

Mekong Basin is considered to have the world's largest inland fisheries (Poulsen et al, 2004; Dugan 2008). Due to the diversity of fisheries and fishers in the region, it is difficult to estimate the actual amount and value of Mekong fisheries; the most recent estimates for the annual value range from US$2 billion up to US$3 billion, with catch estimates as high as 2.5 million tonnes a year (Dugan, 2008).

When discussing the potential impacts of hydropower development, fish is usually considered to be a resource particularly vulnerable to negative impacts. The impacts upon fisheries due to hydropower development can be divided into two main categories:

1 the so-called barrier effect of dams on fish migration; and
2 the impact that hydropower development has upon water quantity and quality, and, consequently, upon fish habitats, for example.

Overall, the fisheries of the Mekong are dependent upon migration over both long and short distances, with many of the commercial species having highly developed migratory patterns (Barlow, 2008). The migratory fish species generally move upstream to spawn, while juvenile fish then move back downstream to feed and grow on the floodplains and wetlands (Poulsen et al, 2002). Dams act as barriers for fish migrating upstream, while the fish migrating downstream must usually pass through hydropower turbines, resulting in mortalities with very low survival rates (Barlow, 2008). Dugan (2008) estimates that over 70 per cent of the total fish catch in the Lower Mekong Basin is dependent upon long-distance migrant species. Dugan (2008) also points out that existing fish passage facilities simply cannot cope with such a large fish migrations and high species diversity that is present in the Mekong, indicating that effective mitigation measures for the barrier effect are not easy – or cheap – to achieve in the Mekong context.

Hydropower dams also affect river flows, causing different kinds of changes in both water quantity and quality. These include, for example, changes in the extent, duration and timing of annual floods, as well as reduction in suspended sediment concentrations due to sediment trapping of the reservoirs (Kummu and Varis, 2007). Reduced floods with shorter duration reduce the available fish habitats in the floodplains, resulting in lower fish production (Barlow 2008; Halls et al, 2008). Changes in the timing of the floods can also disrupt the crucial spawning and migration cues of fishes (Baran, 2006). Overall, this chapter seeks to synthesize what is known about the magnitude and nature of the expected changes – particularly as a consequence of dam-building – to flow regimes, and their consequent potential impacts upon fish, fisheries and livelihoods. Related to this, the chapter discusses the challenges connected to models and impact assessments, as well as the problems in addressing the real value of diverse small-scale use of different water-related resources – most importantly, fish. The geographical focus of the chapter is on the Tonle Sap Lake in Cambodia that is one of the most productive freshwater ecosystems in the world.

MODELLERS AND MODELLING

The chapter draws largely on the findings of the Finnish component of the Water Utilization Programme (WUP-FIN) project, funded by the Ministry for Foreign Affairs of Finland and implemented during 2001 to 2006 as a complementary project to the MRC Water Utilization Programme (MRCS/WUP-FIN, 2007b). The focus of WUP-FIN was on hydrological and hydrodynamic modelling, scenario simulations, socio-economic and policy analyses, and integrated assessment of ecosystem and socio-economic impacts. Several case studies were carried out in different parts of the Lower Mekong Basin, with emphasis being on the Tonle Sap Lake and the Mekong floodplains in Cambodia.

Although essentially a modelling project, WUP-FIN also included strong socio-economic and policy analysis components. The socio-economic analyses carried out in WUP-FIN aimed to increase understanding of social, economic and political factors in water resources management at intermediate and local levels. This increased understanding was used to support other project components – in particular, model development, case study design and impact assessment, both locally and at the basin-wide scale. While the socio-economic and policy analyses ultimately aimed to also address regional basin-wide challenges, this was done by first studying the challenges at the local level, and then putting these into the larger regional context. As it turned out, many of the regional concerns emerged from impacts felt or foreseen at the local level.

The chapter is a follow-up to a review we made (Sarkkula et al, 2007) about the use of mathematical modelling in integrated management of water resources in a previous volume on improving water governance in the Mekong region (Lebel et al, 2007) that carried three important conclusions.

First, modelling projects must link better with the other dimensions of water management, most importantly with social dimensions where its linkages have traditionally been the weakest. This linkage should preferably be created from the very beginning of any modelling exercise, and enough time and resources should be allocated to this multiple and, essentially, interdisciplinary task.

Second, and related to the above, much deeper integration with social sciences is needed. Some progress in integration of research teams connecting natural and engineering sciences has been made, but integration with the social sciences is still only emerging. To date, the approach adopted by modellers to address these more multidisciplinary connections has typically been merely 'to add some social stuff' to their models (Nancarrow, 2005). This is clearly insufficient and can easily just increase the misunderstandings and even prejudice between modellers and non-modellers.

Third, modellers need to focus more on cooperation and communication. This will require more two-way dialogue with decision-makers, planners and other stakeholders on the models as well as on their results and uncertainties. The aim and outcome of these engagements should be increased transparency and intelligibility of the models and their results. In some instances, joint interpretation and assessment of the modelling and analysis results is also desirable and valuable.

The real change is therefore likely to come through the establishment of teams for integrated assessment and modelling with balanced and equal participation by modellers, social scientists, policy experts and other non-modellers. A spirit of mutual appreciation and respect has to be cultivated between the involved individuals, teams, stakeholders and interest groups, requiring good communication skills (Janssen and Goldsworthy, 1996) and genuine aspiration towards truly interdisciplinary work.

MODELLED FLOW CHANGE AND METHODOLOGICAL DEVELOPMENT NEEDS

Credible and accurate basin-wide hydrological and flow modelling is a task of primary importance in assessing the changes in flow caused by water resources developments in the basin. Relatively small flow alterations tend to have proportionally much greater impacts upon the river ecosystem and particularly upon the floodplains. Therefore, basin-wide hydrological modelling alone is not enough for proper flow assessment, but needs to be coupled with a sub-basin-scale floodplain model to understand the consequences of the flow alteration on the flood characteristics, and furthermore, on the ecosystem productivity in the river and its floodplains.

This chapter discusses some of the current challenges in basin-wide modelling in the Mekong Basin, and then provides an example from the WUP-FIN Project on simulated impacts of Mekong development upon the Tonle Sap floodplain system and its productivity, and consequently, upon fisheries.

Basin-wide modelling and its challenges

The Water Utilization Programme (WUP) was set up at the Mekong River Commission (MRC) to build up a knowledge and model base for the Mekong Basin.

The WUP *Project Implementation Plan* (PIP) states:

> *… The broad aim of Component A is to provide enhanced knowledge base and analytical tools to support the WUP, and the BDP [Basin Development Plan of the MRC], that are based on improved understanding of the interaction between the physical and biological features of the basin and their functions with respect to water resources, and the changes in these that may occur due to human activities.*

It is clear from this statement that a comprehensive basin-modelling package is necessary for the Mekong River Commission Secretariat (MRCS) core activities and implementation of the Mekong Agreement. It implies the need for primary data collection and ecosystem process studies, as well as an understanding of the socio-economic functions and impacts in the basin, on top of building an advanced model system to respond to the complexity of the Mekong Basin environment. This concept was included in the terms of reference (ToR) of the WUP-FIN phase 1 (Tonle Sap modelling project; see MRCS/WUP-FIN, 2003) and has been a guiding principle throughout the WUP-FIN work.

Furthermore, the specific objectives of the WUP modelling component were to:

- develop an integrated and comprehensive basin modelling package that provides analytical support needed by the MRC and the riparian countries to implement the Mekong Agreement, prepare the *Basin Development Plan*, and carry out basin and sub-basin planning for sustainable water resources development;
- develop additional model components to analyse and predict transboundary impacts of proposed actions on the aquatic ecosystem and other water uses and functions of social, economic, regional and global importance.

Although WUP plans set the objective of modelling and integrating water quantity and quality with biological, ecological and socio-economic issues, the actual WUP-A work, in practice, focused on water quantities only. The possibility of including water quality and the Tonle Sap lake and floodplain model in the Decision Support Framework (DSF) was lost when the original consultancy work for water quality and lake/floodplain modelling were eliminated during the project inception phase. Consequently, in the DSF there is only a limited 'set of impact analysis tools that enable the prediction of environmental and socio-economic impacts in response to changes in condition of the river system' (MRC, 2004). The set consists of flood properties and a rather schematic saline intrusion description. All of this has led to critical limitations in implementing the DSF and assessing the impacts of development scenarios.

The Model and Knowledge Base (i.e. Decision Support Framework of the MRC) was developed under an approximately US$5 million contract with Halcrow Group during the period of 2001 to 2003. The contract outputs consist of a basin knowledge base and hydrological/hydrodynamic models and impact assessment tools. Model components include catchment hydrology (SWAT), water use (IQQM) and 1D river hydrodynamics (ISIS). The ability of the DSF to simulate hydropower development impacts was critically studied by Adamson (2006), and reported at the Mekong Region Waters Dialogue meeting in Vientiane in July 2006. Adamson found the DSF hydrological model unsatisfactory for this purpose, stating the following:

> *The hydrological model IQQM selected for the DSF by the Halcrow Consultants was principally developed for the Murray Darling Basin in New South Wales in Australia, with major focus on identifying water allocations in a highly managed and regulated river system. The hydropower element had to be added to the Mekong version of IQQM in order to enable simulation of hydropower storage and power plants. However, this is fairly rudimentary and the DSF would not be the modelling system of choice on which to base an assessment of the consequences for the regional hydrological regimes of hydropower expansion in the Mekong Basin. Output would be coarsely indicative of the cumulative impacts of any regulatory storage. Meaningful hydropower simulation needs dedicated simulation models that are much more sophisticated than the relevant modules within the DSF.* (Adamson, 2006)

Consequently, it is clear that the current DSF system needs to be strengthened to meet the requirements set in the WUP *Project Implementation Plan*, most importantly to ensure that analytical tools are based on improved interaction between the physical and biological features of the basin; model components for analysis and prediction of transboundary impacts of proposed actions on the aquatic ecosystem are being developed and actively used; and modelling of water quantity and quality and its linkages with important environmental, social and economic issues, such as wetlands and fisheries, form the core of the system.

The recently established Information and Knowledge Management Programme (IKMP) of the MRC, which continues the work of the Water Utilization Programme, needs to address the existing gaps in the DSF – in particular, to focus on providing more comprehensive views on social and economic impacts that the changes in the flow and water quality of the Mekong system are likely to cause.

Improving the accuracy of change estimates

Plans for large-scale hydropower dams are mushrooming in the Mekong Basin; yet there are no appropriate and commonly agreed tools to make good estimates of their potential impacts. Credible, validated and transparent models are necessary for good decision-making and public acceptance of those decisions. Further work is needed, especially in improving the reliability of the estimates of hydrological, environmental, social and economic impacts, including:

- modelling the basin-wide hydrological impacts of the developments (e.g. by model studies comparative and complementary to the DSF in order to reduce the current uncertainties in the flow change estimates);[1]
- developing further the integrated indicator of the productivity of the Tonle Sap ecosystem (based on Junk, 1997; Lamberts, 2006), and the Lamberts and Koponen (2008) productivity model;

- continuing work on social impact assessment with particular focus on the inclusion of broader social issues, such as vulnerability and poverty; and
- continuing the work of defining the acceptable reverse flow to the Tonle Sap (as stated in the 1995 Mekong Agreement) as part of the national consultations and in dialogue with the stakeholders.

Changes in Tonle Sap flooding and productivity potential

Changes in the flow and water quality regime of the Mekong River have impacts upon flooding, erosion, sedimentation, navigation, fisheries and agriculture, as well as upon consequent social and economic issues. The objectives of the WUP-FIN phase 2 (MRCS/WUP-FIN, 2007b) were set to provide complementary tools and information to approach these questions. In addition to direct impacts upon the river system, Mekong developments have impacts upon the hydrology and ecosystems of the wetlands connected to the river as their behaviour is largely affected and controlled by the mainstream Mekong River. This is the case with some of the important tributaries' floodplains, the Tonle Sap system in Cambodia being the most highlighted example. The Tonle Sap River is, in some definitions, taken as part of the mainstream due to the dominant role that the river has in the functioning of the Tonle Sap Lake ecosystem. Around 60 per cent of the Tonle Sap flood water originates from the Mekong, and the water level in the lake is controlled by the water level in the Mekong mainstream (Kummu and Sarkkula, 2008). Therefore, the possible flow alterations in the Mekong mainstream will directly affect the flood pulse of the Tonle Sap Lake.

The Tonle Sap system and its remarkable levels of aquatic production are clearly of crucial importance for Cambodia and, indeed, for the entire Mekong Basin (e.g. Keskinen, 2006; Kummu and Sarkkula, 2008). The productivity of the Tonle Sap lake-and-floodplain system is driven by the flood pulse of the Mekong and by the rich floodplain biodiversity. The flood pulse transfers terrestrial primary products into the aquatic phase during flooding and creates an extremely rich ecosystem for aquatic life (Junk, 1997). Primary production (phytoplankton, periphyton and plants) fuels the food webs, resulting in one of the world's most productive fisheries grounds. The floodplain of the lake offers ample opportunities and conditions for fish to breed and grow (Lamberts, 2006; Lamberts and Koponen, 2008).

Hydropower development may change the natural flood pulse, directly undermining the productivity of the system by reducing the inundated habitats, delaying the onset of flooding, and shortening its duration (growth period for aquatic organisms). All of these changes are estimated to have a negative impact upon the fisheries productivity of the Tonle Sap system. Hydropower development would probably also reduce the supply of sediments and nutrients to the downstream ecosystems because of sediment trapping in the reservoirs. Fisheries productivity is further likely to be affected by worsening conditions for fish reproduction due to slowly rising flood waters and the associated poor water quality. Lower flow velocities will limit the drift of eggs, larvae and juveniles to the floodplain habitats, while dams will obstruct fish migrations.

A number of models have been used for flow regime simulations with the foreseen hydropower developments (e.g. the high development option with Chinese mainstream dams and Lower Mekong Basin tributaries dams). The DSF flow values were input into the WUP-FIN Tonle Sap model to simulate the flooding change of the Tonle Sap system, such as the change in the extent of the flooded habitat and the change in the dry season water level (two more simulations on the cumulative impact of the hydropower developments that were available in addition to the DSF are referred to below). The results from the different simulations are quite different, especially the dry season water levels. This is mainly due to the different development scenarios and assumptions, as well as the models themselves used in the analysis. An actual analysis and comparison of the models has not been possible due to their different contractual setting. The results require higher confidence levels and this question should be re-examined, with additional models brought into the ensemble (e.g. the Variable Infiltration Capacity (VIC) model and WUP-FIN).

The dry season water-level rise due to Mekong upstream development was, in different assessments, estimated as:

- 0.15m (by DSF, data prepared for IBFM);[2]
- 0.30m (by Henrik Garsdal of the Danish Hydraulic Institute (2004), based on Adamson's (2001) analysis on the mainstream Mekong Basin); and
- 0.60m (by ADB, 2004, using MIKE Basin).

The impact of the water-level rise upon the dry season lake area is presented in Figure 9.1. The 30-day minimum water level during the analysis period of 1997 to 2006 for May was 1.44m above mean sea level (amsl), which was used as a reference level. The bottom of the lake lies at 0.6m amsl; thus, during the low water level the average depth of the lake is only around 0.8m, with a lake area of around 2300km^2. The estimated rise of 0.60m in the dry season water level, as simulated by ADB (2004), would result in the flooding of an area of 3200 km^2, indicating that the permanent lake area would increase by nearly 1000 km^2 or 40 per cent (Kummu and Sarkkula, 2008).

The rise in the dry season water level of the lake would mean an extension of the permanent lake and significant destruction of the flooded gallery forest where it becomes permanently inundated and, consequently, loss of important habitats (Kummu and Sarkkula, 2008). The different cumulative impact assessments (CIAs) also predict that the peak water level would decrease and thus reduce the inundated area of the lake, as presented in Figure 9.2. Thus, the area of the floodplain would decrease, depending upon the CIA, by 7 to 16 per cent. For example, in the case of a CIA carried out by ADB (2004), the total floodplain area would decrease from

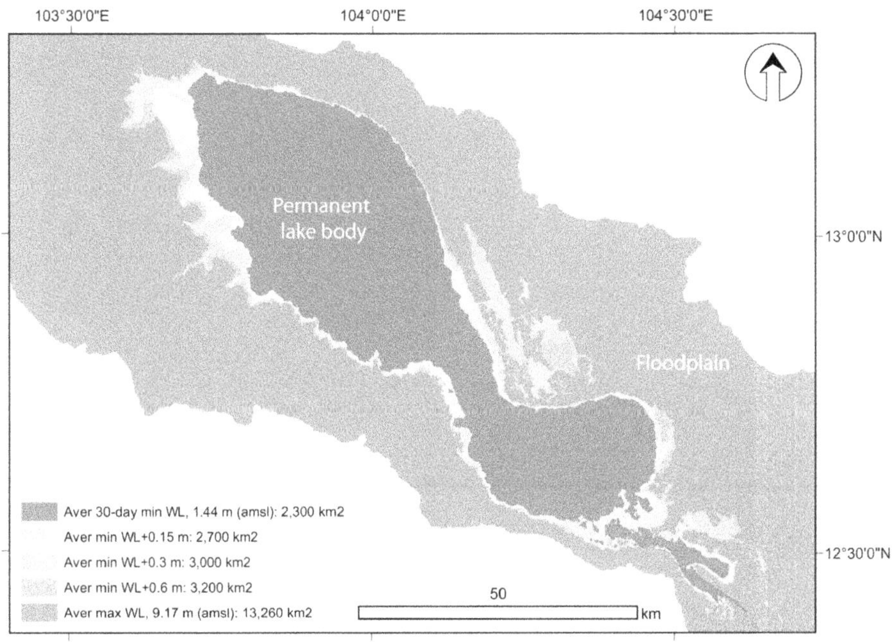

Figure 9.1 *Inundated areas due to the increased dry season water level*

Source: adapted from Kummu and Sarkkula (2008)

the current 10,750km^2 to 9060km^2 by the year 2025, resulting in around a 15 per cent decrease in both cumulative flooded area and volume.

Figure 9.3 shows the change in flood duration over the floodplain during 1997 conditions based on the EIA 3D model results and input of the MRC flow regimes developed for the IBFM project. The period of inundation decreases in most parts of the floodplain by one to two weeks (5 to 10 per cent), while in the lowest parts of the floodplain the inundation is prolonged due to the increase of the dry season lake level. Due to permanent inundation, these areas would be transferred from floodplain habitats to become part of the lake proper.

The tall gallery forest strips around the lake make an important physical barrier between the lake and the floodplain. The strips create favourable conditions for sedimentation within the forested zone where nutrients bound with the sediment, mainly from the Mekong, fuelling primary production. The lake extension would cause permanent submersion – in essence, destruction – of considerable strips of gallery forest surrounding the lake (Kummu and Sarkkula, 2008). The reduction of the flooded forest area could therefore have a significant impact upon the whole Tonle Sap ecosystem, and probably also upon floodplain dynamics. The evolution of the floodplain to its present state and biological functioning has taken several

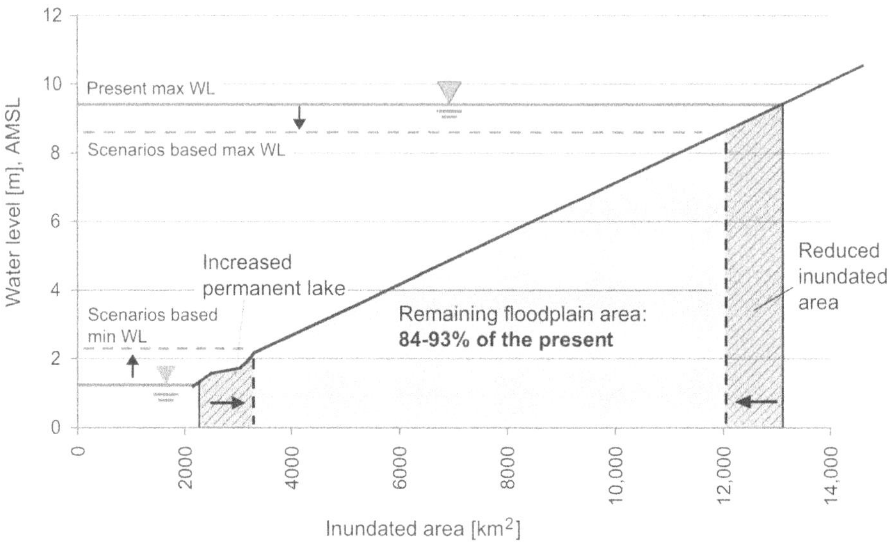

Figure 9.2 *Schematic presentation of the possible impacts on the floodplain extent due to changes in flow regime*

Source: adapted from Kummu and Sarkkula (2008)

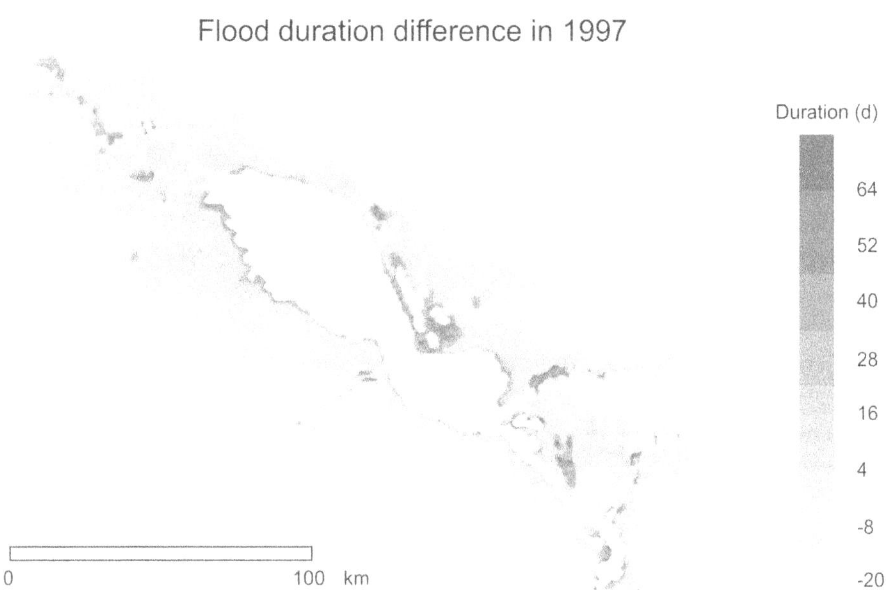

Figure 9.3 *Flood duration difference based on simulation results*

Source: adapted from MRCS/WUP-FIN (2007b)

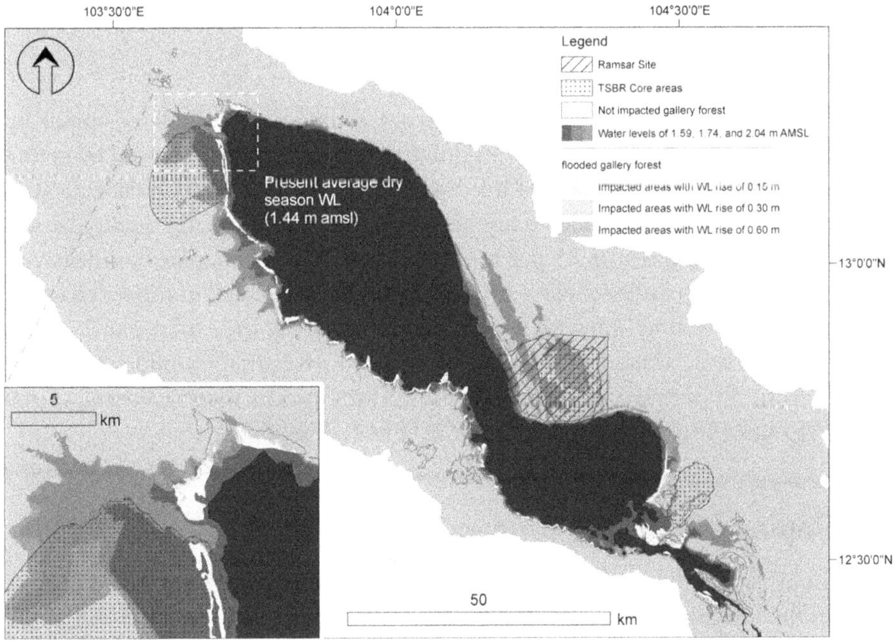

Figure 9.4 *Flooded and remaining tall gallery forest in the case of 0.6m dry season water-level rise*

Source: adapted from Kummu and Sarkkula (2008)

thousands of years (Tsukawaki, 1997), which means that what is lost in its structure and productivity can hardly be mitigated in any way.

Scenario work carried out under the WUP-FIN project, aiming to estimate the cumulative impact of the changing floodplain conditions in the Tonle Sap, focused on comparing the flow changes between Flow Regime FR3[3] (MRCS/IBFM, 2006a) and the baseline in 1997 and 1998 on a number of flood and water quality indicators of the lake and the floodplain. The comparison of the simulation results gave the following results:

- The inundated floodplain habitat would be reduced by 15 to 20 per cent.
- The period of inundation would be shortened by one to two weeks.
- The increased dry season water level would permanently inundate a major part of the flooded forest around the lake (extending the permanent lake).
- Dissolved oxygen conditions would worsen by extending the strongly hypoxic/anoxic period in the floodplain during early flooding due to slowly rising floods.

- Sediment and nutrient input to the lake with the flood waters would be reduced.

A first estimate of the impact of the changing floodplain conditions was made by introducing a cumulative indicator for floodplain productivity potential by giving an estimate for the minimum and maximum value for each individual indicator. The calculation of the cumulative impact of the physical and water quality factors gave a value in the order of 25 per cent reduction in the floodplain productivity potential, even with rather conservative estimates for individual indicator changes. This estimate is well in line with the assessment made by the expert panel within the IBFM phase 2, where it was estimated that Flow Regime 3 would result in an overall 20 to 30 per cent or more reduction in the productivity potential of the Tonle Sap Lake and its floodplain (MRCS/IBFM, 2006a).

Developing a Tonle Sap productivity model

The Tonle Sap flood pulse is largely (60 per cent) driven by the water that is pushed up into the lake by the reversed flow of the Tonle Sap River during the rise of the flooded Mekong River. The remainder is runoff from Tonle Sap's own catchment as well as direct open-water precipitation. The Mekong flood waters not only bring water and floods, but also nutrient-laden sediments that are largely deposited in the floodplain. The flood water integrates the terrestrial vegetation within the aquatic phase of the ecosystem, and this interaction between the terrestrial and aquatic phases is the driving force of ecosystem productivity. Knowledge of the relation between ecosystem productivity and the flood pulse, as ultimately determined by the flows in the Mekong River, is still very limited.

Lamberts and Koponen (2008) have developed a quantitative model of the ecosystem productivity of the Tonle Sap Lake and floodplain, with a particular focus on its response in the function of altered flow regimes in the Mekong River.

The underlying assumption is that the Tonle Sap secondary production (including fish production) is mostly endogenous (i.e. based on primary products generated within the ecosystem rather than imported with the flood waters). While there are no specific data to support this, the assumption is reasonable based on the low organic matter contents of the inflowing rain and Mekong waters and the known migration of fish between the Tonle Sap and the Mekong. This is believed to be largely a net export of fish biomass from the river, with mostly fish juveniles, larvae and eggs drifting into the Tonle Sap ecosystem with the flood waters. Furthermore, the assumption is made that most of the organic matter in the Tonle Sap ecosystem is produced locally by four categories of primary producers: periphyton, phytoplankton, rooted macrophytes and floating macrophytes.

The main result from this research is a spatially explicit model of Tonle Sap ecosystem primary production, which is the basis of the secondary food webs and determines, to a large extent, the overall productivity of the system. The modelled production is based on environmental factors, most of which are directly or indirectly dependent upon the hydrological cycle. The model allows quantitative assessments of the environmental impact (relative or absolute) upon ecosystem primary production. It uses the few data specific to the Tonle Sap and depends upon the hydrodynamic model for making the results spatially explicit. It has demonstrated where the main data and knowledge gaps are, and the model has been developed so that its accuracy can be improved by relatively few field measurements. In this way, it can be constantly refined while already providing the best available quantitative assessments of the impact of flow alterations in the Mekong River upon the primary production of the Tonle Sap ecosystem.

The link between primary production of the ecosystem, its fish production and the fish catches is very complex, and there is no specific information on these links, nor are there sufficient data on fish catches. However, with the reasonable assumption that most of the productivity of the Tonle Sap is located within the ecosystem rather than imported, taking into consideration the high fish production, and given the dynamic character of the ecosystem, it can be assumed that any loss of primary production will directly result in loss of secondary production and, hence, fish catches. The precise nature of this relationship is unclear and may be impossible to establish.

The ecosystem and economic values of resources

The Mekong and its floodplain are rich in natural resources, particularly in fish, supporting local livelihoods in a variety of ways. This was also evident in the socio-economic surveys carried out during the WUP-FIN Project in different parts of the basin (MRCS/WUP-FIN, 2007b). Although the study areas in the Lower Mekong Basin were diverse and, therefore, different in many ways, there were also findings – related, for example, to the value of and the dependency upon resources – that were similar in all studied areas; these are discussed next in more detail.

Estimating the real value

Measuring poverty and the value of resource use in the widely varied conditions present in the Mekong Basin is not simple. For example, the findings from the Tonle Sap, as well as from Nam Songkhram Basin in northeast Thailand, indicate that both of the areas are considered poor in monetary terms, but rich in natural resources. In both areas local people rely on diverse natural resources that provide both food and income, although not necessarily in cash. In Nam Songkhram, the area's wetlands are considered to be 'nature's supermarket' where you need no money to 'shop' for the large variety of different resources provided (MRCS/

WUP-FIN, 2007b). This coexistence of economic poverty and resource wealth raises questions about the validity of current poverty measures, and suggests that macro-economic analyses measuring poverty in pure monetary terms are not able to properly measure the wide and diverse use of natural resources. This leads to an underestimation of the real values of different natural resources for the people living in the Mekong Basin.

Consequently, the close linkage between the viability of river ecosystems and people's livelihoods is often not taken seriously into account in social and, particularly, economic analyses at national and regional levels (see Chapter 12). Everyone agrees that the health of the river ecosystem feeds directly back into supporting the welfare of the people in the basin, particularly those amongst the poorest sections of society. However, the importance of maintaining the river ecosystems – and, consequently, the diverse set of resources and services that they provide – is still neglected in most policy discussions about the development of the basin.

The dependence of poor people upon aquatic resources

The findings from the WUP-FIN case study areas indicate that the poorest villagers are those most directly dependent upon fisheries and related aquatic resources for their livelihoods. At the same time, the poorest have usually less capabilities and resources to adapt and make use of the changes that take place in the availability of these resources due to changes in water flows – for instance, irrigation development or dam construction. When coupled with prevalent governance challenges and problems of unequal access to different resources, this is likely to lead to growing disparities between villagers: while those who are better off are likely to gain, those who are already poorer will lose (e.g. Fox and Sneddon, 2005; Keskinen et al, 2007; Sneddon, 2007).

This finding was particularly evident from the results of the participatory village surveys in the Tonle Sap area (see Keskinen, 2006; MRCS/WUP-FIN, 2003), which concluded that the people living in the villages closest to the lake were in many ways most vulnerable to the changes in natural resources. They are generally poorer, less educated, have fewer livelihood options, do not own agricultural land, and depend strongly upon common-pool resources such as fish and flooded forests for their livelihood. Differences also seemed to emerge within the villages between the capabilities of the poor and the better-off to respond to the changes in natural resources.[4] Similar findings were also apparent in the Cambodian floodplains (MRCS/WUP-FIN, 2006a, 2006b), in Nam Songkhram (MRCS/WUP-FIN, 2007a; MWBP, 2005), as well as in the Mekong Delta (MRCS/WUP-FIN, 2006c), where livelihood developments focusing on intensive agriculture and aquaculture do not seem to take the poor's dependence upon, for example, wild-capture fisheries properly into account.

Small-scale utilization and distributing the benefits of development

Small-scale utilization of Mekong's water resources offers a more sustainable basis for poverty reduction than large-scale projects. As discussed above, we believe that many of the current economic assessments in the Mekong Basin are underestimating the actual value of natural resources for local people. Following on from this, the common justification for water development is the 'underutilization' and 'underdevelopment' of the basin's resources. However, the Mekong basin is – through small-scale fishing, farming, use of wetlands and floodplain resources – already extensively utilized in a variety of ways at the local level. Consequently, most future development options in the basin are focusing on the development of modern sectors such as irrigated agriculture, while a majority of the population in the basin actually depends upon more traditional livelihood sources (MRCS/IBFM, 2006b; Keskinen et al, 2008). As noted by Phillips et al (2006):

> *The key development paradox of the region is that economic growth is necessary to bring many of the populations out of poverty, but the 'classical' route involving the subsidized construction of massive infrastructure is most unlikely to provide the optimal result in this respect for the poorer sections of the populations.*

Indeed, we see that 'classic' large-scale development interventions such as irrigation and hydropower projects are, despite their objectives of poverty reduction, actually often undermining the foundations of the livelihoods of the poorest groups by impacting negatively upon the different common pool resources – most importantly, fisheries.

Consequently, the management and development of Mekong's water resources – if aiming at poverty reduction – should be based much more upon already existing 'decentralized' utilization of the Mekong's resources. In addition to the actual value of this diffuse utilization, its distributional benefits should also be considered. Compared, for example, with the distribution of the benefits from hydropower dams, small-scale utilization usually allows for more equal distribution of the benefits derived from the Mekong's resources, reaching the poorest more easily.[5] Hence, if, for example, the value of hydropower development and of sustaining river fisheries were of the same order of magnitude, in terms of poverty reduction, fisheries would provide a more favourable basis due to its more equitable – and already existing – distribution of benefits.

It is important to emphasize that our findings do not imply that infrastructure should not be built when it is needed for national economic development. What we are highlighting, however, is that in terms of poverty reduction, we believe that much more emphasis should be put on sustaining and developing existing small-scale and local livelihood resources. Overall, small-scale infrastructure reduces risks of unintended impact, better involves local stakeholders, and the distribution of benefits and costs from the project are much easier to monitor and address. This finding is also supported by the results from the ADB-funded Built Structures Project for the Tonle Sap Lake (CNMC and WorldFish Centre, 2007).

The recognition of the actual value of the traditional livelihood sources and their distributional benefits will most probably lead to more balanced discussion about the possible trade-offs required in developing the Mekong's resources. When considering the huge number of different plans for Mekong's development, it is worrying to notice a complete absence of well-informed and transparent discussion about the different trade-offs that are unavoidably required – both within and between the riparian countries – due to changes caused by water resources development. We see that there is an urgent need to acknowledge that water development requires trade-offs, and that the discussion about trade-offs is always highly political. Achieving the best compromises (where possible) on different trade-offs requires open and transparent discussion, access to relevant information by all concerned parties, as well as research focused on socio-economic and livelihood issues (such as small-scale fisheries) that are most likely to experience radical changes.

Peoples' well-being under threat

Up until 2006, only China was actively building and pursuing mainstream dams in the Mekong region. Since then, there has been a sudden surge in mainstream dam plans within the Lower Mekong Basin in Laos, Thailand and Cambodia (see Chapter 2). These intentions further increase the potential for destructive impacts upon fisheries in the Lower Mekong Basin. Apart from alteration of the flood pulse, changes in timing and duration of flooding and water levels, sedimentation in reservoirs, and other flow-related impacts such as changes in larvae and juvenile drift, the dams will block essential fish migration routes and disconnect spawning and living habitats. The impacts will very likely be significantly bigger than in the so-called high-development scenario (Chinese dams and Laos tributary dams) that was discussed in the section on 'Modelled flow change and methodological development needs'. The risk to people's well-being in the basin is consequently magnified.

This chapter has considered several issues related to living conditions in the Mekong region, often focusing on groups of rural poor who are particularly vulnerable to changes in floodplains. A critical question is about their life and future in relation to basin developments. The rural poor are defined as poor as a result of their low cash income, an indicator unable to describe their living circumstances and well-being. A much more important source of livelihood, however, originates from the rich natural resources in the basin, especially the enormously productive fisheries. Fisheries do not only benefit the people living next to the river or the floodplains, but all of the Lower Mekong Basin countries, providing their populations with the main portion of their animal protein needs.[6]

It is paradoxical to talk about the poverty of the population without properly taking into account this enormous natural resource. Basin developments are repeatedly and without critical views being justified by the statement that large-scale infrastructure construction is the solution to reducing poverty, although there is no evidence this will lead to the claimed result. The dams are expected to generate income; but how will the benefits and costs be shared?

Chapter 12 in this volume discusses the hydropower and fisheries trade-off and the storylines that are embedded within it. Friend et al identify the inherent assumption in the approach that society can afford to trade off fisheries for the economic benefits of hydropower. They claim that the notion that fisheries can be traded off rests on the highly questionable hypothesis that what is lost can be replaced. They conclude that an empirically based counter-narrative is required that can provide a counter-scenario where fisheries are not merely a resource of conservation value, but a resource whose management is central to meeting the development challenges of the basin.

The destructive effects of the dams at different timescales must be included in the development equations, as well. The lifetime of the dams is very short compared to the evolution needed for the ecosystem to develop its services. The richest ecosystem in the Mekong Basin, the Tonle Sap floodplain, was created about 6000 years ago when the Mekong and the Tonle Sap rivers were connected as a consequence of an elevated sea level (Tsukawaki, 1997). Since then, part of the Mekong flood waters have entered into the lake and the surrounding terrain through an annual reversal of the flow of the Tonle Sap River, resulting in the ecosystem's high biodiversity and productivity, particularly in the aquatic–terrestrial transition zone.

As discussed in this chapter, the modification of the Tonle Sap flood pulse due to construction of hydropower dams will negatively affect floodplain productivity. Already, the cumulative effects of the Chinese dams and the Lower Mekong Basin tributaries dams[7] have been estimated to have a significant negative effect upon the Tonle Sap ecosystem's productivity. It is important also to realize that permanent reduction in the flood extent and duration is not only likely to reduce fish catch and, thus, threaten livelihoods and food security, but can also threaten the long-term sustainability of the fish populations by reducing their reproductive potential (mean size) and by making fish more vulnerable to capture (Halls et al, 2008). In addition, while the latest plans of building a number of mainstream dams in the Lower Mekong Basin may have a limited impact upon the flood pulse, they will critically block fish migration routes and disconnect spawning grounds and living habitats. Most of the total fish catch in the Lower Mekong Basin is dependent upon long-distance migrant species (Dugan, 2008). There is little doubt that the impacts upon fisheries would be dramatically destructive.

In terms of trade-offs, the distribution of hydropower's benefits and costs is obviously between upstream and downstream, and between different social groups. It would be fair and reasonable to take these concerns as one elementary starting point in the development plans discussion, and to see whether the countries involved and their people are able to bear the consequences of losing an essential part of their ecosystem productivity and services. This also means taking social responsibility of the most affected, and being strict in assessing the impacts regarding general social conditions and food security risks in riparian countries.

The valuation of the fisheries resource has been largely pending and has often been excluded from the basin development equations. For example, Cowx et al (2004) state how poorly the true value of this sector is reflected in official statistics and discussions of food security and livelihoods; as a consequence, fisheries suffer in the face of relatively higher economical priorities such as hydropower. As highlighted by Sokhem and Sunada (2006), one of the key problems in the Mekong Basin is that the value of the fisheries resource is usually ill defined, severely undervalued and poorly represented from both an economic and social perspective.

Recent efforts to truly value the Mekong fisheries (e.g. by the MRC Fisheries Programme) are expected to give a more valid starting point to assess the well-being of people and give improved means to compare development impacts and prices.[8] It is necessary to try to estimate the real value of fish and fisheries: first, to highlight their importance for livelihoods, as well as to draw a more realistic picture of people's well-being and their vulnerability to change. There are many indicators and indices developed and available for assessing people's well-being and life quality that can be used as starting points to make such an evaluation in the Mekong (Henderson, 1996; Prescott-Allen, 2001).

Molle (2006) pointed out several risks in large water infrastructure projects. The challenges, he argues, are how to ensure that:

- projects are not primarily moved by bureaucracies seeking to perpetuate themselves or by the financial and political interests of decision-makers;
- displaced people are fairly compensated (the lack of a voice from poor rural people in many countries suggests that attention to their fate will remain limited);
- benefits are not captured by, and concentrated upon, a few well-off elites, and costs and risks are not borne by poorer people; and
- development alternatives are fairly assessed.

In the case of the Mekong region, the worst scenario – made more likely with the recent boom of dam projects – is that hydropower development will be given priority, with nil or insufficient consideration of, or compensation to, the diverse groups of people undergoing the negative impacts of that development. It is hard to see how the costs and benefits of this development will be fairly balanced – both within and between the riparian countries. Unfortunately, the MRC has so far, despite its mandate and constitution (the 1995 Mekong Agreement), done very little to ensure that a worst-case scenario does not unfold (see Chapter 14).

The conventional economic development path currently pursued in the Mekong is fraught with risks. It is necessary to step back to properly see and assess the plausible alternatives and their implications. We propose a radical strengthening of development dialogues that places emphasis on broad multi-stakeholder participation. We anticipate that this would bring to the fore the uneven sharing of burdens, costs, benefits and risks between upstream and downstream, as well as between different social groups. This kind of finding will most probably draw more attention to the potential of alternative, smaller-scale, development options. In addition, it will demand vastly improved dissemination of information relating to development projects with the result that the public will be better engaged in scrutinizing proposals, promises and potential impacts.

This radical change needs to be supported by research and continuing improvement of the methods for assessing development project impacts, improving their accuracy, transparency and credibility. In this process, the MRC still has an unfulfilled potential. Through credible and comprehensive assessment tools and constructive dialogue with different partners, it can still become a key organization in helping to find an appropriate balance between acceptable levels of hydropower development and the maintenance of fisheries as a vital environmental and social resource in the basin.

NOTES

1 The Mekong basin-wide distributed hydrological Variable Infiltration Capacity (VIC) model provides an existing tool and opportunity to improve the hydrological basin development simulations (Costa-Cabral et al, 2007). Connected with the WUP-FIN Lower Mekong Basin river and floodplain model (MRCS/WUP-FIN, 2007b), the coupled system largely responds to the needs and complexity of the Mekong Basin and its ecosystem. Both the VIC and WUP-FIN models have water quality and ecosystem productivity simulation capability.

2 This refers to the integrated basin flow management process of the MRC.

3 FR3 represents the potential high developments in the basin, as foreseen in 2003 (including the Chinese mainstream and Laos tributaries hydropower dams).

4 This conclusion is also supported by the findings of the ADB-supported Built Structures Project (CNMC and WorldFish Centre, 2007)

5 Distribution of the benefits from different common pool resources (such as fisheries) is naturally not without problems either: the poorest groups, in particular, often have problems getting equal access to these resources.

6 Hortle (2007) summarizes the available information on consumption and yields of inland fish and other aquatic animals (OAA) in the Lower Mekong Basin, reaching an estimate of the total consumption of fish and OAA at about 2.1 million tonnes per year and 0.5 million tonnes per year, respectively. Annual consumption of inland

fish plus OAAs as country averages is in the range of 40kg to 50kg per capita, and its portion of all animal protein consumption values is high across the LMB (as high as 82 per cent in Cambodia).

7 So-called high-development scenario, designed during BDP phase 1.

8 The current estimates for the economic values (first-sale value) of freshwater fish and aquatic products range between US$1 billion and US$1.5 billion (Ahmed et al, 1998; MRC, 2002; Baran, 2005; MRCS/IBFM, 2006a). Including all multiplier effects, the fishery is worth several times more than this figure and its replacement value is far higher (Baran, 2005).

REFERENCES

Adamson, P. T. (2001) 'Hydrological perspectives on the Lower Mekong Basin – the potential impacts of hydropower developments in Yunnan on the downstream flow regime', *International Water Power and Dam Construction*, March, pp16–21

Adamson, P. T. (2006) *Hydrological and Water Resources Modelling in the Mekong Region: A Brief Overview, Mekong Region Waters Dialogue*, Co-convened by IUCN, TEI, IWMI, M-POWER, Vientiane, Laos, 6–7 July

ADB (Asian Development Bank) (2004) *Cumulative Impact Analysis and Nam Theun 2 Contributions*, Final report, Prepared by NORPLAN and EcoLao for Asian Development Bank, Vientiane, p143

Ahmed, M., Hap, N., Ly, V. and Tiongco, M. (1998) *Socioeconomic Assessment of Freshwater Capture Fisheries of Cambodia*, Report on a household survey, MRC/DoF/ Danida Project for the Management of the Freshwater Capture Fisheries of Cambodia, Mekong River Commission, Phnom Penh, Cambodia

Baran, E. (2005) *Cambodian Inland Fisheries: Facts, Figures and Context*, WorldFish Centre and Inland Fisheries Research and Development Institute Phnom Penh, Cambodia

Baran, E. (2006) *Fish Migration Triggers in the Lower Mekong Basin and Other Tropical Freshwater Systems*, MRC Technical Paper no 14, Mekong River Commission, Vientiane, Laos

Barlow, C. (2008) 'Dams, fish and fisheries in the Mekong River Basin', *Catch & Culture*, vol 14, no 2, September, Mekong River Commission, Vientiane, Laos

CNMC (Cambodian National Mekong Committee) and WorldFish Centre (2007) *Influence of Built Structures on Tonle Sap Fisheries – Findings and Recommendations*, CNMC and WorldFish Centre, Phnom Penh, Cambodia

Costa-Cabral, M. C., Richey, J. E., Goteti, G., Lettenmaier, D. P., Feldkötter, C. and Snidvongs, A. (2007) 'Landscape structure and use, climate, and water movement in the Mekong River Basin', *Hydrological Processes*, vol 22, no12, pp1731–1746

Cowx, I. G., Almeida, O., Bene, C., Brummett, R., Bush, S., Darwall, W., Pittock, J. and Brakel, M. V. (2004) 'Value of river fisheries', in Welcomme, R. and Petr, T. (eds) *Proceedings of the Second International Symposium on the Management of Large Rivers for Fisheries*, vol I, FAO Regional Office for Asia and the Pacific, Bangkok, Thailand, RAP Publication 2004/17

Danish Hydraulic Institute (2004) *Study on Natural Reverse Flow in the Tonle Sap River*, Prepared by DHI (Danish Hydraulic Institute) Water and Environment for the Mekong River Commission Secretariat, Financed by World Bank, Vientiane, Laos

Dugan, P. (2008) *Examining the Barrier Effect of Mainstream Dams to Fish Migration in the Mekong, with an Integrated Perspective to the Design of Mitigation Measures*, Conclusions from an Independent Expert Group Meeting, Vientiane, Laos, 22–23 September 2008

Fox, C. and Sneddon, C. (2005) *Flood Pulses, International Watercourse Law, and Common Pool Resources. A Case Study of the Mekong Lowlands*, Research Paper no 2005/20, United Nations University/World Institute for Development Economics Research (WIDER), Helsinki, Finland, p20

Halls, A., Sopha, L., Pengby, N. and Phall, T. (2008) 'New research reveals ecological insights into dai fishery', *Catch & Culture*, vol 14, no 1, June, Mekong River Commission, Vientiane, Laos

Henderson, H. (1996) *Creating Alternative Futures: The End of Economics*, Kumarian Press, West Hartford, CT

Hortle, K. G. (2007) *Consumption and the Yield of Fish and Other Aquatic Animals from the Lower Mekong Basin*, MRC Technical Paper no 16, Mekong River Commission, Vientiane, Laos

Janssen, W. and Goldsworthy, P. (1996) 'Multidisciplinary research for natural resource management: Conceptual and practical implications', *Agricultural Systems*, vol 51, no 3, pp259–279

Junk, W. J. (1997) *The Central Amazon Floodplain: Ecology of a Pulsing System*, Ecological Studies 126, Springer, Berlin

Keskinen, M. (2006) 'The lake with floating villages: A socio-economic analysis of the Tonle Sap Lake', *International Journal of Water Resources Development*, vol 22, no 3, pp463–480

Keskinen, M., Käkönen, M., Prom, T. and Varis, O. (2007) 'The Tonle Sap Lake, Cambodia: Water-related conflicts with abundance of water', *The Economics of Peace and Security Journal*, vol 2, no 2, pp49–59

Keskinen, M., Mehtonen, K. and Varis, O. (2008) 'Transboundary co-operation vs internal ambitions – the role of China and Cambodia in the Mekong region', in Pachova, N. I., Nakayama, M. and Jansky, L. (eds) *International Water Security: Domestic Threats and Opportunities*, United Nations University Press, Tokyo, pp79–109

King, P., Bird, J. and Haas, L. (2007) *The Current Status of Environmental Criteria for Hydropower Development in the Mekong Region: A Literature Compilation*, Consultants Report to ADB (Asian Development Bank), MRCS (Mekong River Commission Secretariat) and WWF (World Wide Fund for Nature), Vientiane, Lao PDF

Kummu, M. and Sarkkula, J. (2008) 'Impact of the Mekong river flow alteration on the Tonle Sap flood pulse', *Ambio*, vol 37, no 3, pp185–192

Kummu, M. and Varis, O. (2007) 'Sediment-related impacts due to upstream reservoir trapping in the lower Mekong River', *Geomorphology*, vol 85, pp275–293

Lamberts, D. (2006) 'The Tonle Sap Lake as a productive ecosystem', *International Journal of Water Resources Development*, vol 22, no 3, pp481–495

Lamberts, D. and Koponen, J. (2008) 'Flood pulse alterations and productivity of the Tonle Sap ecosystem: A model for impact assessment', *Ambio*, vol 37, no 3, pp178–184

Lebel, L., Dore, J., Daniel, R. and Koma, Y. S. (eds) (2007) *Democratizing Water Governance in the Mekong Region*, USER and Mekong Press, Chiang Mai, Thailand

Molle, F. (2006) *Planning and Managing Water Resources at the River Basin Level: Emergence and Evolution of a Concept*, Comprehensive Assessment Research Report 16, International Water Management Institute, Colombo, Sri Lanka, www.iwmi.cgiar.org/assessment/publications/research_reports.htm

MRC (Mekong River Commission) (2002) *Fisheries in the Lower Mekong Basin: Status and Perspectives*, Technical paper no 6, MRC Secretariat, Phnom Penh

MRC (2004) *Decision Support Framework, Water Utilisation Programme Component A: Final Report*, no 1, DSF 100 Main Report, Mekong River Commission, Phnom Penh, Cambodia

MRC (2005) *Overview of the Hydrology of the Mekong Basin*, Mekong River Commission, Vientiane, November

MRCS/IBFM (2006a) *Flow Regime Assessment Integrated Basin Flow Management Report*, no 8, Mekong River Commission, Vientiane, Laos

MRCS/IBFM (2006b) *Social Team – 2006 Final Report, Integrated Basin Flow Management (IBFM) Process*, Water Utilization Programme/Environment Programme, Mekong River Commission, Vientiane, Laos

MRCS/WUP-FIN (2003) *WUP-FIN Phase I – Modelling Tonle Sap for Environmental Impact Assessment and Management Support, Final Report*, Mekong River Commission and Finnish Environment Institute Consultancy Consortium, Phnom Penh, p107

MRCS/WUP-FIN (2006a) *In the Bend of the Bassac River – Participatory Village Survey in Chheu Khmau Commune, Cambodia*, Mekong River Commission and Finnish Environment Institute Consultancy Consortium, Vientiane, Laos, www.eia.fi/wup-fin/wup-fin2/publications.htm

MRCS/WUP-FIN (2006b) *Living in the Mekong Floodplain – Participatory Village Survey in Preah Sdach Commune, Cambodia*, Mekong River Commission and Finnish Environment Institute Consultancy Consortium, Vientiane, Laos, www.eia.fi/wup-fin/wup-fin2/publications.htm

MRCS/WUP-FIN (2006c) *Mekong Delta Socio-Economic Analysis Report*, Mekong River Commission and Finnish Environment Institute Consultancy Consortium, Vientiane, Laos, www.eia.fi/wup-fin/wup-fin2/publications.htm

MRCS/WUP-FIN (2007a) *Nam Songhkram Model Applications, WUP-FIN Phase II – Hydrological, Environmental and Socio-Economic Modelling Tools for the Lower Mekong Basin Impact Assessment*, Working Paper no 5, Mekong River Commission and Finnish Environment Institute Consultancy Consortium, Vientiane, Laos

MRCS/WUP-FIN (2007b) *WUP-FIN Phase 2: Hydrological, Environmental and Socio-Economic Modelling Tools for the Lower Mekong Basin Impact Assessment, Final Report: Part 2 – Research Findings and Way Forward*, Mekong River Commission and Finnish Environment Institute Consultancy Consortium, Vientiane, Laos, www.eia.fi/wup-fin/wup-fin2/publications.htm

MWBP (Mekong Wetlands Biodiversity Conservation and Sustainable Use Programme) (2005) *Thai Baan Research on the Ecology and Local History of the Seasonally-Flooded Forest in the Lower Songkhram River Basin*, IUCN and MWBP, Vientiane

Nancarrow, B. E. (2005) 'When the modeler meets the social scientist or vice-versa', in Zerger, A. and Argent, R. M. (eds) *MODSIM05: International Congress on Modelling and Simulation Advances and Applications for Managements and Decision Making*, Modelling and Simulation Society of Australia and New Zealand Inc, Melbourne, December 2005

Phillips, D., Daoudy, M., McCaffrey, S., Öjendal, J. and Turton, A. (2006) *Trans-Boundary Water Cooperation as a Tool for Conflict Prevention and for Broader Benefit-Sharing*, Phillips Robinson and Associates, Windhoek, Namibia, www.egdi.gov.se/pdf/44699_om_web.pdf

Poulsen, A. F., Poeu, O., Viravong, S., Suntornratana, U. and Tung, N. T. (2002) *Fish Migrations of the Lower Mekong Basin: Implications for Development, Planning and Environmental Management*, MRC Technical Paper no 8, Mekong River Commission, Phnom Penh

Poulsen, A. F., Hortle, K. G., Valbo-Jorgensen, J., Chan, S., Chhuon, C. K., Viravong, S., Bouakhamvongsa, K., Suntornratana, U., Yoorong, N., Nguyen, T. T. and Tran, B. Q. (2004) *Distribution and Ecology of Some Important Riverine Fish Species of the Mekong River Basin*, MRC Technical Paper no 10, Mekong River Commission (MRC), Vientiane, Laos

Prescott-Allen, R. (2001) *The Wellbeing of Nations: A Country-by-Country Index of Quality of Life and the Environment*, IDRC/Island Press, Washington DC

Sarkkula, J., Keskinen, M., Koponen, J., Kummu, M., Nikula, J., Varis, O. and Virtanen, M. (2007) 'Mathematic modelling in integrated management of water resources: Magical tool, mathematical toy or something between?' in Lebel, L., Dore, J., Daniel, R. and Koma, Y. S. (eds) *Democratizing Water Governance in the Mekong*, Mekong Press, Chiang Mai, Thailand

Sneddon, C. (2007) 'Nature's materiality and the circuitous paths of accumulation: Dispossession of freshwater fisheries in Cambodia', *Antipode,* vol 39, no 1, pp167–193

Sokhem, P. and Sunada, K. (2006) *Key Fisheries Issues in the Mekong Region*, Mekong Region Waters Dialogue, Co-convened by IUCN, TEI, IWMI, M-POWER, Vientiane, Laos, 6–7 July 2006

Tsukawaki, S. (1997) 'Lithological features of cored sediments from the northern part of the Tonle Sap Lake, Cambodia', in *Proceedings of the International Conference on Stratigraphy and Tectonic Evolution of Southeast Asia and the South Pacific*, Bangkok, Thailand, 19–24 August 1997, pp232–239

Vörösmarty, C. J. (2000) 'Anthropogenic disturbance of the terrestrial water cycle', *Bioscience*, vol 50, no 9, pp753–765

World Bank (2004) *Modelled Observations on Development Scenarios in the Lower Mekong Basin, Mekong Regional Water Resources Assistance Strategy*, Prepared for the World Bank with Mekong River Commission Cooperation, p142

Part III

Institutions, Knowledge and Power

The 'Greening of *Isaan*': Politics, Ideology and Irrigation Development in the Northeast of Thailand

François Molle, Philippe Floch, Buapun Promphakping and David J. H. Blake

INTRODUCTION

Large-scale and comprehensive development of river basins is a child of colonialism. Massive irrigation development in India, Egypt and Sudan by the British, emulated by the French or the Dutch, heralded a new area of the colonial economy. Large swathes of arid lands were brought under cultivation for the production of industrial crops such as cotton, sugarcane or rice. After gaining independence, national governments took over the colonial model in an attempt to deliver the promises of 'development' and foster economic growth in rural areas, and were influenced by the full basin development model of the Tennessee Valley Authority (Molle, 2006). The worldwide development of water regulation infrastructures and irrigated areas in the period of 1950 to 1980 achieved many benefits, including increased incomes, yields and production, and a global food sufficiency reflected until recent days in long-term declining grain prices (Molden et al, 2007). In the absence of opportunity costs for labour, such rural development projects had large multiplier effects and their economic justification was quite strong.

With time, because of the opposition to dams, declining cost-benefit ratios and – perhaps – the very successes achieved in terms of food production, such projects have lost their economic appeal and funding by leading development banks dramatically dropped during the late 1990s. In the last five years or so, however, a fresh rhetoric of justification for large-scale water projects was observed[1] and calls for renewed investments in agriculture have been made after the recent food crisis. A number of countries have implemented or floated proposals for massive interbasin diversion projects (e.g. China, India, Jordan, Brazil). Whether they are justified by hydropower generation, flood control, urban supply or irrigation, dams and canals and many megaprojects still feature prominently on the agenda of many governments (see Chapter 11).

Planners and politicians in dry countries have frequently been captivated by the 'desert bloom' syndrome, whether this led them to embracing small-scale irrigation or large-scale river engineering. Irrigation is still often seen as a redemptive solution and politicians have long seized the promise of water and the pledge to 'green the desert' as an electoral trump card. It has also been the favoured option of governments seeking to ensure national food security, alleviate poverty and control potential social unrest (Sampath, 1992; Abu Zeid, 2001). The northeast of Thailand although not arid by any standard, is considered the driest and poorest region of Thailand. If rice cultivation, supplied by derivation of small streams, has been practised in valley bottoms for centuries, the expansion of cultivation on higher lands has made irrigation a crucial instrument of control of both climatic and social uncertainty.

This chapter first recounts the chronology of river basin development in northeast Thailand, reviewing the different projects that have been planned, designed, dreamed of, and sometimes implemented during the last 60 years.[2] The ensuing section focuses on the rationale and justifications, the ideological underpinning, and the political and strategic dimensions of these successive projects. We are concerned here with the governance of large-scale project planning and with the justifications brought up by the national and foreign proponents (or opponents) of these projects. We hold that ideology and politics are overarching drivers of water resource development and that the way in which dominant players are able to cast their agenda largely determines outcome. Yet, there is evidence that the political arena where development trajectories are shaped is also conditioned by both supranational evolutions and the growing clout of players from civil society at large.

THE 'GREENING OF *ISAAN*': A RECURRING SYNDROME

Isaan, or the northeast of Thailand, makes up 85 per cent of the Thai territory that drains to the Mekong River (Koontanakulvong, 2006). The main river systems in the northeast are the Mun, the Chi (the main tributary of the Mun) and the Nam Songkhram. The largest sub-basin by far is the Chi-Mun Basin, which roughly covers 120,000km^2 and empties into the Mekong River at Khong Chiam. Rainfall in the northeast is seasonally distributed, with around 85 per cent of the total annual precipitation concentrated in the months from May to October, making irrigation a necessity if year-round cultivation is contemplated. Soils are generally considered of poor quality for agricultural production and yields are much lower than the national averages. The 0.9 million hectares of irrigated land only amounts to 10 per cent of the region's cultivated land.

Northeast Thailand is often identified with underdevelopment and stands out as the poorest region of the country. Although the percentage of the population living below the poverty line has fallen dramatically (from 57 per cent in 1962 to 38.5 per cent in 1976 and 12.7 per cent in 1996), poverty remains higher in rural areas, in general (16 per cent), and the northeast, in particular (26 per cent), where this diminution has been slower (Fan et al, 2004). The region distinguishes itself by a higher degree of specialization in rice farming, a higher rate of subsistence farmers, a lower use of agrochemicals, indebtedness of two farmers out of five, and a low density of industrial units that produce only 4 per cent of the national manufacturing added value (World Bank and NESDB, 2005). As a result of this situation, the development of water resources, in general, and of irrigation, in particular, has always been a top priority of planners and politicians since World War II (e.g. Sneddon, 2000, 2002).

Early development and piecemeal projects

Securing, expanding, intensifying and irrigating agriculture in *Isaan* has been taken as a mission by most decision-makers during the last 60 years. This section briefly recounts the chronology of water resources planning and development in *Isaan*. The ensuing one focuses on justifications and motivations.

Traditionally, irrigation in the northeast was confined to the alluvial soils of the valley bottoms of the secondary rivers, where earthen weirs, locally referred to as *thamnop*, were used to divert streams to the paddy fields (Neawchampa, 1999). Fukui and Hoshikawa (2003) reported that in 1920 as many as 503 earthen bunds could be found in the province of Nakhon Ratchasima alone. They also argue that irrigation of paddy fields around the Chi-Mun Basin was the norm rather than the exception, with cultivated fields located in the alluvial plains and valley bottoms. Additional storage was limited to natural or small village ponds, which catered for a variety of domestic water uses and provided water security in the dry season.

State-sponsored irrigation started in northeast Thailand in 1939, when the Royal Irrigation Department (RID) experimented with pilot tank irrigation projects and river diversion weirs. C. Kambhu, the charismatic head of the RID in the 1950s, was an early advocate of small-scale solutions and vigorously argued for small- and medium-scale reservoirs as the best option for *Isaan* (Kambhu, 1956). Further to these early efforts, small-scale development intensified in 1951 with the 'tank programme', initiated with US assistance (USBR, 1965). In

1963, a total storage capacity of 250,000m^3 had been attained, with 40,000ha of potentially irrigable land that added to 100,000ha potentially served by river diversion schemes.

The hydraulic mission: Large- and medium-scale developments

The difficulty of managing diversions of unregulated flows and the somewhat slow and tedious implementation of the Tank Programme during the late 1950s and early 1960s led the RID to look into possibilities of large-scale storage projects (Floch et al, 2007). Based on a Japanese reconnaissance survey of Mekong major tributaries (EPDC, 1960), the newly formed Mekong Committee proposed to submit two irrigation and five multipurpose dam projects to lending agencies 'after thorough feasibility studies', and listed a total of 16 large-scale projects, which together would potentially store 9.2 billion cubic metres and serve an irrigable area of 278,720ha. In 1965, the United States Bureau of Reclamation (USBR) conducted the first river basin development planning study for the Chi-Mun Basin with the principle objective of recommending a 'programme for the orderly economic development of the Chi-Mun Basin, and to establish an order of priority for undertaking feasibility grade surveys on the potential projects in the basin' (USBR, 1965). Though slightly differing, all of these planning documents pointed to the importance of large-scale irrigation and multipurpose development, and identified the few topographically suitable sites that later would guide planners and decision-makers time and again during the following 50 years.

The first implemented large-scale storage project in northeast Thailand was the Nam Pung hydropower project, which was finalized in 1965, followed by the Ubol Ratana Dam in 1966, the Lam Pao Reservoir in 1968, the Lam Takhong Reservoir in 1969, the Lam Pra Plerng in 1970, the Sirindhorn Reservoir in 1971, and the Chulabhorn Dam in 1972 (see Figure 10.1). The most favourable sites for large-scale construction were developed within a timeframe of only ten years, leaving only the Upper Chi, the Nam Yang and the Lam Dom Yai rivers unharnessed by large-scale infrastructure. At the same time, medium-scale water resources development was also increasingly pursued and totalled close to 400 million cubic metres) of storage by 1978.

Small is beautiful?

In 1975 the government of then Prime Minister Kukrit Pramoj made funding available (through sub-district or *tambon* funds) for small-scale water infrastructure, notably several thousand village ponds and weirs in *Isaan* (Bruns, 1991). In 1978, the Thai government established an Accelerated Water Resource Development Committee and the sub-district funds were transformed by military governments into the Rural Economy Rehabilitation Programme and, in 1980, the Job Creation

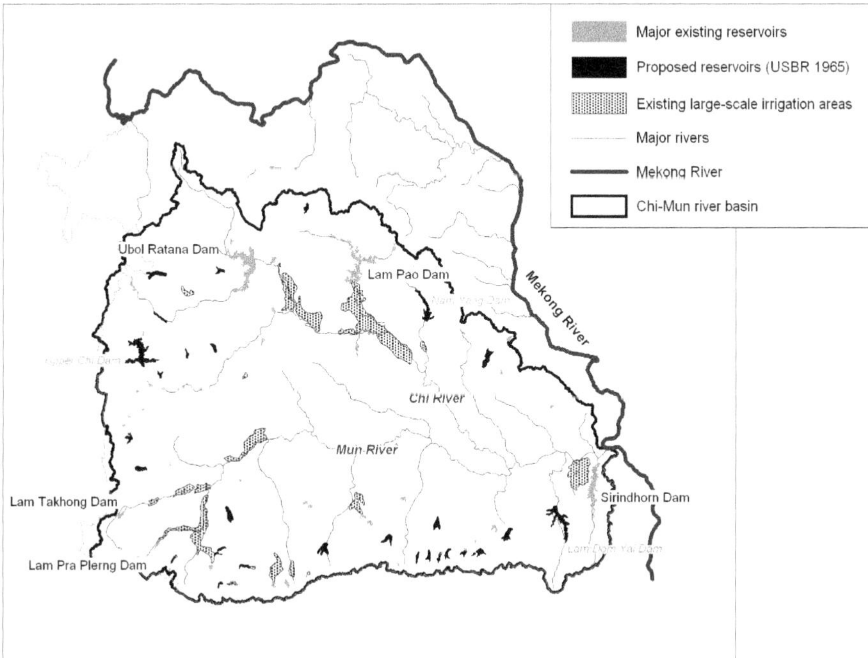

Figure 10.1 *Major infrastructure in the Chi-Mun Basin and the 1965 US Bureau of Reclamation (USBR) vision of full development*

Source: Adapted from USBR (1965)

Programme, both of which primarily included water resources development projects. Implementation was, more often than not, problematic, with reportedly up to 80 to 90 per cent of the weirs constructed under this programme said to have faced technical problems and to have failed (Bruns, 1991).

In 1978, in the wake of a few years marked by a communist insurgency, political turmoil and farmers' protests, the government adopted a two-pronged water policy with a focus on:

- the development of distribution systems from reservoirs and rivers; and
- the development of small-scale resource projects in every village as a means of meeting basic domestic water needs and allowing for minimal supplementary irrigation and for minimal dry season irrigation of backyard gardens (AIT, 1978).[3]

The 1978 study by the Asian Institute of Technology (AIT) found that a maximum of 115,200ha were potentially irrigable in northeast Thailand (i.e. roughly 12 per cent of the arable land suitable for agricultural production). The survey predicted problems of shortages during the dry season and the experience with small-scale projects was deemed 'discouraging'. The Royal Irrigation Department's own efforts to continue the development of the remaining sites that had been identified earlier as suitable for large-scale water resources development were now considered inappropriate by the National Economic and Social Development Board (NESDB), which made it considerably harder to implement them. The Lam Dom Yai and Upper Chi projects had been repeatedly studied; but both were eventually dropped by decision-makers because resettlement difficulties were considered insurmountable. And it was only the relatively smaller Lam Nang Rong Project (1991), the Upper Mun Reservoir (1996) and the Lam Chae Dam (1998) that the RID could implement during this period of time, none of which matched the earlier projects in scale.

Because of these difficulties, the RID increasingly developed small- and medium-scale irrigation projects on basically all tributaries and watersheds in *Isaan*, with over 4000 small-scale irrigation projects, storing over 800 million cubic metres (Boonlue, 2005), built between 1978 and the present day. These projects were paralleled by 600 small-scale projects implemented by the Office of Accelerated Rural Development and numerous additional programmes, including the Thai–New Zealand Project (Hafner, 1987), during the early 1980s, and the German-funded Small Irrigation Projects (SIP).

In parallel with the Royal Irrigation Department's construction efforts in northeast Thailand, a new and increasingly powerful actor – the National Energy Authority (NEA) – emerged during the late 1970s. The NEA started to implement electric pumping stations along the main rivers of the country, each station typically serving an area of 500ha located within 1km of the stream. The NEA had constructed the first hydropower project in northeast Thailand, the Nam Pung, and was now looking for means of promoting the utilization of the energy generated. It is estimated that some 1000 pumping stations have been implemented in northeast Thailand between the late 1970s and the present day (Boonlue, 2005). Considered together, all of these investments make northeast Thailand a region with a diversified and diffuse irrigation infrastructure that started to reshape the land and waterscape of the region.

Water imports and regional water resource developments

It was recognized, early into the reconnaissance surveys of northeast Thailand, that internal water resources were ill suited for the development scenarios envisioned by planners and decision-makers. A low runoff-to-rainfall ratio and a mostly flat and undulating topography (which puts considerable limits on surface water storage and gravity diversions) made planners look into ways to import water from the Mekong River from the onset. In the Mekong dam cascade – a series of dams planned to be built on the mainstream Mekong River considered in early plans – import into, and distribution throughout, the region hinged on the Pa Mong Dam that was to be constructed 20km upstream of Vientiane. As the implementation of the Pa Mong began to appear increasingly distant, Thai authorities and planning partners explored other options to augment water supply in northeast Thailand in the continued effort to 'Green *Isaan*'.

The 'Green Isaan*' Project*

The first regional study that looked into ways to make northeast Thailand bloom was aptly called *'Isaan Khiew'*, or 'Green *Isaan*'. The fifth national economic and social development plan (1981 to 1986) had (for the first time and on the grounds of 'national security') included greater social and economic equity as an objective: a poverty alleviation programme identified the 12,652 poorest villages (60 per cent of which were located in *Isaan*) and showered them with water supply, roads, schools, irrigation, electrification and soil improvement (Baker and Phongpaichit, 2005). In 1987, Thai Army Commander-in-Chief General Chavalit Yongchaiyudh was aiming to become prime minister and, in an attempt to build political support in *Isaan*, undertook to present His Majesty the King with a master plan for the development of the north-eastern region. A severe drought had just hit northeast Thailand, and the project was presented as a response to it (Bruns, 1991). The report, prepared by British Biwater Company, was presented to General Chavalit in late 1987 and was geared towards the accelerated development of water resources, ensuring water supply, increasing reforestation and improving rural incomes (Biwater, 1987). The project met with 'considerable criticism and scepticism' from politicians and academics (Bruns, 1991).

With irrigation seen as an essential input for regional development, the study to detailed strategies for water resources development. Numerous projects of all sizes were identified and it was thought possible to store almost 5 billion cubic metres of additional water (basically the sum of all technically feasible storage sites at full development, regardless of costs), serving an additional 288,000ha. Additionally, Biwater looked into inter-basin transfer options (some of them studied earlier) worth an additional 448,000ha. Even though Chavalit tried to negotiate a loan agreement with the World Bank, the proposed project did not materialize beyond a few eucalyptus or cashew nut plantations and a few failed agricultural projects.

The Khong-Chi-Mun Project

After the failure to implement the 'Green *Isaan*' plans, a new grand project was elaborated upon by the NEA under the banner of the Khong-Chi-Mun Project (KCM). The project largely drew from earlier planning documents that had accumulated over the years and integrated them within one large planning framework. The Rasi Salai Dam, for example, had already been studied in 1982 by Dutch consultant NEDECO (1982), who had earlier assisted the Mekong Secretariat in studying pump irrigation in *Isaan* and the use of floodplains for storage,[4] and became a trademark of the KCM project (RID, 1988).

In 1989, the proposed Khong-Chi-Mun Project received a boost from the government of then Prime Minister General Chatichai Choonhavan (1988 to 1991), whose declared intention 'to turn the battlefields [of Indochina] into marketplaces' soon became the semi-official policy for development plans in north-eastern Thailand (Pednekar, 1997; Kamkongsak and Law, 2001). The feasibility studies completed in 1992 by the NEA (which later became the Department for Energy Development and Promotion, or DEDP) claimed that it was technically feasible to irrigate an area of 796,800ha in 15 provinces, with construction being envisioned in three successive stages over a period of 42 years (ASEAN et al, 1992; see also Figure 10.2 for a general layout of the project at the proposed full development).

Unlike the earlier Green *Isaan* Project, however, the KCM project infrastructure was (only partly) implemented. Some weirs in the Chi and Mun floodplains were constructed and new and larger pumping stations complemented the already impressive number of small-scale electric pumping stations constructed in earlier years by the NEA. Construction of the Rasi Salai Weir/Dam on the Lower Mun River was completed in 1994, followed by the Huana Dam, the largest dam structure within the overall scheme. Both projects triggered land disputes and salinization impacts, sustained protest from the local population whose livelihoods depended upon the services so far provided by the floodplains, and drew heavy criticism from civil society and academics, which pointed to the lack of research, transparency and participation (Sretthachau et al, 2000; *Rasi Salai Declaration*, 2003; Shannon, 2005).

Despite the outcry, in 1997, then Prime Minister General Chavalit gave full support to the KCM project as the only way to ensure sufficient water supply to the 'long-suffering farmers of the northeast' and waved the long-held promise of 'turning the northeast green' in front of an assembly of village and district chiefs gathered in a five-star hotel in the city of Khon Kaen (Sneddon, 2003). With the advent of the financial crisis in 1997, large-scale capital-intensive projects were once again shelved. The KCM remained incomplete, with its cascade of weirs along the Chi and Mun lower reaches challenged on social and environmental grounds, few of the planned pumping stations effectively implemented, and with no additional water imported from the Mekong River.

The Water Grid

It was not until 2003 that the next avatar of the 'greening' syndrome materialized, when the Thaksin government launched the idea of investing US$5 billion in a project supposed to do away with water problems in the country (see full details in Molle and Floch, 2008). Despite the alleged priority given to water demand

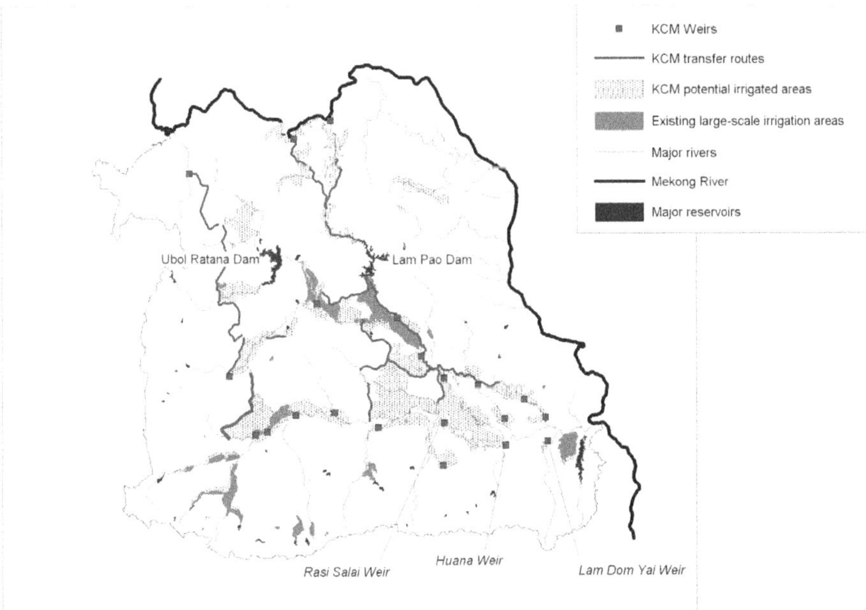

Figure 10.2 *The Khong-Chi-Mun Project: General layout and typical infrastructure*

Source: ASEAN et al (1992)

management (i.e. improving efficiency and reducing demand) proclaimed in the *Ninth National Plan* (2002 to 2006), it was announced that the irrigated land of 29.46 million rai would be incremented by an additional 103 million rai within five years, with the expected benefit of enabling farmers to cultivate and access water all year round. Although project targets announced in the newspapers proved to be fuzzy and contradictory, they all pointed to a dramatic increase in irrigated land (Molle and Floch, 2008). Borrowing from the power-generation sector, the project was dubbed 'Water Grid', to describe a set of interconnected reservoirs and basins allowing for the movement of water from sources to water-deficient areas.[5]

The north-eastern region was to be the major beneficiary of the development project, with projects including the Nam Ngum-Chi-Mun Project, which would transfer over 4 billion cubic metres of water from the Mae Ngum Dam in Laos through a siphon under the Mekong River, with pumping stations allowing transfer to the Chi Basin. Figure 10.3 sketches out this project, as well as other companion projects for the northeast region.

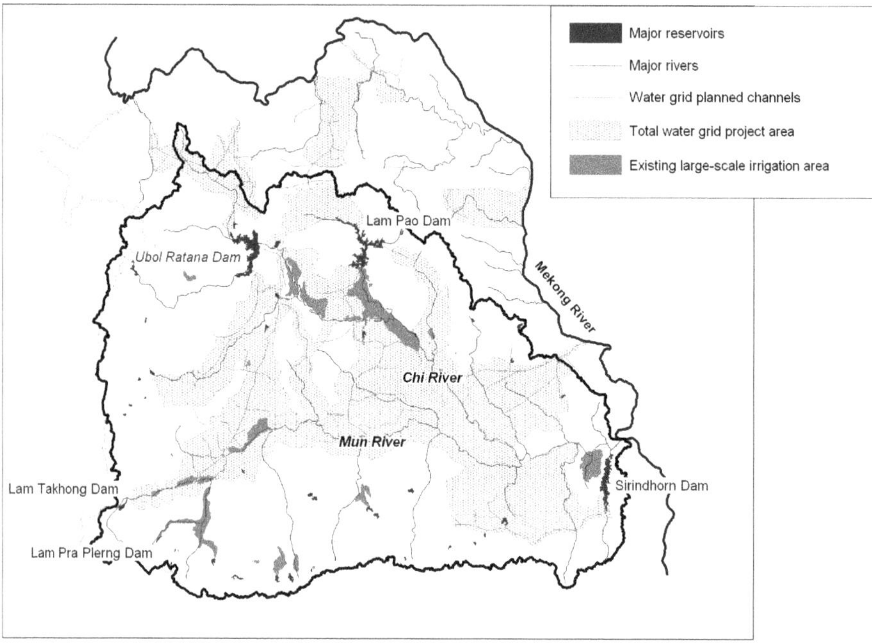

Figure 10.3 *Water Grid in northeast Thailand*

Source: Floch et al (2007)

In early 2004, the project came under fire from several quarters, including academics doubting its economic profitability (*Bangkok Post*, 2004a), as well as environmentalists predicting salinity problems or recalling that earlier pilot projects had failed (*Bangkok Post*, 2004b, 2004c). Water experts such as Senator Pramote Maiklad opined that the 'project is neither cost effective nor feasible in terms of engineering techniques' (*Bangkok Post*, 2004c) and its timetable unrealistic (*The Straits Times*, 2003). In late 2005, the government also planned to spend up to 1.7 trillion baht (US$43 billion) over five years on megaprojects aimed at boosting activity and reducing poverty, including investments in the irrigation sector (MOAC, 2006). All of these plans were largely temporarily set aside following the 2006 coup that ended the administration of Prime Minister Thaksin.

Transfers from Laos and other recent alternatives

Although the grand projects of the Thaksin era seemed to have faded into oblivion, the idea of tapping water from Laos tributaries of the Mekong River, siphoning it under the river and using it in *Isaan*, reappeared in 2007. This idea, also part of the Water Grid, was first tested in 1998 by Sanyu Consultants, who envisaged building two dams on the Xe Banghiang River in Laos, close to the confluence with the Mekong, from which 3.3 billion cubic metres of water could also be abstracted and siphoned under the Mekong into *Isaan* (RiversWatch, 2002). Another plan studied by Sanyu in 2004 considered siphoning water off the Nam Ngum Dam in Laos to the Huay Luang stream. While this option is technically feasible, the expected cost of 0.5 baht for 1 cubic metre of water raises serious doubts about the economic relevance of the project. In 2008, a study financed by the World Bank and the French Agence Française de Développement (AFD) was carried out to examine the possibility of diverting water from the Nam Ngum and Xe Bangfai rivers (in Laos), but did not lend support to further investigation on that matter.

In 2006, a trade journal announced that a newly formed Thai agency, the Department of Alternative Energy Development and Efficiency,[6] was reviewing plans at Pa Mong and Sambor (on the mainstream of the Mekong), and that a private Thai engineering firm (Panya Consultants) would conduct a US$2.4 million study revisiting a total of seven sites first identified by bureau staff in 1952 (Biggs, 2006).

In June 2008, the newly elected Prime Minister Samak announced his intention to spend US$15 billion in megaprojects, with US$5 billion targeted for water diversions (Ekachai, 2008). One plan was a 600 million cubic metre diversion from Huay Luang to Lam Pao Dam, with a second phase expected to bring water from the Nam Ngum River in Laos (a revival of earlier discredited studies). Another plan is linked to the study of three run-of-the-river dams on the Mekong mainstream, with one located in Pak Chom (upstream of Vientiane, close to the site of the Pa Mong Dam) targeted to divert water by gravity to the Lam Pao Reservoir 'through underground tunnels to Loei and Udon Thani, where reservoirs will act as distribution centres to send the water on to farms in other provinces during the dry season. The water will be transported through small pipelines' (Charoenpo, 2008). Most recently, the project to siphon water from the Nam Ngum River won cabinet approval (Wipatayotin, 2008). The desert bloom syndrome is, thus, alive and well.

ASPECTS AND CROSS-CUTTING THEMES

The storyline of the development of water resources in the Chi-Mun River basins presents a number of recurring themes that are analysed in this section. They include the justifications given and the politics of water resource development, the engineering ethos and lopsided governance patterns.

Meta-discourses and the rhetoric of justification

Stigmatizing *Isaan* as 'poor and dry', Thailand's development agencies saw water resources development as the key solution to the problems of the region as early as the 1950s (Sneddon, 2002). Development was the post-war magic wand that would partly come from joint regional development of the Mekong River Basin (see Molle and Floch, 2008). Investment in dams, pumps and tube-wells, but also roads, electrification or eucalyptus plantations, would bring prosperity. In the Green *Isaan* Project, for example, the establishment of agro-industry was the focal point of development and would:

> *... produce the processed goods for regional export, create employment opportunity in the urban areas and create the demand for agricultural products. ... Irrigation, required to produce raw materials for the agro-processing industry, will create wealth and job opportunities in the rural areas.* (Biwater, 1987)

The standard description of *Isaan* ascribes its lack of development to natural causes: unpredictable climate, 'dryness', poor soils, lack of storage, population pressure, or 'traditional', if not backward, farming practices: all putative reasons why 'the Northeast has historically lagged behind other regions' (World Bank, 1975). These perceived deficiencies inevitably lead to calls for increased water storage and irrigation infrastructure, and, secondarily, roads and better links to markets, as well as, occasionally, improved social services. Irrigation is generally justified by positive (desert bloom) or negative (cracked soils during water shortages) images and by the mere observation that farmers in irrigated areas are better off than in rain-fed areas. Then Prime Minister Thaksin, for example, reportedly said that 'it would not be a problem if the [Water Grid] project required a lot of money because it would be worthwhile eventually'; likewise, the deputy prime minister in charge of the project saw the project as 'a worthwhile investment because it will benefit 30 to 40 million people nationwide' (*The Nation*, 2003).[7] Prime Minister Samak's recent proposal is also 'an ambitious water project aimed at helping farmers in the Northeast, the country's poorest region' (Charoenpo, 2008). That 'every farmer, especially those from the 19 provinces in the Northeast, should have access to water' (*The Nation*, 2004) seemed to be taken as an uncontroversial and desirable future, with no reference whatsoever to costs or alternative options.

Another classical means of furthering projects is to propose them under the umbrella of politically charged and overriding meta-justifications (Molle, 2008). Such meta-justifications typically include national goals or priorities such as food self-sufficiency, national security, 'modernization', or the fight against poverty. Justifications for developing the Water Grid, in general, and irrigation, in particular, were based on arguments that merely emphasized expected benefits and were shrouded in a pro-poor rhetoric that magnified the assumed power of the state and attendant benefits. Thaksin 'vowed to eradicate all water-related problems plaguing the country, which he said were major hurdles in the government's war on poverty', and the study, to be completed within a year, would design 'projects to control levels of water in 25 river basins, to help rehabilitate forest and soil resources', helping him to meet his goal of eradicating poverty by 2009 (*The Straits Times*, 2003). The 'war on poverty' was clearly branded as an overriding meta-justification that offered a means to silence opposition since, obviously, nobody is against poverty reduction.

A major meta-justification of water resource development in *Isaan*, well until the mid 1970s, was the threat of communism, used by both the US and the Thai military to justify their objectives (Bell, 1969). The scare of communism was used to legitimate foreign aid, military build-up and suppression of opposition to the regime (Darling, 1965). Such meta-justifications present projects as the result of 'pressing needs' that bear no contestation. In 1988, the then Armed Forces Chief General Sunthorn Kongsompong, for example, was reported to say that it is 'necessary for us to launch a campaign like the Green Northeast project. It is a matter of national security and the Northeast is of much strategic importance' (Labournet, 2004).

The politics of water resource development also often include manipulation of symbolic power. The July 2003 workshop on Sustainable Water Resource Management organized for the launching of the Water Grid was opened with a quote from His Majesty the King:

> *The main point is the need of water for consumption, water for agriculture because water is life. People can't live without water. People can live without electricity. If there is electricity but no water, people can't live.*

Symbolic support from the king is frequently marshalled by recalling his fondness for irrigation and rural development and his support of dams for flood protection. The Green *Isaan* Project was thus aptly billed *Nam Pratan Nai Luang* – that is, Water from the King – while opposition to projects such as the Pasak Cholasit Dam were efficiently silenced by stressing the king's patronage of the dam.

An interview with C. Roongrueng (1999), a former director general of the Royal Irrigation Department, provides a textbook illustration of the range of discursive devices that are mobilized to justify more infrastructure:

> *At present, the quantity of water is not sufficient because of an increase in the population which has led to more demand for water. ... And because many forests have been destroyed, water cannot be retained. So it became necessary that we build a big reservoir to retain water for the dry season. ... The increased population has led to more agriculture and more demand for water. It would be good if people were not born. But since the population has increased, everything has been affected.*
>
> *However, water is a necessity. When there is a water shortage, it is the RID who is responsible for it. We have tried to propose every solution to solve the problems. ... Nowadays, in the IMF [International*

> *Monetary Fund] time, the population in Bangkok has decreased because some people have gone back to agriculture. But they would not have water if the RID did not provide them with water sources. How could we survive? Luckily, the RID has prepared for this.*
>
> *I think we are 'lost'. Many people have imitated foreigners. They want to preserve resources without them being fully developed. It is necessary to develop everything to its full capacity before preserving it. If we preserved our natural resources, what could we use? Would you want to buy them from elsewhere?... Many people ask why we want to do it [inter-basin diversion project]. It is because it is a duty we have been assigned. We have learned to find water for you, not for ourselves. You live well right now because of what we have done in the past.*

The statement borrows from faulty hydrological knowledge, glosses over the fact that urban needs hardly total 15 per cent of water diversions, does not discuss how farmers' 'needs' are themselves related to past irrigation overdevelopment, stresses the ills of population and urban growth that leave no choice to dutiful engineers, contrasts their disinterested mission with the irresponsibility of dissenters and with the foolishness of foreigners, and concludes by closing the debate ('there is no alternative').

Whether out of good intentions or as channels of official statements, the media also often contributes to turning unavoidability into common wisdom. Recently, for example, the *Bangkok Post* (2007) discussed the hypothesis that at some time in the future Thailand would not be able to feed its own people and would depend upon food imports, having 'to fork over a hard-earned foreign exchange advantage to buy ever more expensive food', pointing to the imperative to 'never abandon its determination to maintain food security'. Alarmist discourses are also commonplace, as illustrated by a high official justifying a project because the [Phetchaburi] province ran the risk of 'becoming a 'desert' because the province received less rainfall than the amount of water evaporating from its soil (*Bangkok Post*, 2004b). Clichés of the region as 'the water-starved Northeast' (*Bangkok Post*, 2008), widely resorted to by the KCM and other project proponents (see Figure 10.4), implicitly legitimize supply augmentation projects.

The stigma of drought and flood and the 'naturalization' of poverty in *Isaan* divert attention from other structural aspects of regional development. Extension of upland and fibre crops in the 1960s, or of eucalyptus in the 1980s, has benefited large urban-based entrepreneurs. The region's agricultural surpluses have been tapped for export, the benefits of which are appropriated by the metropolis, with only marginal changes in technology or living standards in the producing areas (Bell, 1969).

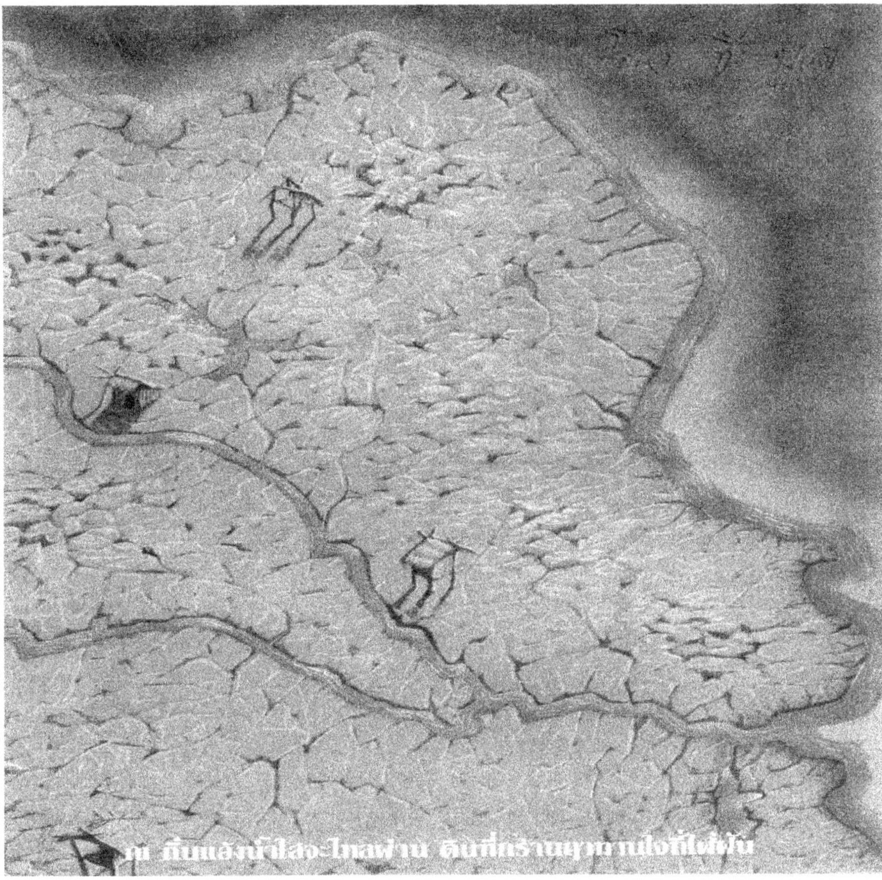

Figure 10.4 *Cover of promotional material for the Khong-Chi-Mun Project*

Geopolitics and politics

Post-war Mekong geopolitics has been a central determinant of government interventions in *Isaan* until the mid 1970s. The political situation dictated that Thailand would be the only country to be able to benefit from US/Western willingness to help develop water resources in the region – a country that would fully embrace the objective of combating the spread of communism by investing in rural infrastructures. Indeed, a major objective of small-scale investment programmes, as well as major efforts to open up so-called 'pink areas' by expanding road networks, was countering insurgency.

Security considerations have been paramount in the composition of the US aid programme in Thailand (which started with the communist takeover in China in 1949, and the spread of internal communist insurrection in Myanmar/Burma, Malaya and the Philippines).[8] As Steinberg (1986) pointed out, US interventions 'at their most naive ... have been justified by the "domino theory", [and] at a sophisticated level they have attempted to help the Thai authorities establish productive sovereignty over their periphery'. By 1973, the US Agency for International Development (USAID) director in Thailand characterized the programme in Thailand as consisting of two types: security with development aspects; and development with security aspects (Hill, 1973).

As mentioned earlier, one of the major security-related programmes, funded by the US and the World Bank (World Bank, 1975), was the Accelerated Rural Development Programme (ARD), which constructed rural feeder roads, potable water systems and small-scale irrigation systems in security-sensitive areas in north and northeast Thailand. It was designed in such a way as to integrate remote and ethically diverse regions and to allow the central government to exert control over these areas (Steinberg, 1986). Jacobs (1971) described it as 'an ambitious direct action, paternalistic, government-service programme, frankly aimed at winning friends for the existing political order', and one that would deliver to the villagers what the central government thought they would need. Later on, the then director of the United States Overseas Mission (USOM) would comment that 'it is a known fact, disputed only in degree because of the inadequacy of the information available, that during those 23 years [of US assistance to Thailand] the poorest segment of the population has benefited least from all those expenditures' (Hill, 1973).

Although concerned by these geopolitical considerations, Thai politicians also saw massive public investments as a means of procuring private political and financial gains. By associating themselves with a large water project and conjuring up images of water abundance in order to dispel precariousness and poverty, politicians expect to establish political support and constituencies. This explains the ever-returning grand development projects reviewed earlier. Announcements of non-credible targets reveal the political motivations of these projects. The Green *Isaan* Project, for example, promised to make the northeast 'green' within five years by improving water resources and raising the percentage of forest areas (Bruns, 1991), while the 1991 regional development plan for the lower northeast region foresaw industrial development in the region, with Korat destined to become the 'Detroit of Thailand' (Bruns, 1991). Likewise, in 1997, General Chavalit reiterated the promise to 'turn the Northeast green', while Thaksin's Water Grid was to triple Thailand's irrigated area in five years with 'a nationwide tap water system ... installed by 2005 so that villagers and farmers throughout the country can enjoy running water all year round' (*Bangkok Post*, 2004c).

Notwithstanding the influence of external factors and the political gains sought by politicians, the Thai administration also pursued its own version of the hydraulic mission enmeshed in local and national politics (Floch et al, 2007). As elsewhere, water resource development stood at the confluence of interest groups motivated by professional, financial or political gains (Molle, 2008). As explained by Bruns (1991):

> *Irrigation projects are large and visible rewards that politicians can offer in exchange for support. Members of Parliament are active in lobbying RID for projects at the request of their constituencies. MPs and representatives in provincial assemblies may be contractors themselves or have links to them and stand to gain from building projects funded by the Job Creation Programme or the provincial administration. At the national level there has been strong political pressure for construction of water resources projects.*

With *Isaan* remaining both the poorest region and the largest 'reservoir' of voters (40 per cent of the population), it is no wonder that, as recalled sympathetically by the *Bangkok Post* (2003):

> *The idea of transforming the Northeast into a 'promised land' where poor farmers can grow rice and other crops and raise livestock to make enough money to sustain a traditional livelihood without having to travel to the city to make a living every dry season has never faded from the minds of some caring north-eastern politicians.*

This vision is shared and promoted by consulting and construction companies more than willing to contribute to greening *Isaan*. As summarized by Samudavanija (1995): 'in the name of "economic development" the military and bureaucratic complex acquired additional financial sustenance through sponsoring infrastructure construction in rural areas. The corruption associated with these projects helped the various patron–client networks maintain their political authority over the rest of the country's population.'

Financing, however, whether from state coffers or through bilateral/international funding, is not always forthcoming. Although Chavalit tried to negotiate a loan with the World Bank for his Green *Isaan* Project (Hewison, 1994), and although the project was bundled into a major arms deal purchase with the British government (*The Nation*, 1994; LabourNet, 2004), the project did not materialize. Thatcher's government was ready to grant US$100 million and provide a loan of US$500 million for the project if agreement was found on a major package of military equipment purchase. Although the Thai government allocated money for the programme in the 1989 budget plan, the joint project foundered, partly because the Americans succeeded in reasserting themselves as the main arms supplier (LabourNet, 2004). Likewise, when the Mun River Basin Water Resource Development Plan (Binnie and Partners, 1995) was completed in 1995, with European Union (EU) funding under the auspices of the RID, the NESDB eventually denied funding to RID, although proposals for further development of water resources and irrigation had been dramatically downsized. Promotion of the KCM project was also allegedly embroiled in corruption linked to the military regime during the early 1990s (Samudavanija, 1995).

Within the administration, the prospect of massive projects and attendant funding also awoke professional and financial interest. The KCM project was developed by the NEA/DEDP, an agency under the Ministry of Science and Technology which succeeded in challenging RID's monopoly on water/irrigation infrastructure. The Water Grid also demonstrated the financial and political attractiveness of such projects for both line agencies and politicians. The project remained delayed as a 'result of a row between Natural Resources Minister Suvit Khunkitti and Agriculture Minister Somsak Thepsuthin over who should oversee the project' as 'both ministers want[ed] to supervise the project because it could be promoted in their election campaigns' (*Bangkok Post,* 2004d).

Corruption in Thailand and its links with politics has been well documented (Ockey, 1994; Phongpaichit and Piriyarangsan, 1996; Phongpaichit and Baker, 1998). Corruption in large projects mainly involves three categories of actors: the state bureaucracy and high-rank officials, politicians, and the business sector (Phongpaichit et al, undated). Politicians intervene in the reshuffle of high-ranking officials in order to ensure that their men hold strategic positions and offices. High-ranking officials, in turn, nominate and assign their subordinates key positions and functions in state agencies in order to ensure their control over projects. If these officials are moved away from their department, control will remain through their subordinates. Third, the business sector seeks high rents, offers bribes or other forms of benefits, and teams up with the other two parties in order to obtain contracts or minimize their costs of operating projects. Imbalance of power between the three parties can create tensions and may cause change in the Thai political regime (Phongpaichit et al, undated).[9]

Much of the water investments in *Isaan* during the military regimes involved high-ranking officers, as shown by the relationships between General Chavalit and Sia Leng, a *jao pho* (godfather) from Khon Kaen who assisted his Green *Isaan* Project (Phongpaichit and Piriyarangsan, 1996). Programmes such as the ARD were also known to be associated with corrupt practices. Contracts for road design and construction supervision were cancelled in 1979 'due to alleged irregularities on the part of some ARD staff and consultants. Charges of inadequate work performance also led to several court actions against contractors' (World Bank, 1985).

This situation is in no way peculiar to Thailand. In post-war Japan, a system of collusion between politicians, businessmen and bureaucrats evolved. They formed a so-called 'iron triangle' of shared benefit and influence which made public works projects the centre of a system of vested interests that encouraged bribery and bid-rigging (Woodall, 1993; Feldhoff, 2002). In the US, too, much of the construction drive of the Bureau of Reclamation and the Army Corps of Engineers has been fuelled by, and linked to, a triangle of shared interests (Reisner, 1986; McCool, 1987). Collusion between business, politics and bureaucrats in the water sector is a commonality shared by virtually all countries (Repetto, 1986).

The development industry: A fixed discourse in a changing world?

Beyond general justifications of development and the pervasiveness of political interests, analysis of the last 50 years shows a remarkable regularity in the promotion of large-scale water resources development in northeast Thailand. Although this vision has gradually been challenged and has somehow evolved with regard to which projects are pushed forward, it has largely adhered to the ethos of the 'hydraulic mission', where the development of water infrastructure is seen as obvious and other considerations – whether economic, social or environmental – are at best treated as externalities to be mitigated.

Although a common feature of the four post-war decades that saw 'modernization' and technology as central to economic development, this ethos has also been linked to, and nurtured by, the wider debate around water resources development in the Mekong Basin (Jacobs, 1995; Friesen, 1999; Nguyen, 1999; Molle et Floch, 2008). Grand plans, modelled after the Tennessee Valley Authority (TVA), to achieve a 'comprehensive development' of the basin, including several major dams on the mainstream, ensured the prominence of the engineering mindset. The engineers of the US Bureau of Reclamation, in particular, transplanted their 'culture of irrigation' to Asia, particularly to Thailand where American influence was greater: a dozen Thais visited the bureau in the US as early as 1946 (Biggs, 2006) and intense exchanges lasted at least two decades. The concept of river basin full development promoted by the TVA informed the reconnaissance report carried out by the bureau in 1965 at the request of USAID, where almost every single tributary to the Chi and the Mun rivers was dammed in its upper course (see Floch et al, 2007).

The Royal Irrigation Department created a culture where floods and drought automatically translated into proposals for more dams and more irrigation schemes (see Chapter 7). This was predicated, as shown above, on the self-defined engineering mission of RID, but also on a disregard for indirect costs and on the argument that 'water flows to the Mekong unused' (Roongrueng, 1999) – a typical argument insensitive to wider ecosystemic functions of the water regime, as well as to pre-existing people's livelihoods, echoed in 1995 by the Foreign Minister of Thailand: 'It has been a pity to let the Mekong River, with its abundance of water resources, just flow to the sea' (cited in Friesen, 1999).

In addition, while the rhetoric of participation and local- and small-scale developments featured prominently in basically all water planning documents after the 1978 AIT report, the underlying understanding of the uses of water and the discourses surrounding the utilization of basin resources have merely been readjusted to comply with the overall policy, with no substantial rethinking of the benefits and costs associated with water resources developments at large.

In retrospect, the above review of water resources development plans in *Isaan* reveals an impressive insistence and ingenuity in finding ways of mobilizing water for the region. We can at least identify six main options that have been explored and/or implemented at different points in times (see Figure 10.5):

1 Small-scale storage or diversion structures (ponds, tanks and weirs) have been an early solution that resurfaced during the late 1970s.
2 Damming the various tributaries of the Chi and Mun rivers has been the hallmark of the 1965 USBR survey and several of these dams have been constructed.
3 The Pa Mong Dam was, for long, the cornerstone of irrigation development based on diversion of the Mekong waters (and gave way to several variants in the KCM project, where water would be pumped from the rivers).
4 Pumping stations along the main rivers were disseminated by the NEA/DEDP, starting in the 1970s up to the mid 1990s.
5 Storing water in the floodplain itself, through a succession of weirs, was first introduced in the 1980s (NEDECO, 1982) and was later incorporated within the KCM project.
6 Last, frustrated efforts to import water directly from the Mekong led to creative plans to siphon water from dams located in Laos under the Mekong into *Isaan*.

This engineering drive and the fixed discourse stressing the 'urgent need to bring water from the Mekong to alleviate the region's water needs' (Interim Committee, 1988) have been gradually challenged on economic, environmental and social grounds. Economic considerations have never featured prominently in *Isaan* project planning. The 1988 *Revised Indicative Plan* of the Mekong Committee reveals that the:

> *… economic returns [of the five* Isaan *projects] are not very attractive. This is due to forecasted low rice prices.… Nevertheless, the economics of the projects may improve considerably if a greater proportion of upland crops is introduced.… A major consideration with respect to irrigation projects in north-eastern Thailand is that economic merits of a project do not capture other critical strategic and political aspects, such as employment generation and risk reduction.… under the 'Green E-sarn Scheme' and for socio-political reasons, irrigation in north-eastern Thailand is likely to be acceptable at somewhat lower economic returns than elsewhere in the country.* (Interim Committee, 1988)

The cost-effectiveness of small-scale projects has also been, at best, dubious, with 'a consensus among government engineers building small reservoirs and other small projects that such projects cannot be justified in economic terms, but are necessary for political reasons or for their social benefits' (AIT, 1978). Some academics have

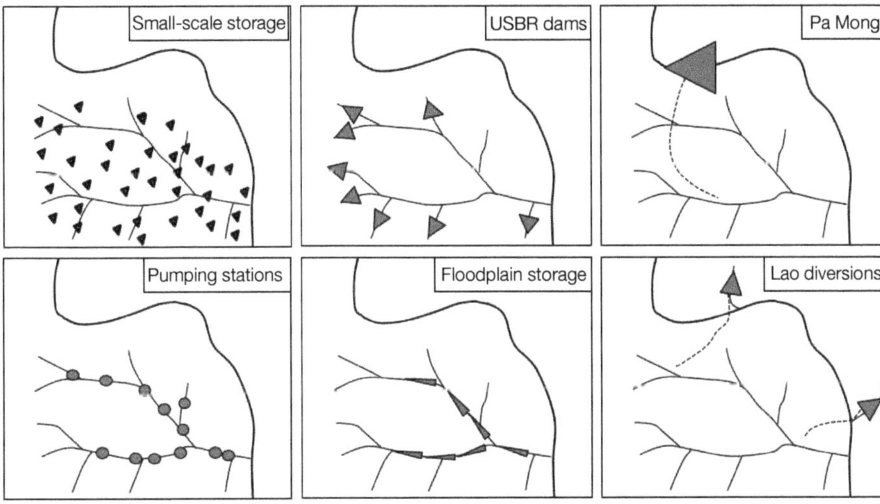

Figure 10.5 *Main options for the mobilization of water resources in* Isaan

also disputed the economic relevance of the later KCM and Water Grid projects; but these arguments have had relatively little effect on official discourse.

Environmental issues have been somehow more prominent. The salinization problems and the conflicts on floodplain management around the Rasi Salai and Huana dams on the Mun River have generated debates about environmental impact assessments (EIAs) and critiques about the ways in which shoddy assessments are used to 'green wash' projects. For example, the first EIA done for the Songkhram Irrigation Project in 1992 by consultants AEC et al was rejected by the National Environmental Board (NEB) after finding that these were exact copies of EIAs that had been done earlier for another large-scale water diversion project, the Khong-Chi-Mun (Breukers, 1999). In 1991, Mark Rentschler of the US Treasury Department advised the US government that 'The environmental impact assessment prepared for the Pak Mun Project did not appear adequate to allow the [World Bank's] board of directors to evaluate the environmental soundness of the project' (Wangpattana, 1996). Nevertheless, the World Bank and the Electricity Generating Authority of Thailand (EGAT) never commissioned a new EIA and continued to state that the Pak Mun Dam would only have minimal impact upon the surrounding natural and human environments (Friesen, 1999). Likewise, in 2002, in the middle of conflicts related to the Rasi Salai and Pak Mun dams, the DEDP submitted a proposal and an EIA for the second phase of the KCM project; but the expert panel which analysed the EIA rejected the proposal on the ground that the comprehensive groundwater study carried out by KKU had not yet been completed (Wiszniewski, 2003).

The Minister of Natural Resources and Environment, Praphat Panyachartrak, attempted to upgrade EIA procedures in order 'to catch up with the rapid economic growth' and to promote participation from the public, who – according to him – should 'be allowed a much bigger say in state development projects, which will also face tougher scrutiny from a new agency' (*Bangkok Post*, 2004e). His efforts were not rewarded and perhaps not unrelated to his removal and replacement by Suvit Khunkitti, the main proponent of the Water Grid. In sum, EIAs are seen as a 'bureaucratic hoop' to be jumped through in order to start construction, not as an authentic mechanism to decide whether or not the dam should be built (Friesen, 1999). This has led local groups to engage in grassroots research in order to mobilize local knowledge, empower local communities, build resource user networks, and produce alternative assessments such as a 'people's EIA' (Manorom, 2007).

The conflicts around the Pak Mun and Rasi Salai dams, let alone earlier occurrences of displacement because of dam construction, have shown abundantly that social impacts have been equally neglected, belittled in the name of national development. Unfortunately, the debate between the state and non-governmental organizations (NGOs) is now polarized. The government distinguishes between 'good' and 'bad' NGOs, and generally either attempts to keep planning secretive[10] or envisions debates in terms of confrontation, as suggested by Prime Minister Samak's recent declaration that 'he did not care about opposition from non-governmental organizations' (Charoenpo, 2008).

The fixity of the development discourse is also demonstrated by the obsessive focus on water resource and irrigation development (see above) and a disregard for alternatives, although reservations surfaced – at times – in official reports. For example, a World Bank report in 1969 noted that 'the northeast is merely the more striking example of the widespread predisposition to disregard the potential for rain-fed agriculture – at least at the official level'. Even if irrigation development was, arguably, a sound public investment option during the 1970s, the lack of re-evaluation is striking. At a time when neighbouring countries such as Malaysia have resolutely moved out of an agrarian economy to higher-value economic activities, the option of reducing the farming population through 'the introduction of substitute job opportunities', as put forth by Apichart Anukularmphai (cited in Le-Huu and Nguyen-Duc, 2003), is little debated. Billions of dollars worth of investment plans in irrigation infrastructure still make the headlines in Thailand despite the long-term decline in rice prices and the fact that agricultural labour is already in short supply and many remaining soils are salinity-prone (in *Isaan*) (Molle and Floch, 2008). This is besides the very low rate of land under dry-season rice cultivation, which remains at about 14 per cent of the total irrigated area (Kamkongsak and Law, 2001), an official figure that is disputed or seen by many as being optimistic. No doubt, the recent food crisis and hikes in price will probably lend support to continued investments.

Conclusions

After decades with the environment considered as a mere physical support to conventional economic growth, international institutions such as the World Bank or the ADB are supposed to now rest with a resource management paradigm, which should be reflected in their practices and, perhaps, might have somehow percolated down to borrowing countries (Colby, 1989). In the case of *Isaan*, or Thailand, in general, evidence of such a shift is suggested by several facts: the greening of legislation; the advent of the Ministry of Natural Resources and Environment; the requirement of EIAs (under certain conditions); the strengthening and maturing of environmental, human rights and/or livelihood-oriented NGOs of different stripes; the organization of civil society around struggles on the Huana or the Pak Mun dams; etc.

At the same time, it is debatable whether the mindsets of engineers and bureaucrats, or politicians' views of water resources, or the core business of consultant companies have really evolved in parallel. The above account suggests that most of the worldviews and interests that underpinned the ideologies of the hydraulic mission or the 'frontier economics' are alive and well. Avoidance, distortion or manipulation of EIAs, attempts to denigrate social movements, continued use of overriding objectives (food security, national security, poverty alleviation, etc.) to close debates, token participation of stakeholders to build legitimacy, and other political devices all show that practice has only been superficially modified. According to Chomchai (2001): 'national environmental policy seems to have been overshadowed and, indeed, supplanted by a development strategy that favours the industrial and commercial sectors and vested interests at the expense of natural ecological balance and overall national interest'.

Although the grand projects of the Cold War era may have reflected both the ideology of full river-basin development and the geopolitical interest of the US, the logic and bundle of political and financial interests underpinning large-scale water projects have not significantly subsided since that time. In developing countries, classic 'iron triangles' give way to influential and lasting 'iron rectangles' that combine politicians, state bureaucracies, private consulting and construction firms, and development banks and cooperation agencies, which all have vested interests in maximizing disbursement of funds (Molle, 2008). Relationships between these four apexes are very fluid and vary with time; but the confluence and coincidence of their interests are extremely strong.

This does not mean that these four groups of organizations are homogeneous. Other segments of the administration (typically the Ministry of Finance or the NESDB) may disagree and successfully oppose projects; development banks may also strengthen their social and environmental criteria and not support particular projects. Yet, the declaration of Prime Minister Samak – less than a week after being appointed – about a megaproject to bring water to *Isaan* provides a fascinating and remarkable continuity to 60 years of water policy centred on the 'desert bloom' promise. Whose dreams and visions are being fulfilled by the river development schemes? 'Who benefits from the projects and who determines what projects are carried out in the name of "progress and development"' (Hudson-Rodd and Shaw, 2003)?

One of the main (unanswered) questions is why, after all, governance shifts are so hard to bring about. Why would it not be possible to do 'good projects', with adequate safeguards, compensations, detailed assessments of future impacts and strict screening of projects? For Grey and Sadoff (2006), who acknowledge the need for improved project governance, 'the world is a different place in the 21st century, and there is no doubt that the costly mistakes of the past can and must be avoided in the future'; investment in hardware should be paralleled by 'investment in institutions', with capacity-building, participation and goodwill supposed to make a difference. The above account of decision-making in water resource development in *Isaan* suggests that this view includes a good dose of wishful thinking and that institutions are not easily swayed by the injection of money or rhetorical calls for 'responsible growth'. Governance shifts are slow and result from the complex interplay of local, national and global dynamics, with democratization more likely to result from hard-fought battles than from the mere desirability of social and environmental sustainability.

NOTES

1 Notably at the World Bank; see Briscoe (2003); World Bank (2005).
2 A more detailed account of the history of water resources development is given elsewhere (Floch et al, 2007).
3 The report estimated that only 20 per cent of the population could benefit from large-scale together with river-pumping schemes and that small projects could go a long way in serving the water needs of the remaining 80 per cent.
4 In 1989, the Mekong Committee reported that it had introduced a new concept in the design of flood control and storage projects by constructing reservoirs in the areas affected by annual flooding (Mekong Secretariat, 1989). Beset by resettlement problems and constrained by the depletion of attractive dam sites in northeast Thailand, the intergovernmental body recommended that the Government of Thailand should adopt a strategy in a consolidated way so that each step would be taken with the firm knowledge that in the event that each project is demonstrated to be economically and technically feasible, the government would wish to pursue further implementation of such a scheme (Mekong Secretariat, 1989).
5 Of course, because of its bulky nature, moving water is a much more expensive venture than moving electricity. Some examples of such pressurized grids, or 'carriers', exist in small arid countries such as Israel, Cyprus or Tunisia; but their costs have generally prevented expanding the concept at a very large scale.
6 This department originated from the earlier DEDP and is now under the Ministry of Energy.

7 The project would also be justified because farmers in irrigated areas earn three times more than those forced to find their own water supplies, said Mr Thaksin, and because 'if the irrigation system was extended, both farmers and the government would reap higher revenues' (*The Straits Times*, 2003).

8 US interventions in the region have been prompted by fears such as the invasion from the People's Republic of China, regional security after the French defeat in Dien Bien Fu, deterioration of conditions in Laos, the inception and the active communist insurgency in Thailand, Thailand as a base for action in the Vietnamese War, the international trade in narcotics, and the Vietnamese invasion of Cambodia (Steinberg, 1986).

9 For instance, the domination of business in Chatchai's 'buffet cabinet' led to a *coup d'état* and greater scrutiny of business interests through political participation (election), resulting in the 1997 Constitutional Law. More recently, the presence of Thaksin in Thai politics characterized the alliance between business interests and politicians who control voters in rural constituencies. This has led to the latest coup in September 2006. The coup leaders gained political support not only from the middle class, averse to 'corrupt politicians', but also from high-rank officials who were manipulated by the Thaksin regime. Such shifts in power illustrate the competition between the three parties (Phongpaichit et al, undated).

10 An article in the *Bangkok Post* (2004) referred to military units conducting 'psychological operations' in order to convince local people to accept the construction of the Kaeng Sua Teng Dam.

REFERENCES

Abu-Zeid, M. (2001) 'Water pricing in irrigated agriculture', *International Journal of Water Resources and Development*, vol 17, no 4, pp527–538

AIT (Asian Institute of Technology) (1978) *Water for the Northeast: A Strategy for the Development of Small-Scale Water Resources*, vol 1, AIT, Bangkok, Thailand

ASEAN, PALCON, SWHP and NIPPON KOEI (1992) *Me Kong-Chi-Mun Detailed Feasibility Report*, Prepared for the Department of Energy Development and Promotion, Bangkok, Thailand

Baker, C. and Phongpaichit, P. (2005) *A History of Thailand*, Cambridge University Press, Cambridge, UK

Bangkok Post (2003) 'Thaksin taps into Northeast dream', *Bangkok Post*, 28 July

Bangkok Post (2004a) 'Suvit in hot water over B400bn "fantasy" project for farmland: Blasted for bypassing panel of water experts', *Bangkok Post*, 30 March

Bangkok Post (2004b) 'Plodprasop proposes new pilot project: Bid to end row over water-grid project', *Bangkok Post*, 14 April

Bangkok Post (2004c) 'Government policy fails to address root causes', *Bangkok Post*, 3 May

Bangkok Post (2004d) 'Water crisis looms, says grid study', *Bangkok Post*, 13 June

Bangkok Post (2004e) 'Public to get more say in state projects: Agency to ensure EIA specifications are met', *Bangkok Post*, 13 February

Bangkok Post (2004f) 'Military units conducting 'psychological operations', *Bangkok Post*, 7 May

Bangkok Post (2007) 'Food security is paramount', *Bangkok Post*, 3 November

Bangkok Post (2008) 'Budget sought for river water diversion studies', *Bangkok Post*, 1 February

Bell, P. F. (1969) 'Thailand's northeast: Regional underdevelopment, "insurgency", and official response', *Pacific Affairs*, vol 42, no 1, pp47–54

Biggs, D. A. (2006) 'Reclamation nations: The US Bureau of Reclamation's role in water management and nation building in the Mekong Valley, 1945–1975', *Comparative Technology Transfer and Society*, vol 4, no 3, pp225–246

Binnie and Partners (1995) *Mun River Basin Water Resources Development Master Plan: Final Technical Report*, Prepared by Binnie and Partners in association with WS Atkins International Ltd, UK, ITC, The Netherlands, ATT Consultants Co Ltd and TA&E Consultants, Report prepared for the Royal Irrigation Department, Ministry of Agriculture, Bangkok, Thailand

Biwater (1987) *Investigation and Preparation of a Water Resource Development Programme for Northeast Thailand (Final Report: Water Resources)*, Bangkok, Thailand

Boonlue, C. (2005) 'The present condition on water resources development in the northeastern region of Thailand', in *Proceedings of the International Symposium on Sustainable Development in the Mekong River Basin*, Japan Science and Technology Agency, Tokyo

Breukers, S. (1999) 'Who defines the "problem", who defines "development"? The case of the Songkhram Irrigation Project', *Watershed*, vol 4, no 2, pp25–32

Briscoe, J. (2003) *Country Water Resources Assistance Strategies*, Presentation at the World Bank Water Week, World Bank, Washington, DC

Bruns, B. (1991) *The Stream the Tiger Leaped: A Study of Intervention and Innovation in Small Scale Irrigation Development in Northeast Thailand*, PhD thesis, Cornell University, NY

Charoenpo, A. (2008) 'Megaprojects "a top priority" for new government: Big plans for diversion of water from Mekong', *Bangkok Post*, 4 February

Chomchai, P. (2001) *Environmental Governance: A Thai Perspective*, Workshop on Mekong Regional Environmental Governance, Regional Environmental Forum (REF) for Mainland Southeast Asia, June 2000, Chiang Mai, Thailand

Colby, M. E. (1989) *The Evolution of Paradigms of Environmental Management in Development*, Policy, Planning, and Research Working Papers, World Bank, Washington DC

Darling, F.C. (1965) *Thailand and the United States*, Public Affairs Press, Washington DC

DWR (Department of Water Resources) (2004) 'Water Grid', PowerPoint presentation posted at www.dwr.go.th

Ekachai, S. (2008) 'Destructive mega dreams', *Bangkok Post*, 12 June

EPDC (1960) *Comprehensive Report on the Major Tributaries of the Lower Mekong Basin*, Mekong Reconnaissance Team organized by the Government of Thailand, Bangkok, Thailand

Fan, S., Jitsuchon, S. and Methakunnavut, N. (2004) *The Importance of Public Investment for Reducing Rural Poverty in Middle-Income Countries: The Case of Thailand*, Discussion paper no 7, Development Strategy and Governance Division, International Food Policy Research Institute, Washington DC

Feldhoff, T. (2002) 'Japan's construction lobby activities – systemic stability and sustainable regional development', *ASIEN*, vol 84, pp34–42

Floch, P., Molle, F. and Loiskandl, W. (2007) *Marshalling Water Resources: A Chronology of Irrigation Development in the Chi-Mun Basin, Northeast Thailand*, M-POWER Working Paper, Unit for Social and Environmental Research, Chiang Mai University, Chiang Mai, Thailand

Friesen, K. M. (1999) *Damming the Mekong: Plans and Paradigms for Developing the River Basin from 1951–1995*, PhD thesis, Faculty of the School of International Service, American University, Washington DC

Fukui, H. and Hoshikawa, K. (2003) 'Earthen bund irrigation in northeast Thailand', in *Proceedings of the 1st International Conference on Hydrology and Water Resources in Asia Pacific Region*,13–15 March 2003, vol 1, pp179–184, Kyoto, Japan.

Grey, D. and Sadoff, C. (2006) 'Water for growth and development', Background paper for the World Water Forum, Mexico

Hafner, J. (1987) 'View from the village: Participatory rural development in north east Thailand', *Community Development Journal*, vol 22, no 2, pp87–97

Hewison, K. (1994) *Greening of Isaan – More than Just a Pinch of Salt*, Thai-Yunnan Project Newsletters, issue 24, Australian National University Coombsarchives, Canberra, Australia, www.nectec.or.th/thai-yunnan/24.html#6

Hill, R. M. (1973) *An Overview of USAID Participation in the Thailand Programs of Security and Development, 1951 to 1973*, United States Overseas Mission (USOM), Bangkok, Thailand (cited in Steinberg, 1986)

Hudson-Rodd, N. and Shaw, B. J. (2003) 'Mekong river development: Whose dreams? Which visions?' *Water International*, vol 28, no 2, pp268–275

Interim Committee (1988) *Perspectives for Mekong Development: Revised Indicative Plan for the Development of Land, Water and Related Resources of the Lower Mekong Basin*, Interim Committee for the Coordination of Investigations of the Lower Mekong Basin, Bangkok, Thailand

Jacobs, J. W. (1995) 'Mekong Committee history and lessons for river basin development', *The Geographical Journal*, vol 161, no 2, pp135–148

Jacobs, N. (1971) *Modernization without Development: Thailand as an Asian Case Study*, Praeger Publishers, New York, NY (cited in Steinberg, 1986)

Kambhu, C. (1956) *Memorandum on Water Condition, Water Requirement, and Water Control Projects in the Northeast Region of Thailand*, Royal Irrigation Department, Ministry of National Development, Bangkok, Thailand

Kamkongsak, L. and Law, M. (2001) 'Laying waste to the land: Thailand's Khong-Chi-Mun Irrigation Project', *Watershed*, vol 6, no 3, pp25–35

Koontanakulvong, S. (2006) *Water Situation in Thailand in the Year 2003*, Faculty of Engineering, Chulalongkorn University, Bangkok, Thailand, www.watercu.eng.chula.ac.th/publications/Present_191006_1.pdf, accessed 5 March 2007

LabourNet (2004) www.labournet.net/biwater/biwater.html, accessed April 2004

Le-Huu, T. and Nguyen-Duc, L. (2003) *Mekong Case Study*, Technical Documents in Hydrology, PC-CP Series, Paris, UNESCO

Manorom, K. (2007) 'People's EIA: A mechanism for grassroot participation in environmental decision-making', *Watershed*, vol 12, no 1, pp26–30

McCool, D. (1987) *Command of the Waters: Iron Triangles, Federal Water Development, and Indian Water*, University of California Press, Berkeley, CA

Mekong Secretariat (1989) *The Mekong Program and Irrigation Development of the North-East*, Briefing Paper prepared by the Mekong Secretariat, Bangkok, Thailand

MOAC (Ministry of Agriculture and Co-operatives) (2006) *Project: Integrated Irrigation System Development*, Conference of Interested Parties, MOAC, 26 January, Government House, Bangkok

Molden, D. J., Frenken, K., Barker, R., de Fraiture, C., Mati, B., Svendsen, M., Sadoff, C. and Finlayson, M. (2007) 'Trends in water and agricultural development', in Molden, D. (ed) *Water for Food – Water for Life: Comprehensive Assessment of Water Management in Agriculture*, Earthscan, London, pp57–88

Molle, F. (2006) 'River basin development: A few lessons to be learned from history', Paper prepared for Mekong Region Waters Dialogue: Exploring Water Futures Together, 6–7 July 2006, Vientiane, Laos

Molle, F. (2008) 'Why enough is never enough: The societal determinants of river basin closure', *International Journal of Water Resource Development*, vol 24, no 2, pp247–256

Molle, F. and Floch, P. (2008) 'Megaprojects and social and environmental changes: The case of the Thai "Water Grid"', *Ambio*, vol 37, no 3, pp199–204

The Nation (1994) 'Isaan aid project linked to UK arms sales bid', *The Nation*, 7 March

The Nation (2003) 'Infrastructure project: Tap water grid planned by 2005', *The Nation*, 23 June

The Nation (2004) 'Irrigation plan will hurt intended beneficiaries', *The Nation*, 24 September

Neawchampa, C. (1999) 'Socio-economic changes in the Mun River Basin, 1900–1970', in Fukui, H. (ed) *The Dry Areas in Southeast Asia: Harsh or Benign Environment?*, Kyoto University, Kyoto, Japan, pp215–235

NEDECO (1982) *Development of the Lower Mun Basin: Feasibility Study Volume 1: Main Report*, NEDECO Netherlands Engineering Consultants for the Interim Committee for Coordination and Investigations of the Lower Mekong Basin, Arnhem, The Netherlands

Nguyen, T. D. (1999) *The Mekong River and the Struggle for Indochina: Water, War and Peace*, Praeger, Westport, CT

Ockey, J. (1994) 'Political parties, factions and corruption in Thailand', *Modern Asian Studies*, vol 28, no 2, pp251–277

Pednekar, S. S. (1997) *Resource Management in the Thai Mekong Basin*, Working paper no 71, Murdoch University, Asia Research Centre, Perth

Phongpaichit, P. and Baker, C. (1998) *Thailand's Boom and Bust*, Silkworm Books, Chiang Mai, Thailand

Phongpaichit, P. and Piriyarangsan S. (1996) *Corruption and Democracy in Thailand*, Silkworm Books, Chang Mai, Thailand

Phongpaichit, P., Piriyarangsan, S., Treerat, N., Wontephakul, N., Juthateph A., Wongsinsirikul, V. and Thanawatanawong, C. (undated) *Business, Government and Corruption*, Unpublished research report of the National Committee of Protection and Suppression of Corruption, Thailand (in Thai)

Rasi Salai Declaration (2003) *Rivers for Life! The Rasi Salai Declaration*, endorsed at the Second International Meeting of Dam Affected People and their Allies, 28 November–4 December 2003, Rasi Salai, Thailand

Reisner, M. (1986) *Cadillac Desert: The American West and its Disappearing Water*, Penguin, New York, NY

Repetto, R. (1986) *Skimming the Water: Rent Seeking and the Performance of Public Irrigation Systems*, Research report no 4, World Resources Institute, Washington DC, p47

RID (Royal Irrigation Department) (1988) *Chi Basin Water Use Study. Final Report: Appendices*, Sir Alexander GIBB & Partners, TEAM Consulting Engineering Co, Ltd., MINSTER Agriculture Ltd., Institute of Hydrology, Royal Irrigation Department, Bangkok, Thailand

RiversWatch (2002) *Laos–Thai Friendship Water Diversion Project*, www.livingriversiam.org/mk/sub_diversion.html, accessed August 2005

Roongrueng, C. (1999) 'There is no other alternative', *Watershed: People's Forum on Ecology*, vol 4, no 2, pp10–16

Sampath, R. K. (1992) 'Issues in irrigation pricing in developing countries', *World Development*, vol 20, no 7, pp967–977

Samudavanija, C. (1995) *Economic Development and Democracy in Thailand's Industrialisation and its Consequences*, St Martin's Press, Canberra, Australia

Shannon, K. L. (2005) 'The social and environmental impacts of the Hua Na Dam and Khong-Chi-Mun Project: The necessity for more research and public participation', Presentation at Water for Mainland Southeast Asia, 30 November–2 December 2005, Siem Reap, Cambodia

Sir Alexander Gibb & Partners, Team Consulting Engineering Co Ltd, Minster Agriculture Ltd, Institute of Hydrology (1988) *Chi Basin Water Use Study: Final Report: Appendices*, Report prepared for the Royal Irrigation Department, Bangkok, Thailand

Sneddon, C. S. (2000) *Altered Rivers: Socio-Ecological Transformations, Water Conflicts and the State in Northeast Thailand*, PhD thesis, University of Minnesota, US

Sneddon, C. S. (2002) 'Water conflicts and river basins: The contradiction of co-management and scale in northeast Thailand', *Society and Natural Resources*, vol 15, pp725–741

Sneddon, C. S. (2003) 'Reconfiguring scale and power: The Khong-Chi-Mun project in northeast Thailand', *Environment and Planning A*, vol 35, pp2229–2250

Sretthachau, C., Nungern, K. and Olson, A. (2000) 'Social impacts of the Rasi Salai Dam, Thailand: Loss of livelihood security and social conflict', Paper prepared as a submission to the Fourth Regional Consultation of the World Commission on Dams, Hanoi, Vietnam

Steinberg, D. I. (1986) *The Role of External Assistance in the Economic Development and Planning of Thailand: Torques and Tensions in the American Aid Program*, USAID Evaluation Occasional Paper no 6, US Agency for International Development, http://pdf.dec.org/pdf_docs/PDAAS326.PDF, accessed 31 July 2008

The Straits Times (2003) 'Thaksin vows to tackle Thailand's water problems', *The Straits Times*, July 26, http://straitstimes.asia1.com.sg/, accessed June 2004

USAID (US Agency for International Development) (1963) *Long Range Assistance Study for Thailand*, Secret (subsequently declassified), USAID, Washington DC (cited in Steinberg, 1986)

USBR (US Bureau of Reclamation) (1965) *Reconnaissance Report of Findings and Recommendations on Mun-Chi River Basin Water Resources*, USBR, Department of Interior, US

Wangpattana, A. (1996) 'Pak Mun: The lessons are clear, but is anyone listening', *Watershed*, vol 1, no 3, pp16–23

Wipatayotin, A. (2008) 'Concern over Laotian water diversion plan: Samak's pet project requires EIA study', *Bangkok Post*, 19 July

Wiszniewski, I. (2003) 'Risking salinity in Thailand and Lao PDR', *Watershed*, vol 9, no 1, pp12–25

Woodall, B. (1993) 'The logic of collusive action: The political roots of Japan's Dango System', *Comparative Politics*, vol 25, no 3, pp297–312

World Bank (1975) *Thailand Appraisal of the Northeast Thailand Rural Development Project*, Report no 721-TH, World Bank, Washington DC, 31 December

World Bank (1985) *Thailand: Northeastern Rural Development Project, Performance Audit Report*, Report no 5990, World Bank, Washington, DC, 27 December

World Bank (2005) *Reengaging in Agricultural Water Management: Challenges, Opportunities, and Trade-Offs*, Water for Food Team, Agriculture and Rural Development Department (ARD), World Bank, Washington DC

World Bank and NESDB (National Economic and Social Development Board) (2005) *Thailand Northeast Economic Development Report*, Joint Report of Thailand's National Economic and Social Development Board and the World Bank, Bangkok, Thailand

11

The Promise of Flood Protection: Dikes and Dams, Drains and Diversions

Louis Lebel, Bach Tan Sinh, Po Garden, Suong Seng, Le Anh Tuan and Duong Van Truc

Introduction

Governments frequently make promises to their citizens; few promises are as powerful as that of safety and security. The promise of flood protection is often absolute and in concrete. Dikes and dams, drains and diversions are widely seen as key to flood protection. Flood management is portrayed as a technical and apolitical exercise of flood prevention. Issues of why some groups are more vulnerable than others and how risks may be reallocated by interventions are ignored (e.g. Lebel and Sinh, 2007).

Many urbanizing areas in the Mekong region face increasingly difficult flood management challenges as cities expand into drained-out wetlands or inundation-prone areas. As valuable property is placed at risk, costs of protection and damages typically rise together (Takeuchi, 2001). Instances where flood protection interventions have made things worse are multiplying. More and more people are becoming aware of the unanticipated side-effects of urban flood protection measures for wetlands, river ecosystems and the livelihoods of people who depend upon them. Across the region people are asking questions about who really benefits, who is paying for protection, and whether there are alternative ways of managing risks from floods (Lebel and Sinh, 2007). This chapter focuses on the origins and consequences of flood protection promises made to urbanizing regions in the Mekong region.

PROMISES OF PROTECTION

Promises of protection from floods are a subset of the different ways in which society can respond to risks from flood waters. The protection approach usually implies prevention through regulation of flows. Promises of protection are often made in earth or concrete: dams built far upstream will regulate river flows; diversions will take the water around and past the city; dikes higher and longer will hold back the flood waters; drains, pumps and tunnels will move water out faster.

Regulating dams

Promises of protection by dams are perhaps the most common. A recent example is when Thailand's former Prime Minister Samak Sundaravej refloated the idea of building the controversial Kaeng Sua Ten Dam on the Yom River 'to protect Bangkok from flooding' (*Bangkok Post*, 2008a). In doing so, Samak was repeating promises of construction made by successive Thai governments since 1981. An official from the Royal Irrigation Department (RID) we interviewed summarized the logic neatly: 'A dam is a long-term project. It may take decades to overcome obstacles [to its building], but it will happen one day.' Promises shift with time and purpose. The Kaeng Sua Ten Dam was first proposed for electricity generation by the Electricity Generating Authority of Thailand (EGAT); but later, in 1989, it became an 'irrigation' project with the RID. In 2008, it was to protect Bangkok from floods.

But in monsoonal Asia, storage dams rarely provide as much flood protection to cities as the promises made about them. There are several reasons.

First, major reservoirs are often located far upstream from the cities that they are said to protect. In Bangkok, the first line of flood protection is often described in terms of regulating water flowing from the northern region through regulating the Chao Phraya River's main tributaries: the Ping, Wang, Yom and Nan rivers. Construction of the Bhumipol Dam in 1964 and the Sirikit Dam in 1972 enhanced the capacities to store about 12 billion cubic metres of runoff each year and thus regulate flows to the lower Chao Phraya Delta and Bangkok (Molle, 2007). The storage capacity of both dams has declined as a result of sedimentation. For Bangkok, the early wet season risks of floods are less than later in the year because monsoon rainfall is further north; by October, rain tends to fall further south and, thus, downstream of the Bhumiphol and Sirikit dams. Recent analysis of pre- and post-construction discharges suggest that effects on peak flows far downstream in Bangkok have actually been fairly modest (Tebakari et al, 2005) relative to the widely held perceptions of their crucial role in preventing serious flooding of Bangkok.

Second, many dams are conceived as multipurpose; in practice, this implies important trade-offs in operations to maintain levels and regular releases for hydropower and maximizing flood protection. Vietnam's Hoa Binh (Da River tributary) and Thac Ba (Lo River tributary) multipurpose reservoirs upstream from Hanoi are normally used to produce electricity; but in the flood season (June to September), they are operated to prioritize flood management in downstream areas. The two objectives are in conflict with each other at the operational level (Ngo et al, 2008). Similar dilemmas apply to the management of Bhumipol and Sirikit dams (Hangsapruek, 2007). In the series of high floods which threatened Bangkok during October to November 2006, the Bhumipol and Sirikit dams were already full and could not help to retain more water; instead, agricultural areas in the central region had to be sacrificed as flood retention areas. Above the Bhumipol Dam, operators of the smaller Mae Ngad and Mae Kuang dams upstream from Chiang Mai City face a related dilemma. At the end of the wet season they must balance the risks of another depression bringing high rainfall against the benefits of as high as possible storage at the beginning of the dry season (Lebel and Garden, 2006).

Third, the benefits of dams for flood protection are often exaggerated and burdens are ignored. Benefits are frequently overstated and much less well understood than promises of protection seem to imply. Independent studies have challenged claims of flood protection benefits of the Kaeng Sua Ten Dam and noted other burdens and risks (Sathirathai and Kittiprapas, 2000; *The Nation*, 2007). A Thailand Development Research Institute (TDRI) study suggests that flood protection benefits to farms downstream in the Phrae floodplain and further into Sukhothai Province were exaggerated in earlier project documents (Grachangnetara, 2005). Even senior staff of RID have grown more cautious: 'We still need to address the issue of whether it will be beneficial in flood mitigation in the lower Chao Phraya River Basin' (Thanopanuwat, 2007). Conversely, the burdens caused by floods are not always as large as claimed. A characteristic of urbanizing regions and their rural hinterlands in the Mekong region is that they include many people still engaged in agriculture. The soils in their fields may benefit from slow seasonal inundation. Others benefit from fisheries products that may still be available from seasonal wetlands and streams (Fox and Sneddon, 2005; Sokhem and Sunada, 2006). In these landscapes, slow-rising, slow-moving and modest peak flood heights are not necessarily a major disaster for households adapted to life along riverbanks and floodplains in monsoon Asia (Manuta et al, 2006; Tuan et al, 2008).

Finally, infrastructure is widely perceived to be fail proof. In reality, dams are also a source of risks: their operations can lead to catastrophes when releases for dam safety add flows to natural ones, as has been the case several times with the Yali Falls Dam in Vietnam (Hirsch and Wyatt, 2004). Moreover, 'once a dam is built, most of the people living downstream of the dam tend to believe that they will be completely safe from floods' (Maiklad, 1999, p107) and build in the floodplains. The rising costs of flood damages and, consequently, costs of protection are largely a result of the increased value of infrastructure being put at risk by current patterns of land development.

The problem, therefore, is largely to whom and how promises of flood protection are made. Details about risks are not shared or well understood by those at risk. As a consequence, people feel safer than they should, investments in preparedness are less than needed, and other ways of reducing vulnerabilities are ignored.

Diversions elsewhere

Promises of protection through diversion bring out these points. Diverting water also redistributes burdens and risks. In 1999, under increasing political pressure to protect Hanoi as a result of concerns about unpredictable flood heights, the Government of Vietnam issued important regulations that include, as a last resort, emergency response actions for when the flood level of the Red River in Hanoi reaches 13.4m (Socialist Republic of Vietnam, 1999b). These would involve upstream diversion of flood discharges primarily to the provinces of Ha Tay, Vinh Phuc, Phu Tho, Ha Nam and Nam Dinh. Diversions into the Day River (which runs parallel to Hanoi and also collects its drainage waters) are decided by the prime minister. It is estimated that as many as 675,000 people would be affected by such diversions (Xuan, 2006). A regulation was also issued about compensation policies (Socialist Republic of Vietnam, 1999a), which, apart from relief, also includes tax exemptions and job creation support in a recovery and rehabilitation phase. Negotiations of compensation are not well documented, but appear to be largely a technical and closed exercise involving scientists in the field and the Provincial People's Committee (PPC) of the potentially affected provinces (DDMFSC, 1999).

Unfortunately, Thailand has been without a clear system of compensation when diversions to protect Bangkok result in crop or property losses. There has also been very little in the way of clear policy or guidelines for RID on exactly which areas should be flooded and which kept dry in choosing potential receiving places of diverted flood waters. The result has been substantial hardship for farmers.

In October 2006, for example, the Thai government diverted flood waters to agricultural fields in order to protect key parts of Bangkok. Sena District in Ayutthaya Province took some of the initial diversions after His Majesty the King allowed the RID to flood a small area of his own land (*The Nation*, 2006a). The RID looked for others to 'volunteer' to do the same, causing some perplexity among many residents: 'What kind of request to be a volunteer was this? We were informed in the evening, and when we got up the next morning we were surrounded by water' (Hongthong, 2006). Many other areas were subsequently flooded with promises of compensation provided that farmers followed the RID's planting and harvesting instructions (*The Nation*, 2006b). Some farmers threatened to sue the RID for damaging their property (*Bangkok Post*, 2006a). The RID argued that they needed a law giving them authority to designate flood water-receiving areas during high-water periods (*The Nation*, 2006b). Officials from the RID also argued that a collapsed dike caused the most serious flood problems, not their intentional diversions.

The promise of protection by diversions repeatedly appeals to a discourse of sacrifice:

> *Bangkok has been saved once again by the heroes of our time, as communities in Ayutthaya, Ang Thong and more than a handful of other provinces in the countryside paid the price for its security. People in these provinces are the ones crying. Losing one's shelter for the good of the country is patriotic, yet painful.* (*Bangkok Post*, 2006b)

A local non-governmental organization (NGO) told us that Nakhon Pathom was 'being drowned to protect Bangkok'.

Higher and longer dikes

In some ways, dikes are the flip side of diversions, but often with even less consideration for receiving areas. Several of the larger cities in flood-prone areas of the Mekong region have resorted to dikes or walls. The promises, as for dams, often turn out to be hard to keep as land uses and river modification alter flood regimes in new ways, and, as for diversion, there are often significant side-effects of interventions. Finally, like all the other promises in concrete, there are also the direct benefits of construction itself.

The Vietnam government has a challenging task of protecting the rapidly urbanizing region around Hanoi given the huge seasonal differences in the water levels of the Red River. A series of major dikes, some as high as 15m, have been developed in the region over many centuries to protect fields and homes in one of the most densely settled delta regions in the world. Water levels in many places are now regularly much higher than where people live and farm. Some dikes are equipped with sluices and others are designed to allow overflow. Altogether, there are over 3000km of river walls in the delta region, of which about one fifth are to directly protect Hanoi City (FLOCODS, 2003). Dike failures have occurred and caused substantial damage and loss of life – for example, following the major floods in 1971. The promise of flood protection in the Red River Delta region is long standing, supported by myths and social norms, which largely remain fixed despite massive political and economic changes (see the section on 'Social-ecological legacies').

To reduce flooding in Bangkok from the Rangsit irrigation area to the east and from the Chao Phraya River, a polder system was developed with assistance from Japan and The Netherlands. The Japan International Cooperation Agency (JICA) project was completed around 1986 (Vongvisessomjai, 2006). After major floods in 1983, a 72km long, 2.5m high dike was built running from the east bank of the

Chao Phraya from Pathum Thani to major built-up areas of Bangkok. The result, evident in the 2006 floods, was better protection for inner Bangkok, but worse conditions for those outside of the dike, such as along the Sam Wa Canal, Nong Jok, Minburi and parts of Lad Krabang. Moreover, much of these outer parts of the Bangkok Metropolitan Authority (BMA) were now largely urban. The BMA then negotiated with the RID to build the 'King's dike' on or close to the BMA boundary and link with an inner dike to form a polder. Phnom Penh has faced a similar trajectory of problems following construction of a second dike that creates stagnant flooding within the 'flood protection' area.

In order to protect the east bank of Bangkok from river inundation, walls were initially built to protect major roads leaving a narrow strip between roads and the riverbank. This left some 80,000 homes within the BMA highly vulnerable to flooding. Public pressure from these affected groups has led to river walls along 80km of the river at a high cost of 0.1 million to 0.3 million baht per metre.[1] BMA officials told us that the project would negatively impact upon the Nonthaburi area; but noted that it is outside their jurisdiction. Bangkok flood protection is planned to handle a 1-in-100-year flood. Officials said, however, that if upstream provinces were also to use walls, the resulting rises in river levels would reduce the effectiveness of current walls to only a five-year flood.

Races like this benefit everyone and turn promises of protection into promises of construction contracts. The first dikes to protect Vientiane against floods were built after major floods in 1940. But in 1960, these were demolished to make way for the growth of the city. Serious floods in 1966 led to the rehabilitation and expansion of the dike system (Oudomchit, 2006). Various donors have since provided loans or aid for restoration works, including the Republic of South Korea (Thu, 2008). In 2001, the Asian Development Bank (ADB) approved a loan that included funds for several major drainage projects around Vientiane totalling a length of 14km (ADB, 2008). The most recent flooding in August 2008 (Samabuddhi, 2008) can be expected to lead to further promises of urban protection. Phnom Penh, located in the central lowland close to the conjunction of four rivers, is also affected by regular monsoonal flooding of the Mekong River. To protect urban residents and infrastructure of the capital, the Boung Tompun Dike was built early in the city's history. Expansion of the city has led to construction of a second dike. But availability of funds appears to drive dike construction more than requirements or priorities suggested by urban planning.

Faster drains

Rain that falls in and near urbanizing regions represents a distinct set of flood management challenges and, thus, flood protection promises compared to that brought in by rivers from farther upstream or from nearby coasts as tidal surges. Urbanization in the larger cities of the Mekong region has transformed land surfaces, making large areas virtually impermeable to rainfall.

Within a decade (1986 to 1996), the original four districts of Hanoi lost 65 per cent of their water areas, including ponds, streams and reservoirs as they were filled for the construction of buildings. Almost all of the land cover became concrete, while grassland areas were also reduced by 12 per cent (Pham, 2006). These sorts of changes often increase runoff yield and concentration and, thus, heighten flood peaks. At the same time, road networks have often been expanded without much attention to drainage needs. Coordination with other water works, such as the Hao Nam Project, which turns drains into irrigation canals, has also been limited (*Viet Nam News*, 2008b). The result is that many inner city areas have major problems with drainage even as the sewerage system expands. Several drainage, sewerage and lake projects were under way in 2008. Nguyen Le, director-general of the private company Hanoi Water Drainage, promised that 'when these projects are completed, the flooding will be totally sorted out' (*Viet Nam News*, 2008b).

Ho Chi Minh City is located in an estuarine area of the Dong Nai River system. The Dong Nai River is adjacent to the Mekong Delta, where the flood season normally starts in the middle of July, peaks in October and starts to decline in November. Floods in Ho Chi Minh City are caused by combinations of high tides, rainfall storm events and riverbank overflow. Incomplete and poor sewerage and drainage systems exacerbate risks. Some streets and residential areas with no storm drains at all quickly become small rivers after even fairly modest rainfall (SGGP, 2007). Houses are often built in place of drainage culverts and sluice gates (SGGP, 2007). The frequency and duration of flooding appears to be increasing. Rainfall in Ho Chi Minh City over the past 52 years has increased approximately 0.8mm per year, and since 1988, about 2.5mm per year. Almost US$1 billion has been invested in urban flooding control projects since 1998 (Phi, 2007). Flood control in Ho Chi Minh City has been based on the logic of draining storm water as quickly as possible. 'Although Ho Chi Minh City officials have poured billions of dong into projects to combat flooding, people in whole sections of the city are left wading in knee-deep waters after downpours' (*Viet Nam News*, 2007).

Changes to land surfaces and maintaining drains are also a major challenge within Bangkok. The most cost-effective way is through city planning regulation that sets aside green belt areas and makes developers set aside water retention areas and better manage their canals; but these soft measures have largely failed, so they resort to increasing pumping capacity and building expensive tunnels. Re-widening the network of historical canals is often not an option because people have encroached on them, creating slower drains that need to be compensated for with new faster drains.

Achieving flood protection with more or faster drainage, whether it is through canals, tunnels or pumps, is not as straightforward as it first appears. Urban drainage is often complex and full of agents that respond to problems in their own areas of jurisdiction. Water gets shunted around, and operations adapt to introductions of new infrastructure interventions. Promises of protection have a way of going around in circles like water around a clogged drain.

ORIGINS AND CONSEQUENCES

Promises of flood protection are made and infrastructure is built with reference to those promises. Across the different kinds of infrastructure (from dams to drains), locations (inland to deltas) and national contexts, are there common mechanisms at play? Are origins and consequences of flood protection promises intelligible in a more generalized sense?

To guide our analysis, we developed a simple initial conceptual model (see Figure 11.1). In this model promises are seen as arising from perceptions about risks, but often influenced by triggering events. Five mechanisms are proposed as key to understanding the origins and consequences of flood protection promises (see Figure 11.1). The rest of this section is organized around each of these in turn.

Knowledge of experts

Society often turns expectantly towards experts following disasters and 'close calls'. Experts are widely perceived as having a major role in identifying options for flood and disaster management (see Figure 11.1). In response to demands about why it flooded, an experienced flood manager responded laconically: 'It rained a lot.' But what experts have to say is not necessarily what leaders want to hear.

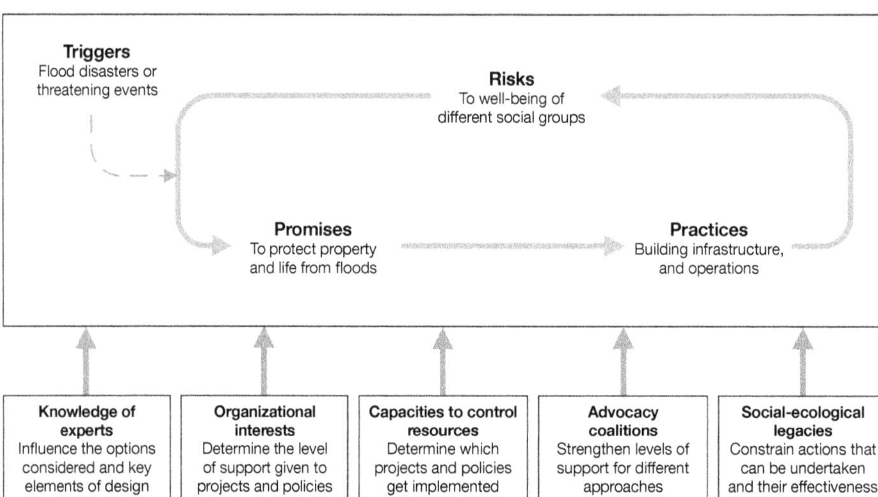

Figure 11.1 *The promise of flood protection: An initial conceptual model*

Source: Louis Lebel

The kinds of experts invited to task forces or to submit project proposals are usually from a relatively narrow set of water management backgrounds. Physical engineers are mostly involved in how to build infrastructure after political decisions have been made. Social and environmental scientists are rare, and are usually only there to meet peripheral requirements in an impact assessment process. As a result, other contributions, apart from infrastructure to reducing flood disaster risks, have been neglected. Countries in the Mekong region still draw on experts from outside their own countries, although their views do not go unchallenged or unqualified.

The new urban development plan of Hanoi City along the Red River, for example, was prepared by Korean urban planning experts, following a model based on Han River in Seoul City. The plan has been criticized by a number of Vietnamese scientists as the plan does not adequately take into account the very different nature of the flood regimes of the Red River and Han River. Hanoi needs dikes. With high sediment flows, any construction preventing river flows, especially during monsoon season, could potentially create a serious flood disaster (Van, 2008).

The development of the scientific knowledge for policy-making around flood protection for Bangkok is founded on past intergovernmental cooperation with Japan and The Netherlands. The Dike System (Vongvisessomjai, 2006), for example, was largely designed by JICA. Much of the expert knowledge is now domestic. The Office of the Royal Development Project Board (ORDPB) increasingly coordinates knowledge inputs and policy links for Bangkok's flood management. As an agency outside and above the RID and the Bangkok Metropolitan Authority, it is able to drive cooperation among different ministries and agencies without resorting to negotiating cabinet resolutions. The RDPD is headed by the prime minister and advised by a member of the Royal Privy Council. The board is joined by permanent secretaries of the ministries, the budget bureau, as well as the armed forces.

Experts in bureaucracy may be starting to accept that the full flood protection promise is implausible. Suwit Thanopanuwat (2007), at the height of the 2006 floods, concluded his assessment for Bangkok as follows:

> *Floods will continue to occur, and the people and property in the floodplains will be at risk. The situation will not improve because the cost of a physical solution is unaffordable. Consequently, although it is challenging, an effective floodplain management needs to be implemented and an integrated flood management programme needs to be prepared.*

Some experts in Vietnamese agencies are drawing similar conclusions. In 1999, Hue City and large parts of Thua Thien Hue Province and surrounding coastal areas in central Vietnam suffered two serious flooding events, first in October and then again in December. The second event struck just as there was major relief from the first event. At least 780 people were killed; about half were in Thua Thien Hue. The total cost of the disaster was estimated at US$364 million (CCFSC, 2006). In its early response to events, the Central Committee for Flood and Storm Control submitted a report to the National Assembly on 22 November 1999 in which it argued for a set of measures, including disaster preparedness, reforestation, strengthening key reservoir and canal infrastructure, as well as building capacities for self-help (Ti, 1999). The promise was nuanced. There is a growing appreciation within Vietnam that a purely structural approach to flood control will not work (CCFSC, 2006) and that vulnerabilities are not a simple function of natural hazard but also of social factors (Tran and Shaw, 2007).

Organizational interests

Ultimately, experts are a diverse group, varying in how closely their interests are steered by the organizational and power structures to which they belong and how broadly they are allowed to frame the problem of flood management – as 'protection' or 'preparation'. Exaggeration of flood damages in past events or assessments of future threats create opportunities to make promises of bigger infrastructure. Fulfilling protection promises is good business. The views of external experts may be supported because they fit interests neatly. Thus, the new urban development plan for the Red River, mentioned above, was strongly supported by the Hanoi Department of Urban Plan and Architecture. Bui Van, deputy editor of the *Vietnam Net* newspaper and others argued that large projects such as this are supported because well-connected individuals can profit from land speculation (Van, 2008).

In Vietnam, there is significant competition with respect to flood protection (and other issues in the water sector) between the Ministry of Agriculture and Rural Development (MARD) and the Ministry of Natural Resources and the Environment (MoNRE). MARD has pursued engineering-oriented approaches, while the more recently created, and substantially less influential, MoNRE has promoted more integrated methods (Molle and Hoanh, 2007).

In Thailand, the Department of Water Drainage of the BMA and RID have fundamentally different flood management objectives (Kamolvej, 2006). Capital city administrations are powerful and can shape flood management policies of national departments and agencies. The BMA essentially pursues a zero-tolerance flood policy despite the implausibility of such a strategy and the difficulties this creates for upstream and surrounding areas managed by the RID. The BMA has an interest in aligning its dike infrastructure with BMA boundaries as it doesn't have to worry about what happens beyond those boundaries.

The BMA infrastructure for preventing flooding includes protecting the new Suvarnabhumi Airport; but the airport has its own hydrological aspiration. It was built on a wetland, although it cannot be flooded at any cost. The airport location is just outside the King's Dike south of the flooded area. It is surrounded by its own polder-blocking canals that could drain flood-prone areas. BMA officials talked about meetings with airport officials with frustration and disappointment. The government policy is to keep Bangkok dry at any cost, and to keep the airport drier. This negatively affects people outside the polder system, particularly the people close to Suvarnabhumi Airport.

The key formal policy tool of the Bangkok governor is the four-year BMA management plan, which becomes effective at the beginning of the governor's term. This plan is negotiated and allocates budgets to BMA agencies and specifies verifiable indicators of success. In flood protection, the harder the objective, the more the investment. The objective is set very specifically – for example, the number of major roads that are flooded in rain events of less than 60mm per hour and the number of times that it could drain flood areas in less than two hours. The Department of Drainage and Sewerage was allocated approximately 5.5 billion baht over the four years to achieve this task and will plan its future investment accordingly. The department is one of the major infrastructure builders in the city, benefiting from flood prevention policies.

The construction of dikes around Phnom Penh has been led by the municipality and its technical working group. Work on the dikes began under a centrally planned economy in which all land belonged to the government. People still have few avenues in which to complain about the impacts of construction upon their livelihoods; if they do, they are forced to move. Instead, they have to take individual action, where they can afford to do so – for instance, raising their houses and plots. Otherwise they are condemned to living in stagnant flood waters. High-ranking officials, military officers and wealthy individuals are able to access planning documents and influence plans so that their land ends up protected by dikes. With good connections and influence, they can also buy land and resell for profits.

Capacities to control financial resources

Floods create political opportunities: new budget lines are allocated for rehabilitation and recovery that can be captured or directed; crises open windows to transfer control of existing resources. Identifying and securing, or otherwise controlling, opportunities to pour concrete, construct walls and roads, or move earth are the fundamentals of project-making. Not surprisingly, allegations of corruption are not unusual around flood protection infrastructure projects (e.g. *The Nation*, 2008; Wongpreedee, 2008). Current leaders use flood events to defend their hold on power and make financial gains; would-be leaders use flood events to garner support for themselves by promising alternative solutions.

In 2005, Chiang Mai experienced a series of major flood events that acted as 'focusing events' for policy change (Garden, 2007). Several project proposals for protecting Chiang Mai from floods included in consultant or RID reports years earlier – ranging from notions of making more space for water upstream of the city, stronger enforcement of river encroachment regulations, through to flood walls and removal of traditional weirs – resurfaced as the flood waters receded (Garden, 2007).

The crisis created opportunities to implement projects. By February 2006, then Deputy Prime Minister Suwat Liptapanlop, who had been put in charge of flood prevention, had received approval for funding of projects worth approximately 13 billion baht. Other projects were led by then Minister Newin Chidchob of the Prime Minister's Office. Almost immediately, local communities organized opposition to the larger dike and weir-levelling projects (Phanayanggoor, 2006). The opposition was largely not against more infrastructure of any sort *per se*, but on the way in which such decisions were being made: which options, who should manage and how they will be maintained (Garden, 2007).

The politics of floods changed direction but did not lose momentum following a military coup. Campaign images for the mid 2007 municipal elections for Chiang Mai pictured most candidates standing waste deep in flood waters. Debate continues over whether or not to remove several traditional weirs from near the centre of the city, as these also raise water levels and distribute flood waters, whether high walls should be built to protect the city from peak flows, or whether efforts should focus on restoring river channel width, banks and floodplains (Garden, 2007). Removal of traditional weirs and replacement with upstream dams and gates operated by the RID would transfer control of water resources back to a state agency.

Advocacy coalitions

Advocacy coalitions emerge around flood protection promises. On the one hand, there is a group of core agencies and a coalition of associated consultants, firms and banks whose primary interest is in capital-intensive construction works. Projects that make something tangible and create work will also be subsequently owned and controlled by an organization. Systematic patterns of corruption may reinforce the alignment of interests. On the other hand, there is also a range of actors ready to mobilize against many forms of modifying rivers – in particular, the construction of dams. Coalitions are rarely a neat separation of state and non-state, but bring diverse elements of local and international, public, civil and private interests.

The kinds of actors coalescing around particular promises depend to some extent upon the types of infrastructure involved. Groups opposing dams, for instance, are large and well organized in the Mekong region. But those that scrutinize more 'within-city' issues related to riverbank walls and dikes or drains, tunnels and pumping systems are much more locally organized and, consequently, less visible to the mass media. A good example of the latter was the formation of opposing coalitions in Chiang Mai: one in favour of removing old irrigation weirs and the other in favour of their retention as part of the city's cultural heritage.

Municipal and local interests are not always aligned when it comes to the purpose of new infrastructure and, thus, rationale for flood protection. This is well illustrated by the recurrent (and understandable) opposition to projects that require resettlement and compensation of people to make way for new or expanded canals, dikes or flood walls. A recent example is the controversy within Ho Chi Minh City over a US$18.5 million plan to expand the Ba Bo Canal in Binh Chieu ward for water quality and volume reasons. Local opposition has been strong (Trieu, 2008).

Advocacy coalitions in the Mekong region have been quick to challenge the protection benefits of dams. The higher-than-usual flood levels that occurred in the Mekong River in August 2008 triggered a rapid response from a coalition of local and international organizations that are typically opposed to mainstream dams. The Thai People's Network on Mekong, including the NGOs Foundation for Ecological Recovery and Living River Siam, were quick to assert that the serious flood conditions were, in part, a result of operations of dams in China's Yunnan Province. Dams, they argued, were a cause of flooding, not a source of protection, as has been frequently claimed. The Mekong River Commission was quick to defend China, stating that there was no evidence that dam operations had any impact upon the severity of the flood (Wipatayotin, 2008). The print media in the region closely followed the debates, continuing to give substantial space to dam critics. Many other related articles appeared in the media in the following days, reporting on the perceptions of people along the banks about river-level change and its likely causes (Samabuddhi, 2008), as well as the performance of early warning systems (Charoenpo, 2008). By 1 September, the MRC had followed up with a detailed situational report backing its initial claims of no significant effect of dams in China on flood conditions (MRC, 2008).

Social-ecological legacies

Social-ecological legacies constrain and influence perceptions of risk, how floods are talked about, and how and why promises of protection are made and pursued. The floodplains around several major cities in the Mekong region have been transformed by human actions over centuries. Hanoi has perhaps the most obvious legacy in its extensive system of dikes.

When promises are largely kept, and no major floods have occurred to remind people of risks, those living under the protection of dikes and dams easily misjudge risks. Current perceptions of low flood disaster risk appear to be creating a dangerous situation around Hanoi (Hung et al, 2007). Rapid urbanization around Hanoi has led to settlements in riverside floodplain areas outside the protective system of dikes, first by poor people and migrants, and, more recently, by wealthier people (Hung et al, 2007). Riverside settlements have reduced floodplain widths and, consequently, may also be increasing flood vulnerability for the entire city (Hung et al, 2007).

The legacies of successful dike control of flood waters in the north have also affected flood protection promises and practices in the entirely different flood regimes of southern Vietnam (see Chapter 8). On the other hand, the flood protection promises made for urban residential areas such as Ho Chi Minh City have been difficult to keep for other reasons – in particular, difficulties with runoff, infiltration and drainage. Natural legacies might be better incorporated within flood protection plans rather than, as is typical, contested. Doan Canh from the Tropical Biology Institute believes that in order to solve Ho Chi Minh City's ongoing flood problems, 'any solution for flood prevention should give priority to protecting the natural ecology. This would help to decrease pollution, control floods and improve the urban landscape' (Kinh, 2006).

Across the monsoon region, major cities and city states were ringed by agricultural fields that supported them with food and taxable harvests. This social-ecological legacy has meant further expansion of cities, which, in turn, has resulted in the conversion of frequently prime agricultural land into residences and streets. The traditional *muang-fai* irrigation system using canals and weirs is a legacy that the Chiang Mai municipality is struggling with. On the one hand, there is a nostalgic interest in preserving aspects of local culture; on the other hand, some of the weirs and canals contribute to flooding parts of the city. While, formerly, diversion of abundant wet season flows in the Ping River through these local irrigation systems for growing rice made sense, it no longer fits with new suburban housing and lifestyles. In Bangkok, the struggles between the BMA and RID are legendary; with the involvement of the ORDPB, negotiations have sped up and coordination has improved. Even so, controversies over flood management in eastern Bangkok remain serious.

Bangkok, more than any other city, has acquired its flood protection system largely by fragmented accumulation. After major events, different parts of the city take action, acquiring pumps and building canals, river walls or dikes. Adding new measures in such a complex system invariably creates side-effects for others on the wrong side of the wall, end of the tunnel or receiving end of a drain. Each new intervention triggers a series of compensatory responses, both operational and infrastructural. Some are undertaken by state agencies over larger areas, while others are done piecemeal as individual properties lift themselves higher above the floodplain.

Current approaches to flood protection inadvertently introduce new vulnerabilities. The BMA system, for example, is electricity intensive because of high dependence upon pumps. If there is a problem with electricity during a flood episode, the Department of Drainage and Sewerage's drainage power is drastically reduced. Other aspects of land development further reduce resilience. Canals and rivers that were once used for transport are now walled storm drains. The canal system, which had supported a lifestyle that fitted the monsoonal pulse, has been partially converted to allow for road expansion (Ross et al, 2000). Loss of resilience to seaward flooding, as a result of high tides and sea-level change, as well as to upstream and runoff flooding, has been compounded by land subsidence as a result of groundwater extraction from deep wells (Dutta et al, 2005; Babel et al, 2006). Legacies such as these which reduce resilience have helped to drive a vicious cycle of flood protection promises that cannot be kept; responses involve building more infrastructure rather than changing the way in which development unfolds in the first place.

Discussion

Explaining promises

Promises to protect urbanizing regions are made and pursued to different degrees for a variety of reasons. Five mechanisms help to explain this range: the knowledge of experts; organizational interests; capacities to control resources; advocacy coalitions; and social-ecological legacies (see Figure 11.1).

Society often turns expectantly towards experts following disasters and 'close calls'. Experts – especially specialists in drainage, hydrology and climatology – are widely perceived as key to flood and disaster management. Expert knowledge helps to explain why certain kinds of promises are made and others are kept. The knowledge of experts is influential in terms of the authority that experts have in project documentation, planning and implementation. Such fugitive power (e.g. Farrell, 2004) is crucial to actual practices. But it is not limitless.

The interests of individuals and specific agencies involved in the construction and operation of infrastructure are an important factor in making decisions about investing in particular interventions (see Figure 11.1). Interests do not just trump expertise: they also enslave it. This can be seen in how the promise of flood protection is used by organizations in their proposals despite evidence contradicting claims about benefits or pointing out new burdens and risks. Projects have a functional value (Ferguson, 1994) beyond their impacts upon on risks of flooding *per se*. This is why they keep resurfacing in the pursuit of individual and compatible organizational interests. Floods and disaster are political opportunities to produce projects. Projects do not have to be new.

Interests, however, are unlikely, in practice, to be sufficient to secure projects or to implement changes. This requires a capacity to control both financial and human resources. Much of the politics behind flood protection is bureaucratic, including the power to control resources. The protection promise is easy to make because nobody is against safety; but it is difficult to keep because everybody is in favour of retaining a budgetary advantage for themselves. One of the ways in which actors gain more control is by forming alliances with others.

Advocacy coalitions emerge around major infrastructure projects as promises are developed and declared. Support for, and resistance against, the modification of rivers by dams has generated the strongest disputes among actors and led to extensive coalition-building. Coalitions are rarely a neat and static separation between state and non-state actors; rather, they are a dynamic mixture, with many only active for short periods of time – for example, around flood threats or crises.

Finally, the social-ecological legacies associated with particular places are also needed to explain promises and practices. For instance, a recent history of reliance on dikes, dams or diversions strongly shapes current perceptions of risk. Past infrastructure constrains, by its impact upon the hydrological features of the basin or floodplain, what interventions can later be made. Infrastructure solutions of different kinds, in turn, are supported and maintained by various social institutions and norms that arise around them. These may be very resistant to change.

Form, complexity and novelty

Promises made using different forms of intervention – dikes or dams, drains or diversions – probably involve different combinations of the five mechanisms we introduced in Figure 11.1. One reason is that the scale and, hence, the set of actors that are drawn in by a promise of protection can be quite different. Upstream dams affect and involve people in locations that are far removed from the places for which promises of protection are made. Drains and pumps are usually about much more immediate redistributions of burdens and risks. Dikes and diversions are somewhat intermediate, often playing out in the peri-urban fringe. As a consequence, the kinds of expertise and coalitions likely to be formed are somewhat distinct. Another reason is that the physical form of infrastructure has implications regarding its control: infrastructure that is small or compact is likely to be easier to manage than that which is large or sprawling. Long dikes and diversion canals pass through many different jurisdictions, whereas a pumping station inherently includes possibilities of greater local control.

Promises made in simple and complex settings are also likely to have different dynamics. First, consider complex situations where the interactions between rainfall, land and water uses, and physical interventions are not well understood and there are major uncertainties regarding how people will respond to changes in flood risks or particular kinds of flood events. In these circumstances, different groups can easily argue for different forms of interventions, drawing on alternative expert opinions to back their options. Diverse organizational interests can be pursued as each uniquely frames problems and solutions. Coalition-building, especially with actors capable of controlling key resources, is crucial to securing favourable flood protection projects. In very simple settings, in contrast, knowledge issues are small, so experts have little power. Multiple coalitions are pointless as there is only one course of action under consideration and the choice is primarily support or not. Promises of protection may still benefit some groups more than others; but there is not much less scope for hidden transactions and a real 'danger' that the problem could be solved 'once and for all'.

Another way of thinking about the role of complexity is to consider the number of options involved in making decisions and the number of agents that would need to be coordinated in order to keep a protection promise. This, for example, depends upon the degree to which the river basin falls within a jurisdiction – say, an individual national territory. Options for physical interventions in the name of flood protection for Vientiane or Phnom Penh, lying on mid-stream banks of the Mekong, a large international river, have many more constraints than those for Bangkok. Hanoi with much, but not all, of the Red River Basin within Vietnamese territory falls somewhere in between.

Climate change, insofar as it represents new risks associated with altered flood regimes, appears to be easily accommodated by current modes of making protection promises. Indeed, climate change is rapidly becoming part of the vocabulary of flood experts. Thus, Professor Nguyen Sinh Huy, who directs a flood project for Ho Chi Minh City, points to sea-level rises as high as 0.9m by 2070 to argue that more dikes needed to be built in order to improve the flood protection of Ho Chi Minh City (*Viet Nam News*, 2008a). The proposed dikes are 2.5m high and 8.5m wide, and when complete would add 165km to the total length of dikes (*Viet Nam News*, 2008a). Massive infrastructure projects have also been proposed to protect Bangkok from sea-level rises caused by climate change. A recent example, launched by disaster experts and politicians, was a proposal to build an 80km long wall 300m offshore and 3m higher than moderate sea-level to protect Bangkok and two surrounding provinces (Wipatayotin, 2007). Smith Dharmasorjana, chairman of the National Disaster Warning Committee of Thailand, believes that a system of dikes offshore is needed to protect areas of Bangkok from sea-level rises caused by climate change, as well as intrusions of sea water into major rivers (*Bangkok Post*, 2008b).

A deeper treatment of these three themes – infrastructure forms, complexity and climate change – are beyond the scope of this chapter but are worthy of follow-up. A study that looked more carefully at the links between physical infrastructure and social systems could also address in much more detail questions of performance. In this chapter, moreover, we have not attempted to systematically assess the consequences of different flood protection promises for risks and burdens to different stakeholders.

In summary, flood protection promises are made and kept, or not, for a variety of reasons that largely can be understood in terms of knowledge, interests and power. But a deeper understanding also requires appreciation of social-ecological legacies.

Governing practices

The consequences of why and how promises are made are important because they involve allocating burdens and risks from floods and flood interventions. We saw many examples of the side-effects of flood protection interventions in this chapter. From these concerns we draw several implications for initiatives and other efforts aimed at improving decision-making and practices with respect to flood management in the Mekong region.

First and foremost, flood management needs to be deliberated in the public sphere. To date, the input of experts has been too narrow, while the proposals of government agencies with strong interests, and of consultant and construction firms with vested interests, have been insufficiently scrutinized. The promise of flood protection requires wider deliberation (Lebel and Sinh, 2007). Multi-stakeholder processes are recommended. Even if these do not immediately alter organizational strategies or the design and pursuit of particular projects, they should at least lead to a better, shared, understanding of risks. Done well, they should produce flood management strategies that are much better integrated with the changes in land use and water use that accompany urbanization.

Second, much more attention must be paid to the adverse side-effects of intervention made in the name of flood protection for particular cities. Too often there is insufficient information about, and institutional mechanisms to offset, adverse impacts upon 'other' peoples and places. The promise of flood protection should be accompanied not just with analysis of the costs of construction and operations, but also with full consideration of compensation, insurance and other financial instruments that may be needed to fairly reallocate burdens and risks.

Third, the notions that flood disasters can be eliminated and complete protection afforded should be discarded. The risk of flood disasters may be minimized, and some control of flood water is possible; but there are always, in practice, real and important limits on what magnitude and types of floods can be dealt with effectively. Full protection through infrastructure is rarely a realistic option for rapidly urbanizing regions in developing countries (Schultz, 2002). Understanding risks is a better strategy than ignorance. A likely outcome is greater emphasis on integrating complementary strategies within flood management, including extending early warning systems and improved disaster preparedness.

Fourth, flood protection strategies need to pay much more attention to controlling land use, not just planning and describing it. This means attention must be given to institutional development to ensure that responsible agencies have adequate authority, that administratively there is capacity to assign and reassign rights, and that there is a functional, independent, legal system in place where disputes can be settled and cases of unfair treatment by authorities brought forward (Lebel et al, 2006). Multi-stakeholder processes that inform negotiations around more contentious land-use plans should help to ensure that all voices are heard and not just those of powerful landowners or agencies with vested interests in construction or operation of new infrastructure.

Fifth, and in line with the first four points above, individual urban dwellers and commercial property owners are often keen on high levels of protection, whereas farmers and others in the peri-urban periphery do not wish to become victims of flood protection measures. These actors have agency that cuts across formal institutions and coordination efforts. The cumulative, largely self-organizing, responses to efforts to move water from one place to another cannot be discounted. People and firms build higher, use pumps and organize at local levels (Few, 2003; Molle et al, 2003). Within the bureaucracy of larger metropolitan areas, a similar phenomenon is observable: different districts 'game' each other's flood protection operations. The result in both cases is that actual movement of flood waters is often complex, and the ultimate effectiveness of individual flood protection measures is often much more uncertain than assessments based on linear no-reaction analyses. The contributing causes to flooding in urbanizing regions are both local and more distant, and often interact (Tran and Shaw, 2007; Chang et al, 2008).

The high costs, unintended side-effects and fallibility of infrastructure-centred approaches inherent in pursuing promises of flood protection imply that flood management needs to be redirected. Much greater emphasis should be placed upon land-use planning, both in terms of the use of floodplains and the way in which different vegetation draws upon water resources. Ecosystem-informed approaches that make room for water outside and within urban areas should enable people to live better with more natural flood cycles. Greater investments in early warning systems and community-oriented disaster preparedness may do more for reducing vulnerabilities and risks than a narrow adherence to a logic of control, prevention and protection.

Such alternatives are only likely to emerge as part of wider flood management strategies with changes to current governing practices. Taken together, the five points above emphasize the need to make those in power, with responsibility and authority, more accountable for the promises they make. Much greater opportunity for public scrutiny of proposals and deliberation of alternative approaches to flood management is needed.

CONCLUSIONS

The promise of flood protection is easy to make, but hard to keep. Promises of protection and how they are pursued, it turns out, are better explained in terms of beliefs, interests and power. Promises played out in concrete – as dikes and dams, drains and diversions – regularly exaggerate benefits and ignore the often unanticipated side-effects on people living elsewhere and on the environment. Efficacy in reducing risks of flood-related disasters is often downplayed to a side-effect of projects pursued for other reasons. This does not imply that infrastructure has no role in flood management; but it does underline how frequently the promise of flood protection rings false.

ACKNOWLEDGEMENTS

This chapter presents findings from PN50 Enhancing Multi-Scale Mekong Water Governance, a project of the Consultative Group on International Agricultural Research (CGIAR) Challenge Programme on Water and Food. We thank our colleagues in M-POWER (www.mpowernet.org) for sharing findings with us over the years.

NOTE

1 34 baht = US$1.

REFERENCES

ADB (Asian Development Bank) (2008) *Lao People's Democratic Republic: Vientiane Urban Infrastructure and Services Project*, Completion Report for Project Number 33432, ADB, Manila

Babel, M. S., Gupta, A. D. and Domingo, N. D. S. (2006) 'Land subsidence: A consequence of groundwater over-exploitation in Bangkok, Thailand', *International Review for Environmental Strategies*, vol 6, no 2, pp307–327

Bangkok Post (2006a) 'Farmers to sue irrigation agency over flood diversion', *Bangkok Post*, 3 November

Bangkok Post (2006b) 'Is disaster inevitable?' *Bangkok Post*, 26 November

Bangkok Post (2008a) 'Samak to revive old dam project: PM on collision course with environmentalists', *Bangkok Post*, 6 June

Bangkok Post (2008b) 'Dikes planned to prevent city flooding', *Bangkok Post*, 18 June

CCFSC (Central Committee for Flood and Storm Control) (2006) *National Strategy and Action Plan for Disaster Prevention, Control and Mitigation in Viet Nam – 2001 to 2020*, CCFSC, Hanoi

Chang, H., Franczyk, J. and Kim, C. (2008) 'What is responsible for increasing flood risks? The case of Gangwon Province, Korea', *Natural Hazards*, doi: 10.1007/s11069-008-9266-y

Charoenpo, A. (2008) 'Warnings on floods faulted: MRC bulletins "not reaching" villagers', *Bangkok Post*, 17 August

DDMFSC (Department of Dike Management, Flood and Storm Control) (1999) *Suggestions on Emergency Measures for Extreme Floods Mitigation in Northern Viet Nam and Ha Tinh Province*, Socialist Republic of Vietnam, DDMFSC, Vietnam

Dutta, D., Babel, M. S. and Gupta, A. D. (2005) *An Assessment of the Socio-Economic Impacts of Floods in Large Coastal Areas*, Asian Institute of Technology, Bangkok

Farrell, K. N. (2004) 'Recapturing fugitive power: Epistemology, complexity and democracy', *Local Environment*, vol 9, no 5, pp469–479

Ferguson, J. (1994) *The Anti-Politics Machine: 'Development', Depoliticization, and Bureaucratic Power in Lesoth*, University of Minnesota Press, Minneapolis, MN

Few, R. (2003) 'Flooding, vulnerability and coping strategies: Local responses to a global threat', *Progress in Development Studies*, vol 3, no 1, pp43–58

FLOCODS (2003) *Modelling Embankment Failures in the Red River Dike System: Volume 1, Dikes and Failures – Characterization and State-of-the-Art*, Proc. 0605/1714403, DHA/NRE, Lisbon

Fox, C. and Sneddon, C. (2005) *Flood Pulses, International Watercourse Law, and Common Pool Resources: A Case Study of the Mekong Lowlands*, Research paper no 2005/20, World Institute for Development Economics Research, United Nations University, Helsinki, Finland

Garden, P. (2007) *The Chiang Mai Floods of 2005*, USER Working Paper WP-2007-19, Unit for Social and Environmental Research, Chiang Mai University, Chiang Mai

Grachangnetara, S. (2005) 'Thailand's Kaeng Sua Ten Dam revisited: The flood protection benefit of a controversial agricultural project', *TDRI Quarterly Review*, vol 20, no 3, pp20–27

Hangsapruek, S. (2007) *Flooding Situation and Mitigation Measures in Thailand*, Report on Japan–Thailand–China Joint Symposium on Current Policy and Directions on Flood Damage Mitigation, Cooperative Research Report of Public Works Research Institute no 366, Public Works Research Institute, Japan Science and Technology Agency, Tsukuba

Hirsch, P. and Wyatt, A. (2004) 'Negotiating local livelihoods: Scales of conflict in the Se San river basin', *Asia Pacific Viewpoint*, vol 45, no 1, pp51–68

Hongthong, P. (2006) 'Farmers upset over huge floodwater diversions', *The Nation*, 6 December, Bangkok

Hung, H. V., Shaw, R. and Kobayashi, M. (2007) 'Flood risk management for the RUA of Hanoi: Importance of community perception of catastrophic flood risk in disaster risk planning', *Disaster Prevention and Management*, vol 16, no 2, pp245–258

Kamolvej, T. (2006) *The Integration of Intergovernmental Coordination and Information Management in Response to Immediate Crises: Thailand Emergency Management*, PhD thesis, University of Pittsburgh, Pittsburgh

Kinh, T. B. (2006) 'City looks to solve drainage woes', *Viet Nam Economic Times*, 6 September, Ho Chi Minh City

Lebel, L. and Garden, P. (2006) *Bureaucrats, Consultants and Stakeholders: An Emerging Knowledge System for the Upper Ping River Basin, Northern Thailand*, USER Working Paper WP-2006-11, Unit for Social and Environmental Research, Chiang Mai

Lebel, L. and Sinh, B. T. (2007) 'Politics of floods and disasters', in Lebel, L., Dore, J., Daniel, R. and Koma, Y. S. (eds) *Democratizing Water Governance in the Mekong Region*, Mekong Press, Chiang Mai

Lebel, L., Nikitina, E., Kotov, V. and Manuta, J. (2006) 'Assessing institutionalized capacities and practices to reduce the risks of flood disasters', in Birkmann, J. (ed) *Measuring Vulnerability to Natural Hazards: Towards Disaster Resilient Societies*, United Nations University Press, Tokyo

Maiklad, P. (1999) 'Development and achievements in flood control and management in Thailand', in UN-ESCAP (ed) *Regional Cooperation in the 21st Century in Flood Control and Management in Asia and the Pacific*, United Nations, New York, NY

Manuta, J., Khrutmuang, S., Huaisai, D. and Lebel, L. (2006) 'Institutionalized incapacities and practice in flood disaster management in Thailand', *Science and Culture*, vol 72, no 1–2, pp10–22

Molle, F. (2007) 'Scales and power in river basin management: The Chao Phraya River in Thailand', *The Geographical Journal*, vol 173, no 4, pp358–373

Molle, F. and Hoanh, C. T. (2007) *Implementing Integrated River Basin Management: Lessons from the Red River Basin, Vietnam*, Working paper, Institut de Recherche pour le Développement, Mekong Programme on Water Environment and Resilience, International Water Management Institute, Marseille, Chiang Mai and Colombo

Molle, F., Shah, T. and Barker, R. (2003) *The Groundswell of Pumps: Multilevel Impacts of a Silent Revolution*, ICID-Asia Meeting, Taiwan

MRC (Mekong River Commission) (2008) *Flood Situation Report, August 2008*, MRC Technical Paper no 21, 1 September, Mekong River Commission Secretariat, Vientianne

The Nation (2006a) 'King's move helps save capital', *The Nation*, 11 October

The Nation (2006b) 'Large areas "needed for overflows"', *The Nation*, 12 October

The Nation (2007) 'Shelved Kaeng Sua Ten Dam project may be given green light', *The Nation*, 17 January

The Nation (2008) 'DSI to investigate the BMA bribery allegation', *The Nation*, 16 July, Bangkok

Ngo, L. L., Madsen, H., Rosbjerg, D. and Pedersen, C. B. (2008) 'Implementation and comparison of reservoir operation strategies for the Hoa Binh Reservoir, Vietnam using the Mike 11 model', *Water Resources Management*, vol 22, pp457–472

Oudomchit, B. (2006) 'Flood forecasting system and flood management in Lao PDR', in IFNET (ed) *IFNET Action Report 2006*, International Flood Network, Tokyo, www.internationalfloodnetwork.org/AR2006/AR13Oudmicht.pdf

Pham, N. D. (2006) *Những thách thức môi trường của Hà Nội hiện nay và các giải pháp khắa phục* [*Current Challenges to Ha Noi Environment and Solutions*], Ministry of Science and Technology, www.moc.gov.vn/Vietnam/Management/ScienceTechnology/6270200612121700330/', accessed 28 September 2007

Phanayanggoor, P. (2006) 'Local against dike-levelling', *Bangkok Post*, 6 February

Phi, H. L. (2007) 'Climate changes and urban flooding in Ho Chi Minh City', Paper presented to the Third International Conference on Climate and Water, Helsinki, Finland

Ross, H., Poungsomlee, A., Punpuing, S. and Archavanitkul, K. (2000) 'Integrative analysis of city systems: Bangkok Man and the Biosphere programme study', *Environment and Urbanization*, vol 12, no 2, pp151–161

Samabuddhi, K. (2008) 'Mekong shows its new face', *Bangkok Post*, 25 August

Sathirathai, S. and Kittiprapas, S. (2000) 'Conflict management for environmental and social sustainability', Paper presented to the Second Annual Global Development Conference (GDN 2000), Tokyo, Good Governance for Social Development and the Environment Institute, Bangkok

Schultz, B. (2002) 'Role of dams in irrigation, drainage and flood control', *Water Resources Development*, vol 18, no 1, pp147–162

SGGP (2007) *Mưa lớn gây ngập nhiều tuyến đường và kẹt xe* [*Ho Chi Minh City: Big Rain Caused Flooding and Traffic Jams in Many Streets*], TPHCM, 30 July 2007, accessed 29 September 2007, www.sggp.org.vn/moitruongdothi/2007/7/113022/, Sao Gon Giai Phong

Socialist Republic of Vietnam (1999a) *Quyet Dinh So 185/1999/TTg cua Thu Tuong Chinh Phu Ve Chinh Sach Ho tro doi voi vung phan lu, cham lu thuoc he thong Song Hong* [*The Decision Number 185/1999/QD-TTg on 14 September 1999 on the Compensation Policy for Flood Diversion and Flood Retard Areas in Red River System*], Socialist Republic of Vietnam, Hanoi

Socialist Republic of Vietnam (1999b) *Nghi Dinh So 62/1999/ND-CP cua Chinh Phu ve viec Ban Hanh Quy che ve Phan Lu, Cham Lu thuoc he thong Song Hong de bao ve an ninh thu do Hanoi* [*The Decree Number 62/1999/ND-CP on 31 July 1999 on the Regulation for Flood Diversion and Flood Retard in Red River System to Protect the Capital of Hanoi*], Socialist Republic of Vietnam, Hanoi

Sokhem, P. and Sunada, K. (2006) 'The governance of the Tonle Sap Lake, Cambodia: Integration of local, national and international levels', *Water Resources Development*, vol 22, no 3, pp399–416

Takeuchi, K. (2001) 'Increasing vulnerability to extreme floods and societal needs of hydrological forecasting', *Hydrological Sciences Journal*, vol 46, no 6, pp869–881

Tebakari, T., Yoshitani, J. and Suvanpimol, C. (2005) 'Effects of large-scale dams on the hydrological regime: A case study in Chao Phraya river basin, Kingdom of Thailand', *Journal of the Japan Society of Hydrology and Water Resources*, vol 18, no 3, pp281–292

Thanopanuwat, S. (2007) *The Current Situation (October 2006) on Flood Management in Chao Phraya River*, Report on Japan–Thailand–China Joint Symposium on Current Policy and Directions on Flood Damage Mitigation, Cooperative Research Report of Public Works Research Institute no 366, Public Works Research Institute, Japan Science and Technology Agency, Tsukuba

Thu, P. (2008) 'Lao capital to restore its Mekong dike', *VNDaily*, Hanoi, 20 May

Ti, L. H. (1999) *Experiences of Flood Control and Management in Asia Relevant to Economic and Social Development Strategies for Central Viet Nam in the Twenty-First Century*, United Nations Economic and Social Commission for Asia and the Pacific, Bangkok

Tran, P. and Shaw, R. (2007) 'Towards an integrated approach of disaster and environment management: A case study of Thua Thien Hue province, central Viet Nam', *Environmental Hazards*, vol 7, pp271–282

Trieu, N. (2008) 'Officials say canal needs to be cleaned, not widened', *Thanh Nien News*, Hanoi, 28 June

Tuan, L. A., Hoanh, C. T., Miller, F. and Sinh, B. T. (2008) 'Floods and salinity management in the Mekong Delta, Vietnam', in Be, T. T., Sinh, B. T. and Miller, F. (eds) *Challenges to Sustainable Development in the Mekong Delta: Regional and National Policy Issues and Research Needs*, The Sustainable Mekong Research Network (SUMERNET), Stockholm, Sweden

Van, B. (2008) *Challenging Issues Related to New Urban Development Plan of Hanoi City*, www.tuanvietnam.net/vn/sukiennonghomnay/3424/index.aspx, accessed 12 April 2008

Viet Nam News (2007) 'After billions spent flighting floods, HCM City knee-deep in water', *Viet Nam News*, Hanoi, 5 November

Viet Nam News (2008a) 'Officials step up efforts to prevent floods in HCM City', *Viet Nam News*, Hanoi, 10 March

Viet Nam News (2008b) 'Parts of Ha Noi to face floods', *Viet Nam News*, Hanoi, 21 May

Vongvisessomjai, S. (2006) *Flood Risk Reduction Plan for Chao Phraya River*, TEAM Group of Companies, Bangkok, www.teamgroup.co.th/attach/article4.pdf (in Thai)

Wipatayotin, A. (2007) 'Group floats plan to keep rising sea out', *Bangkok Post*, 25 November

Wipatayotin, A. (2008) 'MRC defends China over Thai floods', *Bangkok Post,* 16 August

Wongpreedee, W. (2008) 'BMA asks Japan to aid bribery probe', *Bangkok Post*, 9 July

Xuan, S. P. (2006) 'Cac Giai phap chong lu cho Dong Bang Song Hong' ['Flood control measures in the Red River Delta'], *Tap Chi Nong Thon Moi* [*The New Countryside Journal*], vol 178, pp3–4

12

Songs of the Doomed: The Continuing Neglect of Capture Fisheries in Hydropower Development in the Mekong

Richard Friend, Robert Arthur and Marko Keskinen

INTRODUCTION

Since early 2007 there has been a rapid acceleration in hydropower development in the Mekong Basin. The value and importance of capture fisheries in the Mekong Basin and the threat of hydropower development to their productivity and sustainability is now well established, widely cited and rarely challenged (MRC, 2003; ADB, 2004; MRCS/WUP-FIN, 2007). Yet, despite this, on the rare occasions in which fisheries enter public debate on hydropower development, they do so in a constrained manner, as something of an afterthought and as an unavoidable, slightly unfortunate, cost of the inevitable march of progress and development.

Fisheries are being downplayed rather than completely denied. There are several factors that appear to be at play. Some of these relate to the nature of policy-making processes, knowledge production and power (e.g. Hirsch 2003; Sneddon and Fox, 2006), and, of course, to the enduring potency of a hydropower-based regional development narrative that traces its roots to the 1950s (Bakker, 1999). These dimensions of policy neglect have been discussed elsewhere.

There is something more deeply ingrained in this neglect. For even when debates about capture fisheries do emerge – whether within the fisheries and hydropower sectors or more broadly – the discussion reverts to a 'conventional wisdom' that the capture fisheries are doomed, facing a very bleak future under practically any circumstances. While similar narratives about the future of fisheries can be observed elsewhere in the world (Pauly, 1990; Wilson et al, 1994; Béné, 2003; World Bank, 2004; Thorpe et al, 2005), it is important to consider how they endure in the Mekong given their significance in terms of global fisheries production, and the special historical, cultural, social and economic importance of capture fisheries in this region. The aim of this chapter is to place this 'conventional wisdom' of doom under closer critical scrutiny. In doing so, this chapter builds on recent work addressing policy that combines the critical analysis of the arguments, assumptions and narratives that underpin policy approaches (Roe, 1995; Shore and Wright, 1997; Hajer and Versteeg, 2005; Johnson, 2006; Verweij et al, 2006). Our concern is that the scale of these fisheries impacts and the implications are so far-reaching that it is important that these narratives and assumptions are explored in the public domain. So far, this has not happened.

Development policy is an attempt to shape the world by making complex problems identifiable and situations of uncertainty manageable. In legitimizing a certain course of action, narratives play a central role in development policy by creating simple storylines of how a 'problem' has arisen and will unfold, and, hence, what the necessary course of action should be. Development narratives are the 'conventional wisdom' that are so deeply embedded that they are rarely challenged, or even considered to be an area that is necessary to be critiqued (Roe, 1991; Johnson, 2006). Narratives legitimize certain types of knowledge and exclude others, and are the means by which actors and institutions make claim to action and ownership over resources (Fairhead and Leach, 1997).

The more complex the situation, the more such narratives endure (Roe, 1991). This is precisely what we see in development policy both within the worlds of fisheries and of hydropower in the Mekong region. Fisheries are complex, diverse and dynamic in many different ways (Wilson et al, 1994). Yet, fishery discussions and policies seem to be dominated by gross simplifications of this complexity, wrapped up in a narrative of doom that leads to their marginalization and neglect in development policy.

Perspectives on the 'importance' of fisheries in the Mekong

The story of the Mekong presents some unique dimensions to what appears to be a global tale. While the neglect of capture fisheries, particularly inland fisheries, in policy arenas is a phenomenon that is not exclusive to the Mekong Basin (see Thorpe et al, 2005), there are few regions in the world that have seen such a concerted and largely successful effort to raise the profile of fisheries and to conduct extensive research on a range of fisheries issues covering biology, ecology, livelihoods and nutrition. This is what makes this story so interesting. We see that while this research has highlighted the importance of fisheries, particularly in terms of employment and nutrition, fisheries have increasingly slipped off the development agendas.

Certain themes have endured in development discourse in the Mekong region. In most of the plans for Mekong development from the early 1950s until the 1980s, fisheries were recognized as 'valuable' in some way. The large numbers of people engaged in fishing, and fisheries' central importance in nutrition, has been widely recognized for many decades (see Tubb, 1966). For example, the Indian Mekong Tonle Sap Team (1962, p1) start their report by saying:

> *The importance of fish in the economic life of Cambodia is too well known to need any special emphasis here. Apart from the fact that it constitutes one of the most vital ingredients in the nutritional content of the people, it is also an important commodity of export.*

Yet, despite this recognized value, the role of fisheries in basin development visions during this period was minimal. For example, when describing the *Master Plan for the Mekong*, the executive agent of the Mekong Coordination Committee (Schaaf, 1966, p5) presented a vision that excluded fisheries, stating:

> *The project seeks the comprehensive development of the water resources of this lower basin, including mainstream and tributaries, in terms of hydroelectricity, irrigation, flood control, drainage, navigation improvement, water management and water supply, along with related far-flung economic and social growth, for the benefit of all the people of the area without distinction as to politics or nationality.*

Even among fisheries experts themselves, a narrative of the limited future of fisheries can be detected dating back several decades. For example, United Nations Food and Agriculture Organization (FAO) Regional Fisheries Officer J. A. Tubb (1966) highlights the importance of Mekong fisheries, but also the limitations of scientific knowledge about the fisheries. He also draws attention to the 'almost cataclysmic changes in the ecology' (Tubb, 1966, p63) that will result from basin development plans, but concludes that such development could and should not be avoided:

> *Water is required and will be used for many other things other than fish production, for irrigation, hydroelectric power, domestic supplies, and these on the whole are likely to have a greater economic value than the mere maintenance of areas of water for the production of fish.* (Tubb, 1966)

We see here the emergence of a central theme – that the developmental potential from capture fisheries will be less than from other development options, and that these losses can be managed even at a time in which the full value of fisheries is openly acknowledged to have not been adequately assessed.

A decade later, there was a further effort to understand the scale and importance of capture fisheries and potential impacts from basin development under the Mekong Basin-Wide Fishery Studies. Assessments, including estimates of the numbers of people involved in fishing, suggested that a quarter of the total population of the Lower Mekong Basin was involved in fisheries in one way or another (University of Michigan, 1976). Additionally, while recognizing the potential of basin development impacts upon the production and value of capture fisheries, similar conclusions were reached. Ultimately, it was suggested that improvements, particularly in the area of aquaculture, could increase the fishery yields so significantly that these possible losses to the capture fishery should not be a cause for concern (University of Michigan, 1976).

Capture fisheries started to have more of a public profile during the mid 1990s. This was driven partly by civil society concerns over impacts of hydropower projects, such as the controversial Pak Mun Dam in Thailand (Roberts, 1993; Bakker, 1999; WCD, 2000; see also Chapter 3 in this volume). At this stage, non-governmental organizations (NGOs), activists and independent researchers were generating important information on capture fisheries (Roberts, 1993; TERRA, 1993; Claridge et al, 1996). It was also influenced by the emergence of a revitalized Fisheries Programme within the Mekong River Commission (MRC) (Sverdrup-Jensen, 2002; Sneddon and Fox, 2006).

A central component of the MRC Fisheries Programme was an attempt to assess and value the productivity of the capture fishery, and from this to identify likely impacts as a result of water resource development, including hydropower (MRC, 1996). This approach had been influenced by an MRC-commissioned report in the mid 1990s, which argued for data and information on the potential impacts upon fisheries related primarily to 'main stem dam developments', but also 'related to irrigation, flood protection, agriculture development, navigation and other changes (Hill and Hill, 1994). The MRC Fisheries Programme of the 1990s can be seen to be a response to this long-established hydropower agenda, and crafted largely in its shadow.

Additional core elements of the MRC's Fisheries Programme have been similarly shaped to focus on managing reservoir fisheries (that would be created as a result of hydropower development) and on promoting aquaculture (with a later emphasis on indigenous species) as a strategy to cope with degradation of capture fisheries. The initial MRC interest in capture fisheries was thus very much framed in terms of understanding the potential impacts of water resource development. Originally, it was less driven by an interest in the fisheries for their own values and potential for economic development. However, it has grown to be the main source of information on the importance of capture fisheries in the region, and an innovative influence in areas such as participatory management and indigenous knowledge (Sneddon and Fox, 2007). Most significantly, the information from the MRC has become widely established and widely cited (Baran, 2007; MRC, 2007).

This renewed interest in capture fisheries in the Mekong since the 1990s generated wider recognition of the hydrological and ecological drivers of fisheries production and a greater understanding of fish species diversity, migration and spawning patterns (Baran et al, 2006; Lamberts, 2006; ADB, 2004; van Zalinge et al, 2004). As a result of this research effort, certain 'facts' have been more widely established, although it is also recognized within the research community that there are still some uncertainties.

These efforts have led to dramatic revisions of estimates of production. For example, during the mid 1970s, total production was put in the region of 48,000 tonnes with a value of US$6 million (University of Michigan, 1976). In contrast, the overall production of the Mekong capture fisheries is now widely cited as being in the region of between 2 and 2.5 million tonnes per year, with an estimated value of around US$2 billion (MRC, 2003; Baran, 2007; MRC, 2007). This level of production is said to constitute 2 per cent of total global fish production and even up to 17 per cent of global inland fisheries production (Baran et al, 2006). The basin's fisheries are also recognized as being based on a system with high species diversity (see Coates et al, 2003), with over 1300 species estimated to occur in the Mekong Basin. As with earlier descriptions of the fisheries, aquatic resources are also widely recognized as central to nutrition and food security, with current estimates of between 27 and 78 per cent of animal protein in diets of rural people across the Mekong Basin (Hortle and Bush, 2003; Meusch et al, 2003). Fisheries are also acknowledged to be of particular importance for poor people partly due to their nature as common property resources, and a safety net and coping strategy (STREAM, 2001; Dixon et al, 2003; Smith et al, 2005; MRCS/WUP-FIN, 2007).

However, even these broad, accepted 'facts' reveal only a partial picture as they merely provide an aggregate representation at the basin level. As a result, much of the Mekong fisheries remain largely invisible. One of the main challenges in talking about 'the fishery' of the Mekong is that, in reality, there are many different fisheries in the region. Across the Mekong Basin, fisheries range from the larger, more visible, examples, such as Tonle Sap and the large river basin-floodplain systems, to the diverse and diffuse small-scale fisheries in ponds, streams and rice fields (Gregory and Guttman, 1996), as well as to small-scale yet productive upland fisheries (e.g. Degen et al, 2005). While some attention has been paid to these larger-scale fisheries, particularly Tonle Sap, there has been far less attention directed towards the smaller-scale fisheries that, when combined, constitute the greater part of the overall picture. As a result, the full scale of involvement in fisheries and its significance for local livelihoods has usually been underestimated (Coates, 2002; Keskinen, 2003; MRCS/WUP-FIN, 2007). Even though the broad evidence concerning capture fisheries is acknowledged, more detailed, nuanced representations remain largely overlooked (Arthur and Sheriff, 2008).

Two main problems appear with the 'facts' of the Mekong region's fisheries. The first is that the accepted truths and facts are incomplete. When the term 'fisheries' appears in hydropower debates, it tends to conjure particular images that are not representative of the full complexity of the Mekong. The second problem is in the interpretation of these facts and their implications. Even though the established broad facts are rarely questioned, the interpretation of their meaning and significance is very much being challenged. As we discuss, the current importance of capture fisheries becomes diminished by an enduring narrative. Even though production estimates have gone up from 48,000 tonnes to 2.5 million tonnes in the last 30 years, this narrative of doom has endured.

Songs of the doomed: Fisheries in the Mekong

In this section we consider how the evidence and arguments generated over the last decade have failed to shift a narrative of doom and despair that runs so deep, and is so heavily institutionalized in the states and departments of fisheries of the region, that it continues to shape the way in which issues, problems and debates are framed.

This enduring narrative of the Mekong contains within it two main elements, one of doom and one of simple solutions, with four distinctive yet overlapping storylines (cf. Roe, 1991; Hajer and Versteeg, 2005):

1 the inevitable decline of an open access resource in the face of population growth and development;
2 fishing as a marginal activity with limited potential for generating economic development;
3 aquaculture as the replacement for wild-capture fisheries; and
4 trade-offs between fisheries and development.

We will now critically examine each of these storylines in turn.

The inevitable decline

The storyline of inevitable decline is the most pervasive of all, underpinning arguments related to fisheries, while also appealing to an intuitive logic that somehow escapes the need for scientific or empirical evidence. In meetings, workshops and consultations, it is often alluded to but rarely addressed directly. It is a storyline that reveals the common sense of received wisdom.

The lynchpin of this storyline is a 'tragedy of the commons' (see Hardin, 1968; Roe, 1991). Capture fisheries are assumed to be 'open access' resources and decline is thus seen as inevitable in the face of rising populations, and – along with such demographic changes – the overwhelming pressures of changes in land use and infrastructure development. For example, it is argued that as a result of 'unprecedented pressure from overexploitation and environmental change, there has been a cumulative decline in total volume of fish caught and average size of fish' (Bush, 2008, p332). Essentially, it is a storyline of unavoidable decline that draws on both the internal nature of capture fisheries as a commons resource, but also includes an implicit recognition of the threats to this resource that come from external developments.

The storyline is simple and appealing. By their very nature, fisheries have no clear boundaries, covering river systems and floodplains that straddle villages, provinces and nation states. Fish themselves are migratory and the habitats upon which they depend are highly dynamic. Whatever happens in one part of the fishery has implications for some other. Capture fisheries are thus easily presented as a classic common property resource facing the inevitable pressures of open access in the face of weak, ineffective management and with no barriers to entry for newcomers (e.g. World Bank/ADB, 2006).

These pressures are argued to be exacerbated by growing populations. Although there is no evidence of a serious discussion of whether populations of fishers are actually increasing, it is inferred by reference to general population increase and assumptions of population increase among poor people. While the arguments that fisheries, like other common pool resources, are of particular importance for poor people have been widely accepted (Smith et al, 2005), these have also become an explanation for the overwhelming threats that fisheries now face. For the story continues that as the numbers of people (and particularly the poor) increase, the threats on fisheries will intensify. For example, Wong et al (2007, p38) single out the 'huge scale of subsistence fishing … [that] is heavy and destructive and there is evidence of declining fish populations as a result'. This is very much what Pauly (1990) describes as Malthusian overfishing. While acknowledging the importance of fisheries for poor people, it presents those same poor people as the greatest threat to their sustainability.

A further dimension of this storyline is the inevitability of threats to the capture fishery that arise from the demands for development. This aspect implicitly recognizes that economic development will have an impact upon capture fisheries. But placed in the context of population growth and subsequent economic demands, these impacts are seen as unavoidable. Management of fisheries can only hope to minimize such impacts as best they can and that fisheries will become a subject for conservation in the face of development (Wong et al, 2007).

These arguments of impending doom have a long history in the Mekong Basin. For example, McCormick Smith (1925) expressed dismay at the apparent overwhelming pressure on the capture fisheries of Thailand, and documented local fishers' perceptions of a widespread decline in production. His prognosis for the future was gloomy, arguing that the combined pressures of population growth and economic progress would inevitably undermine the continued sustainability of the capture fisheries. These kinds of prognoses later appeared during the 1960s amid the early considerations of basin development (e.g. Pantulu, 1966; Tubb, 1966) and were influential in laying out the central elements of regional inland fisheries policy – conservation of fish stocks combined with the expansion of aquaculture.

The notion of the inevitable decline also suggests that even if capture fisheries are important today, the threats that they face are so insurmountable that, in the future, they will not be able to provide the benefits that they are acknowledged as generating now. A recurring theme is that fisheries production has peaked and, as in the case of the Tonle Sap, has potentially 'exceeded the optimum supported by its ecosystem productivity base' (Lamberts, 2006, p489). Without even needing to address the social, economic and cultural acceptability of the impacts of hydropower upon fisheries, the debate can easily be shifted towards future scenarios that no longer include the capture fisheries as a viable option. As we discuss, such a shift requires the combined notions of economic limitations to fisheries and the potential of viable alternatives.

Fishing is an economically marginal activity for poor people

The second storyline addresses the economic importance and developmental potential of capture fisheries. The fact that many people engage in some form of fishing becomes less of a concern if this engagement is somehow marginal and with only limited potential for economic development.

Across the Mekong Basin the vast majority of people fish as part of a diversified household livelihood portfolio (e.g. Dixon et al, 2003; Smith et al, 2005). There are relatively few people who engage as full-time professional fishers. In Laos, for example, over 90 per cent of the catch may be attributed to rural people for whom fishing is not a primary activity (Lorenzen et al, 2000). Capturing the significance of this fishing activity can be a challenge (Keskinen, 2003). There are no reliable figures on the numbers of people engaged in fishing (see Coates, 2002), but there are frequent suggestions that a majority of rural people across the basin engage in some form of fishing activity (Gregory and Guttman, 1996; Sverdrup-Jensen, 2002; Baran et al, 2006; World Bank/ADB, 2006). Where this kind of fishing activity is identified, it is most frequently referred to as a secondary or supplementary occupation (Ahmed et al, 1998). This terminology has important connotations: that fishing is of less importance than other activities; the numbers of people dependent upon fisheries and for whom fishery is important, and the extent of this importance, can be downplayed.

Fishing can also be presented as an activity to which people turn when other options are not viable – for example, as a coping strategy or activity of 'last resort' for the marginalized poor. This can be presented to reinforce arguments that fishing is essentially unimportant except for those who have no other choice. It acknowledges the importance of fishing for the poor and the potential implications of degradation of their main natural resource. As Béné (2003) observes, this is a persistent argument in which the story goes, that poor people fish, and people are poor because they fish.

The combination of these arguments is that capture fisheries have limited opportunities for economic development. In the case of Thailand with a growing industrial marine fleet, inland capture fishing has been regarded as an occupation for the poor and having limited potential for development other than through the introduction of aquaculture, or through the promotion of alternative employment. In general terms, inland fishery is regarded as having no real future in its present form and no real prospects for economic development (Masae and McGregor, 1996). State policy on capture fisheries across the region has focused on a conservation strategy of minimizing degradation, but, as we discuss below, has concentrated efforts on increasing production through aquaculture and stocking (Bush, 2008).

An additional dimension of development policy targeting fishers themselves has been to promote alternative livelihoods, moving fishers into other more productive economic activities. Picking up on the storylines of both the tragedy of the commons and limited opportunities for economic development, the World Bank and Asian Development Bank (ADB) recently stated the case for hydropower development: 'It can be argued that the best basis for intervening in these "common property" problems is provided by the existence of alternative sources of income (as provided by irrigated agriculture) and development generally (as facilitated by the availability of power)' (World Bank/ADB, 2006). This basic perception drives much of regional fisheries policy, with an emphasis on promotion of aquaculture and of alternative livelihoods (World Bank, 2004). Moreover, this serves as a convenient justification for the development of hydropower since the compelling and influential storyline is that through the generation of electricity to spur economic growth, the underlying cause for people's dependence upon fisheries – namely, poverty – can be addressed.

Aquaculture can and should replace the fishery

The substitution of capture fisheries with aquaculture is a global storyline. For example, the World Bank (2007) suggests: 'As production from capture fisheries stagnates, aquaculture is changing the face of our waters.' In the Mekong it has also been argued that, whatever fisheries have provided in the past, the future lies largely in the technology of aquaculture. For example, it has appeared in the press: 'Increasing the amount of fish consumed by Lao people is necessary, but is unsustainable without an expanded aquaculture programme' (*Vientiane Times*, 2008a). In this way, aquaculture has emerged as a hegemonic discourse (Bush, 2008). Throughout the Mekong Basin, aquaculture dominates state-led fisheries policy (see Bush, 2008). Opportunities for increasing fisheries production have been presented almost exclusively in terms of aquaculture production.

These arguments for aquaculture have combined notions of the inevitability of fisheries decline, largely as a result of unavoidable population growth and development. For example, Edwards and Demaine (1997, p11) argue:

> *The need for aquaculture to provide increased supplies of fish should be considered in relation to capture fisheries. These currently dominate production, but are static or in decline globally and in most countries. As wild fish stocks are threatened by human population growth through overfishing and environmental degradation, the stimulus and need for aquaculture are greatest in developing countries where at least 90 per cent of the global increase in population is predicted to take place before the world population stabilizes at a level at least double that of today.*

The promotion of aquaculture reflects development strategies that have focused on modernization and technocratic solutions. Mirroring the agricultural Green Revolution, the fisheries sector has attempted to generate a Blue Revolution based on the development of fish farming (see Coull, 1993). While these strategies have not been limited to the Mekong Basin, this part of Asia has been regarded as having great potential for aquaculture development; Thailand and Vietnam have witnessed dramatic growth in aquaculture production, both inland and coastal.[1]

Recognizing the importance of the capture fishery, one ex-chief executive officer of the MRC reaffirmed the notions of peaked production and the threats of rising populations:

> *Capture fisheries are utilized at its maximum possible level already, and there are only limited possibilities for expanding it in reservoirs and other artificial water bodies. It is more than likely that the development of other sectors may lead to some decline in the overall fish production in the basin in the future.*
>
> *With strong population growth in the Mekong Basin and a natural capture fishery that can hardly be expanded, aquaculture has an important role to play in food security as a whole. It is the most important source for an increase in fish production required to cope with the population increase.* (Kristensen, 2001, p15)

This is a significant assessment as it comes from the one institution credited with raising the profile of capture fisheries and during a period in which there was an explicit commitment to the rhetoric of sustainable development within the MRC. Even within this historical and institutional context, these assumptions remained strong.

The notion that aquaculture is an inevitable response to the decline of the capture fishery has several dimensions. This creates a strange interdependence between aquaculture and capture fisheries in which the uptake of aquaculture by large numbers of fishers and farmers (as opposed to commercial activities) actually depends upon a degraded capture fishery or reduced access to the fishery. Or to put it another way, the degradation of capture fisheries actually becomes an opportunity for the promotion of aquaculture.

The promotion of aquaculture can also be seen as going hand in hand with the development of hydropower. The development of dams affects the wild fishery, contributing to the creation of a degraded fishery which is itself one of the preconditions for the uptake of aquaculture. At the same time, the change to river habitat by the creation of reservoirs provides the very types of water bodies that aquaculture desires. During the 1960s this led to recommendations for government agencies to focus their efforts on promoting aquaculture to take advantage of these opportunities. For example, Pantulu (1966, p65) argues that 'human intervention can, through scientifically based developmental measures, increase fish production to levels probably impossible to achieve in unharnessed rivers' and by taking advantage of the changed environment (Tubb, 1966).

More recently, the lack of suitable water bodies is argued to be one of the major constraints to aquaculture expansion. By creating reservoirs, hydropower provides an additional resource for aquaculture. As Costa-Pierce (1998) writes: 'there are vast areas of new inland waters "locked up" in hydropower and irrigation reservoirs ... hydropower and irrigation reservoirs may be Asia's final "aquatic frontier".' More recently, in talking of dams proposed for the Sekong River in Laos, a Lao government official was quoted as suggesting that 'hydropower not only supplied water for irrigation, but also provided a regular long-term source of fish for local people' (*Vientiane Times*, 2008b). In this way the impacts of hydropower upon an already doomed resource are compensated for by the creation of new aquaculture opportunities.

Trade-offs and tough choices

The notion of trade-offs is increasingly being brought into play in the hydropower debates as a means of framing decisions about the costs and benefits of alternative development pathways as unavoidable choices that can be identified, assessed and mitigated against. In this narrative, the possibility of potential impact upon the capture fisheries is not necessarily denied, nor even downplayed. With connotations of balance, trade-offs have an important resonance. This marks an important shift in the discourse of hydropower development in the Mekong. It is acknowledged that dams do, indeed, have impacts upon capture fisheries and that these can be significant. But the issue becomes one of being in a tough situation where choices between fisheries and dams have to be made, however unpleasant such choices might be. The challenge is to address poverty, although what constitutes 'poverty' and the options for reducing poverty are rarely discussed. For example, the MRC itself noted in 2007 that:

> *Further pressure on wild fish stocks will also come from the increasing exploitation of the Mekong's natural resources by other development sectors. But these developments must proceed, if we are to tackle poverty and underemployment in the basin.* (MRC, 2007, p2)

In this context of poverty, the situation is said to be one in which a choice needs to be made between conservation and development, stagnation and progress, between fish and people (Chapman and Daming, 1996).

It is the framing of this situational context that is most significant. The starting point for making trade-offs is in terms of the unavoidable drive for development, particularly hydropower, legitimized by a poorly defined poverty reduction agenda. The demand for hydropower is presented as being driven by 'a pent-up demand for development' (World Bank/ADB, 2006, p17), and so the choices of trade-offs are framed in terms of what must be sacrificed for the overall good. It is in this context that the sacrifice or loss of capture fisheries becomes a regrettable but necessary trade-off. But this raises critical issues – whether the choices can be reduced so simply and crudely as between fish and people, whether these choices reflect the developmental challenges, whether there are alternative means of framing the challenges and options, how these decisions are made, and who has the power to make them. For as much as it is a choice between fish and people, it is also a choice between food and air conditioning, and to who these benefits accrue.

Of course, this storyline of trade-offs depends upon viewing hydropower as necessary and unavoidable, and upon the associated storylines of the capture fisheries discussed above. This is legitimized by recourse to the challenge of poverty, and the potential of hydropower to meet this challenge. The combined effect of notions of inevitable decline of capture fisheries, the marginal value of fishing and limited potential for economic development, and the viability of aquaculture to replace the degraded capture fishery, all contribute to a narrative in which capture fisheries can, indeed, be traded off for hydropower. Even though by doing so there may be unpleasant impacts, these are not overwhelming or unmanageable. Framing hydropower and capture fisheries in terms of trade-offs closes the space for alternative development pathways.

PROSPECTS FOR A FISHERIES-BASED COUNTER-NARRATIVE: CRISIS UNDER SCRUTINY

A counter-narrative to hydropower development has begun to emerge among a diverse range of civil society groups at various levels, including local communities, NGOs and academics. In this counter-narrative, capture fisheries and fisheries-based livelihoods have come to represent alternative values of development in which the natural dynamism of the waters are entwined with local culture and economics. This combination of concern for impacts and representations of alternative development pathways lies at the heart of attempts to reconstitute the current debate on water resource development.

The emerging counter-narrative presents the river as having value beyond that of water as a commodity. Notions of nature, wisdom and culture are joined around three key arguments:

1 the fundamental ecological, social, cultural and economic importance of capture fisheries;
2 the wealth of local knowledge of fisheries and river ecology; and
3 the capacity of local fishers to manage the Mekong region's resources sustainably and equitably (e.g. Claridge et al, 1996; Shoemaker et al, 2001; Missingham, 2003; Sretthachau and Deetes, 2004; Baird and Mean, 2005).

The counter-narrative also poses a powerful critique to the narratives of open access resources and notions of inevitable decline. Fisheries are argued to have been managed according to traditional rules and norms that have emphasized both sustainability and equity, with rural people acting as the custodians of river ecology (Shoemaker et al, 2001; Missingham, 2003). The degradation of resources is not a result of their perceived status as open access, but actually arises from the 'enclosure of the commons through power-based relations' (Béné, 2003, p965) where political influence, corruption and mismanagement are leading to illegal fishing and overfishing. The threat to managing fisheries is argued to arise from the encroachment of market forces and values, the failure of state-led management initiatives to recognize existing traditional practices, and the incompatibility of state-led fisheries management polices with local management regimes. As pre-existing custodians of river ecology, local fishers are argued to deserve preferential rights in river basin management ahead of other resource users (TERRA, 1993).

This section is concerned with placing the underlying arguments upon which this crisis narrative is based under a more critical examination.

Is there a decline and is this inevitable?

Underpinning the claims of inevitable decline is the argument that capture fisheries production has already peaked. While this argument has some intuitive appeal, it is very difficult to establish the actual status of stocks and production in the Mekong (Coates, 2002). This argument traced back to the 1960s is based on two seemingly contradictory elements: an assertion that fisheries production has peaked and, alongside it, the recognition that there is not enough evidence to determine whether this is so. For example Tubb (1966, p64) argues, on the one hand, that 'production may even now be approaching the maximum', then immediately acknowledges that 'reliable and comprehensive statistical data on production is entirely lacking'. Despite such apparent contradictions, these arguments have continued.

From a purely biological perspective, it is difficult to assess the natural productivity of the Mekong fisheries, which include a huge number of different species with different life cycles, a high seasonal abundance, the ability to migrate over large distances, and are largely invisible. Traditionally, a good deal of the information about a fishery is derived from data taken from what people actually catch. People fish with different gear, targeting a range of habitats, with different levels of intensity at different times of the year. Much of the fish catch is consumed within the household, and as such is invisible to outsiders. As a result, getting a picture of the status of stocks and production across the basin that can be compared from year to year remains a huge challenge (Coates, 2002; Lamberts, 2006). In addition, there is a high natural variability from year to year, particularly for some species. Gathering data and assessing trends is thus extremely difficult (Coates, 2002).

Information remains insufficient to determine whether stocks or production are in decline (Tubb, 1966; Hill and Hill, 1994; Baran and Myschowada, 2008), and certainly, if there is a decline, whether this is inevitable. An intriguing aspect of this argument of fisheries being in decline is that although it can be traced back several decades, it has endured even through a period in which the estimates of fisheries production have dramatically increased. For example, Baran (2007) summarizes the shifts in production estimates through the 1990s. Assumptions of decline have endured during a period in which official production figures have increased almost sevenfold.

This is not to say that concerns about declines of the fishery are not warranted; but this is essentially a management issue and, as such, it is crucial that the factors leading to any decline are identified correctly. Portraying decline as ongoing and inevitable takes the concern out of the sphere of fisheries management.

The tragedy of the commons rests on assumptions of a lack of management, and threats from rising populations of fishers. Considerable effort has gone into documenting the wealth of traditional local management practices in the region that have aimed to ensure sustainability and equity (e.g. Claridge et al, 1996; Degen et al, 2005; Garaway et al, 2006). In contrast to the perception of an unmanageable open access resource, the Mekong Basin provides a wealth of local management regimes, many of which are highly adaptive to changing social and natural environments. Even in cases that are supposedly open access, there is no free for all. There is evidence from across the basin of such management for a range of fishery resources, including river fisheries, floodplains and rice fields providing a range of benefits. Where management regimes are undermined, this is a factor of weak governance rather than due to the intrinsic nature of the fisheries (Thuon, 2004; Keskinen et al, 2007).

The assumption that fisheries face unavoidable pressures from rising populations of fishers also does not hold. Overall, there is no evidence that numbers of fishers are increasing or that where numbers of fishers are increasing that this is a result of population change rather than other socio-economic factors which draw people to the fishery. Again, evidence from the Mekong suggests that the concern is not so much about aggregate numbers of fishers increasing, but about commercial encroachment and use of large-scale destructive gear, privatization of common resources and weak rule of law.

The misrepresentation of fishing within rural economies

The claim that fishing is somehow of secondary importance to local livelihoods or an 'activity of last resort' is also unconvincing. From our perspective, fishing is central to rural economies and fundamental to household livelihood strategies.

There have been several attempts at identifying the numbers of people engaged in full-time or part-time fishing. For example, a comprehensive and often-cited assessment carried out in Cambodia during the late 1990s distinguishes between fishing as a primary occupation or as a part-time occupation (Ahmed et al, 1998). This research suggests that from eight provinces surveyed, 10.5 per cent of the households are engaged in fishing or related activities as a primary occupation, with an additional 34.1 per cent engaged on a part-time basis, indicating a total of 1 million people engaged in fisheries in one form or another. For some areas of Cambodia, the involvement in fisheries is argued to be even higher, up to 90 per cent (Thouk and Sina, 1997).

Yet, these distinctions between primary and secondary occupations can themselves be misleading. The majority of people engage in fishing as a component of diversified household livelihood strategies. For example, in Laos it has been reported that almost everyone who has access to water, fishes (Claridge et al, 1996). The need is then to assess fisheries in this context (Heady et al, 1995; Friend, 2001; Shoemaker et al, 2001; Meusch et al, 2003; Garaway, 2005; Smith et al, 2005; Resurreccion, 2006). Looking again at southern Laos, fishing has an important role in the livelihoods of almost all rural households, and not just the poor, with fishing accounting for up to 70 per cent of household fish consumed and sold across different wealth groups within the same villages (Garaway, 2005). In addition, rather than representing an activity of particular significance in poorer households, people of all socio-economic classes fish and consume fish, with poorer households catching only slightly more on a per household basis (Garaway, 2005).

The majority of rural people across the Mekong Basin tend to refer to themselves as rice farmers (see Luco, 1997; Lorenzen et al, 2000). Yet, although rice farming holds a special place in people's own imaginations and the rural culture of the region, in many cases the low value of rice production and its limited contribution to household economies compared to fisheries is such that it could be argued that they are fishers who farm, rather than farmers who fish (Gregory and Guttman, 1996). As Keskinen (2003) argues for the situation in Cambodia, agriculture and fishing are so intertwined it is impossible to separate them; but the approach of census surveys framed in terms of primary and secondary occupations fails to capture the interdependence of household multiple livelihood strategies.

By considering the amount of time people invest in fishing-related activities, the importance relative to these other activities becomes clearer, even in the smaller-scale fisheries of upland areas (Degen et al, 2005). Other authors have argued that the fundamental importance of fisheries and agriculture (mainly rice production) and water management is such that rural regions of the Mekong can be characterized as comprising 'river-based livelihoods' (Shoemaker et al, 2001) or 'wetland livelihoods' (Friend, 2007).

Can aquaculture really expect to replace capture fisheries?

For proponents of aquaculture, there was some dismay that there were situations in which farmers proved reluctant to take up aquaculture, or if they did, remained reluctant to continue with aquaculture. Despite the efforts of aquaculture extension in many parts of the basin (e.g. the lower northeast of Thailand and southeast Cambodia), the uptake has remained disappointing (Pushpalatha, 2001) and yields and recapture rates differ from those expected (Lorenzen and Garaway, 1998; Garaway et al, 2001).

This led to a reassessment of aquaculture. It became apparent that in these specific areas, the capture fishery that had been assumed to be no longer productive was far more vibrant than had been appreciated and remained an attractive livelihood activity for local people. This in turn led to some important shifts in how aquaculture began to be promoted. It was no longer to be presented as a replacement to the capture fishery, but rather as a supplement, particularly for those engaged in diversified livelihoods (Garaway et al, 2006). Increasingly it was recognized that involvement in both aquaculture and capture fisheries could change from year to year, with people moving in and out of one or the other depending upon a range of factors, including availability of labour and credit, as well as the natural productivity of the capture fishery (Friend and Funge-Smith, 2002). There was also growing interest in the types of aquatic resource activities that combined elements of aquaculture and capture fishery, rather than seeing the two as competing activities.

The most widespread example of this aquaculture–capture fisheries interface is the stocking of capture fisheries (often referred to as 'enhanced' or 'culture-based' fisheries), where natural capture fisheries and reservoirs are stocked with farmed fingerlings and juveniles. This is a key part of government policy throughout the Mekong (Claridge et al, 1996; Warren, 2000; Welcomme and Vidthayanon, 2003). Experiences with village-managed culture-based fisheries in southern Laos indicated that stocking increased the potential biological production. However, the low levels of effort applied as a combined result of restricted access that accompanied the stocking and selected harvesting of larger stocked fish often meant that while the efficiency of harvesting increased, yields were no different from similar unstocked fisheries (Lorenzen and Garaway, 1998; Garaway, 1999; Arthur, 2004; Garaway et al, 2006).

A smaller-scale example can be seen in the management of refuge traps in floodplains. These traps are dug in areas of the floodplains that maintain water in the dry season and attract certain species of fish. While these trap ponds and similar management approaches in rice fields were identified as traditional practices in many parts of the basin (Angporn et al, 1998), it was recognized that in some areas the use of trap ponds was growing and in some cases there was additional stocking of fish in the ponds, constituting a bridge between aquaculture and capture fisheries.

In terms of uptake, it was recognized that while aquaculture involved fish, it did not necessarily fit with fishers' livelihood portfolios, assets, skills and interests. In fact, there was a recognition that in many cases, aquaculture was more akin to farming than fishing, and that the most enthusiastic uptake was often with farmers rather than fishers. As Payne (2000, p2) notes: 'farmers tend to make better fish farmers than people who primarily fish for a living'.

An alternative argument suggests that aquaculture can never expect to replace the capture fishery. In this light, aquaculture is one element of aquatic resource management, but not a substitute for wild fishery (Friend and Funge-Smith, 2002).

When we consider current production levels of capture and culture fisheries in the Mekong Basin, the challenge for aquaculture to replace capture fisheries simply in aggregate production terms becomes clearer. Current estimates suggest that of total production in the Lower Mekong Basin, approximately 80 per cent comes from capture, with 10 per cent from aquaculture and 10 per cent from reservoir-stocking (Sverdup-Jensen, 2002). The imbalance of production figures illustrates the magnitude of required increases in aquaculture to compensate for losses in capture fisheries, even without considering the investments and technical support that would be needed to achieve this and the distribution of benefits.

There are important distributional dimensions in terms of producers and consumers. Aquaculture is currently restricted to farming a few species, as opposed to the great diversity found in local catches. Aquaculture producers are not always fishers, and with growing commercialization, this is becoming even more so. Aquaculture production tends to target urban markets with specific commercially viable species. Typically, this type of production does not represent the daily aquatic resource consumption of the majority of rural people and, as such, is not a substitute.

The notion that hydropower reservoirs constitute the 'last aquatic frontier' (see Costa Pierce, 1998) is also misleading. Despite the gold-rush appeal of such a representation of reservoirs, experience has shown that stocking reservoirs has rarely succeeded in compensating for capture fisheries lost in purely aggregate terms. Reservoirs change the biophysical and institutional nature of the fishery. Stocked reservoirs support different fish assemblages from natural rivers, typically with less species diversity. In terms of how people fish, reservoirs represent a new type of fishery for fishers, requiring new skills and resources. Stocking reservoirs (and other public water bodies) can end up displacing the people who had previously relied on these fisheries. The few cases in the Mekong in which reservoir fisheries have become highly productive are associated with a commercialization of the fishery and the granting of high-investment concessions, such as in the Nam Ngum Reservoir in Laos.

Trade-offs revisited

Framing hydropower and fisheries in terms of trade-offs has an immediate appeal of reasonableness and balance. But at the same time, it reduces complex societal choices based on values of what development means to simplistic choices to be determined by an inferred technical, neutral decision. It simplifies complex options to a set of polar choices, as if these were the only options available, and as if the choice between dams and fish were comparable. At its most crude, the use of the term trade-offs reduces complex economic, environmental, social and cultural values to aggregate economic values. In addition, the trade-offs usually fail to address the distributional aspects of different resources (MRCS/WUP-FIN, 2007).

Inherent in the trade-offs approach is an assumption that society can afford to trade off fish for the benefits of hydropower. But it is not clear what exactly is being traded off. As many authors have argued, fisheries are central to rural livelihoods, providing a range of social and cultural values and benefits. Putting aside these values and benefits, and only focusing on the value of fisheries in terms of food security and nutrition, still highlights what is at stake. The notion that fisheries can be traded off rests on an assumption that what is lost in terms of food can be replaced. As we have discussed above for aquaculture, this assumption is highly questionable. Moreover, the implicit assumption is that there are undifferentiated beneficiaries, rather than considering who and where the winners and losers are.

Additional concerns emerge when this issue of trade-offs is placed in the context of current food production challenges, as well as growing climate change uncertainties. With declining global food reserves, changing agricultural practices towards non-food crop production and rising populations, this notion of trading off a viable, renewable natural resource that sustains large portions of the Mekong region's population appears more risky, and is not a simple choice of trading like for like.

CONCLUSIONS

The narrative of doom surrounding the capture fisheries of the Mekong is persistent and deeply entrenched. It is a narrative that is institutionalized in the fisheries departments, policies and programmes of the region. It shapes the research agendas that fisheries science undertakes, and determines where, when and how fisheries issues enter development debates. As such, it reinforces the current drive for hydropower rather than providing alternatives.

As with other development narratives, this crisis narrative of fisheries in the Mekong simplifies a set of complex issues. The many different fisheries of the Mekong cover a wide geographical area, with significant diversity in all of the key characteristics of a fishery – fishers, gear, habitats, species and fishing practice. Fisheries are characterized by complexity and uncertainty, particularly in large river basins driven by complicated and dynamic ecological processes. Understanding this diversity and complexity is far from straightforward. Developing management and policy in this context of complexity is even more challenging.

The major problem with the ways in which fisheries debates have been framed in the Mekong is that the complexity and diversity of fisheries have not been captured adequately, or have been lost completely. All that has endured is a gross simplification legitimizing a narrow set of management and policy options.

Fisheries science and research needs to seek ways in which they can be more influential. The evidence that has been generated has not been challenged directly – and yet seems to have had so little influence on the course of hydropower development. The problem seems to be the failure to generate compelling arguments that challenge this narrative of doom. Ultimately, the fate of the fisheries of the Mekong region just does not seem to matter.

In this chapter we have attempted to provide an initial critique of the crisis narrative and the assumptions upon which it is based within the context of hydropower development in the Mekong River Basin. We have suggested that the assumptions and arguments embedded in these storylines can be challenged based upon empirical evidence. Yet, even this only takes us so far. Deconstruction alone will not influence policy outcomes until alternative pathways can be demonstrated and articulated. Because of this, there is an urgent need to reframe the evidence and arguments of capture fisheries in the Mekong Basin. Essentially, this requires a move away from simply highlighting the socio-ecological, economic and cultural importance of capture fisheries to creating a counter-narrative (see Roe, 1995) that reverses established thinking, and demonstrates the complexity and multiple realities of fisheries, fishery livelihoods and the fishers themselves across the Mekong Basin. This needs to set out a future scenario of how fisheries and the people who depend upon them can contribute to setting development objectives. Such a rigorous and empirically based counter-narrative should seek to provide a future scenario in which fisheries are not merely a resource of conservation value, but a resource whose management is central to meeting the varied developmental challenges of the Mekong River Basin.

ACKNOWLEDGEMENTS

The authors would like to thank the anonymous reviewers for the helpful comments and suggestions, and also François Molle, Mira Käkönen, Eddie Allison and Rajesh Daniel for their comments on earlier drafts. All errors remain the responsibility of the authors. The title for this chapter has been borrowed from a collected volume of the Gonzo Papers by Dr Hunter S. Thompson: *Songs of the Doomed: More Notes on the Death of the American Dream* (Summit Books, 1990). Furthermore, Richard Friend and Robert Arthur recognize and appreciate the support for this collaboration that was provided by the Swedish International Development Agency (Sida) through the Wetland Alliance. The views expressed here are not necessarily those of Sida.

NOTE

1 Aquaculture, like farming or fisheries, is a term that is used to cover a huge diversity of activity, in terms of technology, investment, scale and intensity, and with a variety of objectives from subsistence to export.

REFERENCES

ADB (Asian Development Bank) (2004) *Environmental Atlas of the Greater Mekong Sub Region*, ADB, Manila, the Philippines

Ahmed, M., Navy, H., Vuthy, L. and Tiongco, M. (1998) *Socioeconomic Assessment of Freshwater Capture Fisheries in Cambodia: Report on a Household Survey*, Mekong River Commission, Phnom Penh, Cambodia

Angporn, P., Guttman, H., Turongruang, D., Mingkano, P. and Demaine, H. (1998) *Survey of Trap Pond Owners in Sisaket and Roi Et Provinces*, Thailand AIT Aqua Outreach Working Paper T-6, AIT, Bangkok

Arthur, R. I. (2004) *Adaptive Learning and the Management of Small Water Body Fisheries: A Case Study in Lao PDR*, Centre for Environmental Technology, T. H. Huxley School for the Environment, Earth Sciences and Engineering, Imperial College of Science, Technology and Medicine, London

Arthur, R. I. and Sheriff, N. (2008) 'Fish and the poor', in Briones, R. M. and Garcia, A. G. (eds) *Poverty Reduction through Sustainable Fisheries: Emerging Policy and Governance Issues in Southeast Asia*, Institute of Southeast Asian Studies (ISEAS) with the Southeast Asian Regional Center for Graduate Study and Research in Agriculture (SEARCA), Laguna, Philippines

Baird, I. G. and Mean, M. (2005) *Sesan River Fisheries Monitoring in Ratanakiri Province, Northeast Cambodia: Before and After the Construction of the Yali Falls Dam in the Central Highlands of Viet Nam*, Sesan Protection Network, Ban Lung, Ratanakiri, Cambodia

Bakker, K. (1999) 'The politics of hydropower: Developing the Mekong', *Political Geography*, vol 18, pp209–232

Baran, E. (2007) *Fish Migration Triggers in the Lower Mekong Basin and Other Tropical Freshwater Tropical Systems*, MRC Technical Paper no 14, Mekong River Commission, Vientiane, Laos

Baran, E. and Myschowada, C. (2008) 'Have fish catches been declining in the Mekong', in Kummu, M., Varis, O. and Keskinen, M. (eds) *Modern Myths of the Mekong*, Water Resources Development Group, University of Helsinki, Helsinki, Finland

Baran E., Jantunen, T. and Chong, C. K. (2006) *Values of Inland Fisheries in the Mekong River Basin*, WorldFish Centre, Phnom Penh, Cambodia

Béné, C. (2003) 'When fishery rhymes with poverty: A first step beyond the old paradigm on poverty in small-scale fisheries', *World Development*, vol 31, no 6, pp949–975

Bush, S. (2008) 'Contextualising fisheries policy in the Lower Mekong Basin', *Journal of Southeast Asian Studies*, vol 39, no 3, pp329–353

Chapman E. C. and Daming, H. (1996) 'Downstream implications of China's dams on the Lancang Jiang (Upper Mekong) and their potential significance for greater regional cooperation, basin-wide', in Stensholt, B. (ed) *Development Dilemmas in the Mekong Subregion*, Monash Asia Institute, Australia

Claridge, G., Sorangkhoun, T. and Baird, I. (1996) *Community Fisheries in the Lao PDR: A Survey of Techniques and Issues*, Natural Environment Consulting Ltd, Benalla, Australia

Coates, D. (2002) *Inland Capture Fishery Statistics of Southeast Asia: Current Status and Information Needs*, FAO, Bangkok, Thailand

Coates D., Poeu, O., Suntornratana, U., Thanh Tung, N. and Viravong, S. (2003) *Biodiversity and Fisheries in the Lower Mekong Basin*, Mekong Development Series no 2, Mekong River Commission, Phnom Penh

Costa-Pierce, B. A. (1998) 'Constraints to the sustainability of cage aquaculture from resettlement from hydropower dams in Asia: An Indonesian case study', *Journal of Environment and Development*, vol 7, no 4, pp333–363

Coull, J. R. 1993 'Will a blue revolution follow the green revolution? The modern upsurge of aquaculture', *Area*, vol 25, no 4, pp350–357

Degen, P., Piseth, C., Swift, P. and Mary, H. (2005) *Upland Fishing and Indigenous Punong Fisheries Management in Mondulkiri Province*, Cambodia FAO/WCS, Phnom Penh, Cambodia

Dixon, P. J., Sultana, P., Thompson, P., Ahmed, M., Lorenzen, K. and Halls, A. S. (2003) *Understanding Livelihoods Dependent on Inland Fisheries in Bangladesh and Southeast Asia*, Synthesis Report, Fisheries Management Science Programme Final Technical Report Project R8118, London

Edwards, P. and Demaine, H. (1997) *Rural Aquaculture: Overview and Framework for Country Review*, FAO RAPA, Bangkok

Fairhead, J. and Leach, M. (1997) 'Webs of power and the construction of environmental policy problems: Forest loss in Guinea', in Grillo, R. D. and Stirrat, R. L (eds) *Discourses of Development: Anthropological Perspectives*, Berg Publishers, Oxford and New York, pp35–59

Friend, R. M. (2001) 'A people-centred fishery: Socio-economic perspectives', in *Siphandone Wetlands, Champasak Province, Lao PDR*, CESVI, Bergamo, Italy

Friend, R. M. (2007) *Securing Sustainable Livelihoods through Wetland Management: Reflections on Experience under the Mekong Wetlands Biodiversity and Sustainable Use Programme (MWBP)*, Technical report, IUCN/MWBP, Vientiane, Laos

Friend, R. M. and Funge-Smith, S. (2002) *Focusing Small-Scale Aquaculture and Aquatic Resource Management on Poverty Alleviation*, FAO and NACA, Bangkok

Garaway, C. J. (1999) *Small Waterbody Fisheries and the Potential for Community-Led Enhancement: Case Studies in Lao PDR*, Centre for Environmental Technology, T. H. Huxley School for the Environment, Earth Sciences and Engineering, Imperial College of Science, Technology and Medicine, London, p453

Garaway, C. (2005) 'Fish, fishing and the rural poor: A case study of the household importance of small-scale fisheries in the Lao PDR', *Aquatic Resources, Culture and Development*, vol 1, no3, pp131–144

Garaway, C. J., Lorenzen, K. and Chamsingh, B. (2001) 'Developing fisheries enhancements in small water bodies: Lessons from Lao PDR and Northeast Thailand', in De Silva, S. S. (ed) *Reservoir and Culture-Based Fisheries: Biology and Management*, Proceedings of an international workshop held in Bangkok, Thailand, ACIAR, Canberra, pp227–234

Garaway, C. J., Arthur, R. I., Chamsingh, B., Homekingkeo, P., Lorenzen, K., Saengvilaikham, B. and Sidavong, K. (2006) 'A social science perspective on stock enhancement outcomes: Lessons learned from inland fisheries in southern Lao PDR', *Fisheries Research*, vol 80, pp37–45

Gregory, R. and Guttman, H. (1996) *Poor in All but Fish: A Study of the Collection of Ricefield Foods from Three Villages in Svay Teap District, Svay Rieng*, Working Paper 5, AIT AquaOutreach Project Asian Institute of Technology, Bangkok, Thailand

Hajer, M. and Versteeg, W. (2005) 'A decade of discourse analysis of environmental politics: Achievements, challenges, perspectives' *Journal of Environmental Policy and Planning*, vol 7, no 3, pp175–184

Hardin, G. (1968) 'Tragedy of the commons', *Science*, vol 162, pp1243–1248

Heady, C. J., McGregor, J. A. and Winnett, A. B. (1995) *Poverty, Equity and Sustainability in the Management of Inland Capture Fisheries in South and South East Asia*, End of Project Report, Centre for Development Studies, University of Bath, Bath

Hill, M. T. and Hill, S. A. (1994) *Fisheries Ecology and Hydropower in the Lower Mekong River: An Evaluation of Run-of-the-River Projects*, Mekong Secretariat, Bangkok, Thailand

Hirsch, P. (2003) 'The politics of fisheries knowledge in the Mekong River Basin', in R. L. Welcomme and T. Petr (eds) *Proceedings of the Second International Symposium on the Management of Large Rivers for Fisheries: Volume 2. Sustaining Livelihoods and Biodiversity in the New Millennium. 11–14 February 2003, Phnom Penh, Cambodia.* FAO, Bangkok

Hortle, K. G. and Bush, S. R. (2003) 'Consumption in the Lower Mekong Basin as a measure of fish yield', in FAO (ed) *New Approaches for the Improvement of Inland Capture Fishery Statistics in the Mekong Basin*, FAO-RAP Publication 2003/1, FAO, Bangkok, Thailand, pp76–82

Indian Mekong Tonle Sap Team (1962) *Increased Fresh Water Fish Production in Cambodia by the Proposed Tonle Sap Barrage*, Central Water and Power Commission, Ministry of Irrigation and Power, Government of India, India

Johnson, D. S. (2006) 'Category, narrative and value in the governance of small-scale fisheries', *Marine Policy*, vol 30, pp747–756

Keskinen, M. (2003) *Socio-Economic Survey of the Tonle Sap Lake, Cambodia*, MSc thesis, Water Resources Laboratory, Helsinki University of Technology, Espoo, http://users.tkk.fi/u/mkeskine/dippa.pdf

Keskinen, M., Käkönen, M., Tola, P. and Varis, O. (2007) 'The Tonle Sap Lake, Cambodia: Water-related conflicts with abundance of water', *Economics of Peace and Security Journal*, vol 2, no 2, pp49–59

Kristensen, J. (2001) 'Food security and development in the Lower Mekong River Basin: A challenge for the Mekong River Commission', Paper presented at the Asia and Pacific Forum on Poverty: Reforming Policies and Institutions for Poverty Reduction, Asian Development Bank, Manila, 5–9 February 2001

Lamberts, D. (2006) 'The Tonle Sap Lake as a productive ecosystem', *International Journal of Water Resources Development*, vol 22, no 3, pp481–495

Lorenzen, K. and Garaway, C. J. (1998) 'How predictable is the outcome of stocking?', in Petr, T. (ed) *Inland Fisheries Enhancements*, FAO Fisheries Technical Paper no 374, FAO, Rome

Lorenzen, K., Nguyen Khoa, S., Garaway, C., Arthur, R. I., Kirkwood, G., Chamsingh, B., Litdamlong, D., Innes-Taylor. N. and Siebert, D. (2000) *Impacts of Irrigation and Aquaculture Development on Small-Scale Aquatic Resources*, Final Technical Report, DFID Project R7235, Imperial College London

Luco, P. F. (1997) *The Community Management and the Traditional Cambodian Society: Villages of the Inundated Forest of the Tonle Sap Lake, Socio-Anthropological Study*, Summary of the original in French, Participatory Natural Resource Management in Tonle Sap Region, FAO, Siem Reap, Cambodia

Masae, A. and McGregor, J. A. (1996) *Sustainability under Conditions of Rapid Development: A Freshwater Fishery in Southern Thailand*, Occasional Paper 01/96, Centre for Development Studies, University of Bath, UK

McCormick Smith, H. (1925) *A Review of the Aquatic Resources and Fisheries of Siam, with Plans and Recommendations for the Administration*, Conservation and Development, Reprinted in 1983 by SEAFDEC, Bangkok

Meusch, E., Yhoung-Aree, J., Friend, R. and Funge-Smith, S. (2003) *The Role and Nutritional Value of Aquatic Resources in Livelihoods of Rural People: A Participatory Assessment in Attapeu Province, Lao PDR*, FAO, Bangkok

Missingham, B. D. (2003) *The Assembly of the Poor in Thailand: From Local Struggles to National Protest Movement*, Silkworm Books, Bangkok

MRC (Mekong River Commission) (1996) *Assessment of Mekong Fisheries – Fish Migrations and Spawning and Impact of Water Management*, MRCS/DANIDA, Vientiane, Laos

MRC (2003) *State of the Basin Report: 2003, Executive Summary*, MRC, Phnom Penh

MRC (2007) *Annual Report 2007*, MRC, Vientiane, Laos

MRCS/WUP-FIN (2007) *Final Report – Part 2: Research Findings and Recommendations, WUP-FIN Phase 2 – Hydrological, Environmental and Socio-Economic Modelling Tools for the Lower Mekong Basin Impact Assessment*, Mekong River Commission and Finnish Environment Institute Consultancy Consortium, Vientiane, Laos, www.eia.fi/wup-fin/wup-fin2/publications.htm

Pantulu, V. R. (1966) 'Study of fisheries aspects associated with the development of the Lower Mekong Basin', *Indian Journal of Power and River Valley Development, The Mekong Project Number*, vol XVI, pp65–67

Pauly, D. (1990) 'On Malthusian over-fishing: Naga', *ICLARM Quarterly*, vol 13, no 1, pp3–4

Payne, I. (2000) 'The changing role of fisheries in development policy', *ODI Natural Resource Perspectives*, no 59, June 2000

Pushpalatha, K. P. C. (2001) 'Community based freshwater fish culture in Sri Lanka', in De Silva, S. S. (ed) *Reservoir and Culture Based Fisheries: Biology and Management*, ACIAR Proceedings no 98, ACIAR, Australia, pp266–273

Resurreccion, B. P. (2006) 'Rules, roles and rights: Gender, participation and community fisheries management in Cambodia's Tonle Sap region', *International Journal of Water Resources Development*, vol 22, no 3, pp433–447

Roberts, T. R. (1993) 'Just another dammed river? Negative impacts of Pak Mun Dam on fishes of the Mekong Basin', *Natural History Bulletin of the Siam Society*, vol 41, pp105–133

Roe, E. (1991) 'Development narratives, or making the best of blueprint development', *World Development*, vol 19, no 4, pp287–300

Roe, E. (1995) 'Except Africa: Postscript to a special section on development narratives', *World Development*, vol 23, no 6, pp1065–1069

Schaaf, C. H. (1966) 'The Master Plan', *Indian Journal of Power and River Valley Development, Mekong Project Number*, vol XVI, pp5–8

Shoemaker B., Baird, I. G. and Baird, M. (2001) *The People and Their River: A Survey of River-Based Livelihoods in the Xe Bang Fai River Basin, in Central Lao PDR*, Canada Fund for Local Initiatives, Vientiane, Laos

Shore, C. and Wright, S. (eds) (1997) *Anthropology of Policy: Critical Perspectives on Governance and Power*, Routledge, London

Smith, L. E. D., Khoa, N. and Lorenzen, K. (2005) 'Livelihood functions of inland fisheries: Policy implications in developing countries', *Water Policy*, vol 7, pp359–383

Sneddon, C. and Fox, C. (2006) 'Rethinking transboundary waters: A critical hydropolitics of the Mekong Basin', *Political Geography*, vol 25, pp181–202

Sneddon, C. and Fox, C. (2007) 'Power, development and institutional change: Participatory governance in the Lower Mekong Basin', *World Development*, vol 35, no 12, pp2161–2181

Sretthachau, C. and Deetes, P. (eds) (2004) *The Return of Fish River Ecology and Local Livelihoods of the Mun River: A Thai Baan (Villagers') Research*, SEARIN, Chiang Mai, Thailand, November

STREAM (2001) *Poverty and Aquatic Resources in Vietnam: An Assessment of the Role and Potential of Aquatic Resource Management in Poor People's Livelihoods*, STREAM, Bangkok, Thailand

Sverdrup-Jensen, S. (2002) *Fisheries in the Lower Mekong Basin: Status and Perspectives*, MRC Technical Paper no 6, Mekong River Commission, Phnom Penh, Cambodia

TERRA (1993) 'Editorial: Mekong fish, Mekong culture', *Watershed*, vol 4, no 3, pp2–4

Thorpe, A., Reid, C., van Anrooy, R. and Brugere, C. (2005) 'When fisheries influence national policy-making: An analysis of the national development strategies of major fish-producing nations in the developing world', *Marine Policies*, vol 29, pp211–222

Thouk, N. and Sina, L. (1997) *Review of the Fisheries and Aquaculture Sector in Cambodia*, DOF, Phnom Penh, Cambodia

Thuon, T. (2004) *Making Space and Access in Fisheries Resource Management for Local Communities in Stung Treng Province, Cambodia*, RCSD's International Conference on the Politics of the Commons, 11–14 July 2003, Chiang Mai, Regional Centre for Social Science and Sustainable Development (RCSD), Faculty of Social Sciences, Chiang Mai University, Thailand

Tubb, J. A. (1966) 'A consideration of the fisheries problem of the Lower Mekong Basin', *Indian Journal of Power and River Valley Development*, Mekong Project Number, vol XVI, pp62–64

University of Michigan (1976) *Fisheries and Integrated Mekong River Basin Development*, Terminal Report of the Mekong Basin-Wide Fishery Studies, Executive Volume, Committee for Coordination of Investigations of the Lower Mekong Basin, University of Michigan, Michigan

van Zalinge, N., Degen, P., Chumnarn, P., Nuov, S., Jensen, J., Nguyen, V. H. and Choulamany, X. (2004) 'The Mekong River system', in Welcomme, R. L. and Petr, T. (eds) *Proceedings of the Second International Symposium on the Management of Large Rivers for Fisheries*, vol 1, FAO, Bangkok, pp333–355

Verweij, M., Thompson, M. and Engel, C. (2006) 'Clumsy conclusions: How to do policy and research in a complex world', in Verweij, M. and Thompson, M. (eds) *Clumsy Solutions for a Complex World*, Palgrave McMillan, Basingstoke, pp241–249

Vientiane Times (2008a) 'Aquaculture set for expansion', *Vientiane Times*, 31 January

Vientiane Times (2008b) 'Xekong 4 dam aims to alleviate poverty', *Vientiane Times*, 2 February

Warren, T. J. (2000) *Indigenous Mekong Fish Species with Potential for Aquaculture, Stocking or Translocation Management of Reservoir Fisheries in the Mekong Basin*, Component Report no 1, MRC, Vientiane, Laos

WCD (World Commission on Dams) (2000) *Dams and Development: A New Framework for Decision-Making*, Report of the World Commission on Dams, Earthscan, London

Welcomme, R. L. and Vidthayanon, C. (2003) *The Impacts of Introductions and Stocking of Exotic Species in the Mekong Basin and Policies for Their Control*, MRC Technical Paper no 9, May, Mekong River Commission, Phnom Penh

Wilson, J. A., Acheson, J. M., Metcalfe, M. and Kleban, P. (1994) 'Chaos, complexity and community management of fisheries', *Marine Policy*, vol 18, no 4, pp291–305

Wong, C. M., Williams, C. E., Pittock, J., Collier, U. and Schelle, P. (2007) *World's Top 10 Rivers at Risk*, WWF International, Gland, Switzerland

World Bank (2004) *Saving Fish and Fisheries: Toward Sustainable and Equitable Governance of the Global Fishing Sector*, World Bank, Washington DC

World Bank (2007) *Changing the Face of the Waters: The Promise and Challenge of Sustainable Aquaculture*, World Bank, Washington DC

World Bank/ADB (Asian Development Bank) (2006) *Future Directions for Water Resources Management in the Mekong River Basin*, Mekong Water Resources Assistance Strategy, World Bank/ADB, Bangkok, June

13

The Anti-Politics of Mekong Knowledge Production

Mira Käkönen and Philip Hirsch

Introduction

Despite the enormous efforts of planning agencies and significant amounts of time and money spent on feasibility studies for water resource infrastructure, the Mekong River has remained one of the world's least developed of the world's major rivers and is thus now perceived by national decision-makers and many international donor organizations as having 'underused potential'. This is at a time when there is a major concern that most large rivers have been *over*developed. There are clear signs that countries in the Mekong River Basin are striving more aggressively after 'modern development' in forms of large-scale dams, irrigation and hydraulic controlling structures. But at the same time, many in the region are aware of the failures of the modernist projects elsewhere, and in those parts of the region where dams and irrigation have been installed, the uncritical belief in human mastery over the forces of nature has been seriously questioned. There are also evolving domains of knowledge that actively contest the scientific and expert knowledge usually used for policy and development plans of the basin. It is thus important to look at how these modernist plans and aspirations, epitomized by large-scale dams and diversions, are being justified and legitimized in relation to competing knowledge domains.

Knowledge produced by experts in the form of models, impact assessments and scenarios dealing with risk play a crucial role in the legitimization process. This chapter provides a perspective on the production of knowledge around major development issues in the Mekong River Basin. At a time when large-scale development plans are being justified and rationalized, it is relevant to ask in what ways scientific knowledge is being deployed, and whose visions of future waterscapes the dominant models and scenarios favour. Important questions also relate to the openness of expert knowledge: is it really open to the public, how is it framed, and what are the possibilities and manifestations of civil contestations and public participation in the current societies of the Mekong region? The focus of this chapter is on the Mekong River Commission (MRC), which is one among the main knowledge brokers in the region. The MRC is mainly foreign funded and, at least as perceived by those who fund it, the MRC has a central role in the water resources management of the Mekong Basin (on the MRC, see also Chapter 14).

Much discussion on the politics of knowledge in the Mekong River Basin and elsewhere draws lines of tension between expert knowledge that employs scientistic discourses, on the one hand, and, on the other, local knowledge of farmers, fishers and other ordinary people who have a living understanding of, and dependence upon, the myriad natural resources of the land, forests and water bodies of the basin. In Thailand, and latterly in other countries of the Mekong, initiatives such as Tai Baan (e.g. Foran, 2006; Sretthachau, 2007) have promoted registers associated with quite different development preferences of the people on behalf of whom such knowledge is asserted. So successful has the new knowledge production been that relatively mainstream organizations, such as the International Union for Conservation of Nature (IUCN) have accepted it as a valid methodology and supported Tai Baan research in the region.

Ironically, however, the participatory turn in mainstream institutions such as the MRC has another side to it. It is at constant risk of being far from a counterbalance to the expert knowledge. Especially as abstracted from its societal context in an international agency such as the MRC, participation tends to mirror a type of development-driven participation that can contribute to the depoliticization of knowledge in support of a particular governance agenda.

This chapter addresses the process and problems of depoliticization of knowledge, or what we call the 'anti-politics of knowledge production'.[1] We focus on two seemingly contrasting aspects of the MRC's work. In the first part the focus is on the MRC's hydrological models, which form the backbone of the MRC's knowledge production and the main source for the MRC's estimations of development impacts. The results of the MRC's hydrological models have also been the main source of justification for future development plans by the Asian Development Bank (ADB) and the World Bank. This domain of expertise is still relatively closed and exclusive. We discuss the knowledge produced by models not only in terms of accountability and validity, but also in relation to the social shaping of interpretations.

The second part of the chapter covers the MRC's response to contestations of the closed nature of its expertise and associated shortcomings of the assessment work. One component of the demands for more participation in knowledge production is that there should be more local knowledge and expertise included in the assessment processes, as well as more diversified and representative interests guiding the knowledge production. The embryonic participatory turn of the MRC is also part of the more general shift that the field of development has gone through. Because the MRC depends heavily upon international donors, its public discourse needs to be in line with the current international discourses, where the new orthodoxy includes participatory decision-making. Hence, for the MRC, it is now not enough to provide only scientific bases for policy-making concerning new development plans. The MRC also has to prove that its expertise is being opened up and democratized, or at least shared more widely. The chapter thus seeks to discuss the rationale of this new participatory turn of the MRC, its promises and limitations in relation to knowledge production, and the role of mainstreamed participation, which tends to *reinforce* rather than defuse the depoliticization of knowledge.

Theoretical dimensions

Knowledge production related to the probable impacts of new development projects, mainly in the form of large mainstream dams, is closely related to the estimations of risk and to assumptions on risk management. Beck's (1992) theory of risk society and his related modernization narrative offer one way of seeing the Mekong region's current situation. Through Beck's lenses, the region appears to be a hybrid of premodern, modern and reflexive modernity. As the Mekong still flows relatively freely, the unpredictable disasters that it brings in some years in the form of exceptionally strong floods or droughts are, in many parts of the basin, often still natural and thus not risks characteristic of modern society. Rather, they are hazards common to premodern societies.[2]

Since the establishment of the Mekong Committee in 1957, there has been an enormous amount of planning effort and expenditure to materialize the 'underused potential' of the river and to engineer out the floods and droughts associated with its monsoon-derived hydrology. For the dam and irrigation plans, whose developers included former Tennessee Valley Authority engineers, the common denominator was the will to control and manipulate nature in order to trigger the economic growth of the region and to fight back the spreading communism (see also Chapter 1). These could be seen as failed attempts to shift the area to the first stage of the era of modernization in which wealth production is driven by a will to control and manipulate nature under the imperatives of economic growth.

After the establishment of a new Western donor-driven MRC in 1995, the emphasis shifted from ambitious plans for a series of large-scale dams to knowledge production on the impacts and risk assessments of proposed development interventions. New domains of knowledge are also evolving that actively contest the scientific and expert knowledge claims used for policy and development plans of the basin. This situation comes close to Beck's key concept for the second stage of modernization: the risk society or 'reflexive modernization' at the heart of which are issues of ecological crisis and contestations between citizen initiatives and formal authority. The key issue, thus, is how the status of knowledge begins to be contested, and how the risks that used to be a matter for experts and scientific legitimation become subject to public debate so that 'modern' scientific knowledge loses its monopoly over truth. The risks that were once presented as being calculable now start to seem incalculable. But still the aspirations of striving after modern development by the governments of the Lower Mekong Basin have not ceased, and the region is now closer than ever to the realization of several large-scale dams on the mainstream Mekong River. The hybrid nature and, even, dissonance of the basin and the MRC in relation to Beck's modernization narrative thus manifests itself in that the underlying current of policy-making is still pursuing the modernization of the first stage, while the knowledge production in the MRC simultaneously needs to face the expectations of the risk society.

Beck's concept of risk is, however, not very beneficial when analysing more profoundly the current knowledge production in which the MRC takes, most importantly, the form of basin flows analysis. For Dean (1999, p177), 'risk is a way – or rather, a set of different ways – of ordering reality, of rendering it into calculable form. It is a way of representing events in a certain form so they might be made governable in particular ways, with particular techniques and for particular goals.'[3] The analysis of how the concept of risk is used in the knowledge production and how it is part of political technologies of the MRC would deserve a deeper and more detailed discussion than is possible in this chapter. But one of the important aspects of the political technologies, also related to risk, is 'technical rendering', a concept developed by Nicolas Rose (1999). This is discussed in more detail in relation to the MRC's hydrological models.

According to Dean (1999), increased numbers of assessments and participatory processes do not necessarily mean that the decisions made are better informed, but that the central target and objective of the governments becomes the reform of the performance of the existing governmental institutions and techniques. This comes hand in hand with the processes where governments are challenged by their capability to control the risks that they produce (Dean, 1999). The analysis of reforms in knowledge production and policy-making in MRC-like organizations benefits from this perspective of seeing them as ways of securing the mechanisms of government.

A central idea of Dean (1999) and Rose (1999) is that the relations of power and truth are inseparably interwoven. In this chapter, the point is not to search only for the obvious '*realpolitik*' dimension in knowledge production.[4] To some extent, there have, indeed, been situations where undesirable assessment results are sidelined and participation often appears to be just a necessary condition for government and lender approval. But there are also less visible yet pervasive power relations that are not simply reducible to '*realpolitik*' or to 'conspiracies'. The assessment tools, like hydrological models, are already shaped by values and power relations. And similarly, even the recent participatory approaches can be perceived as new forms of depoliticizing power relations (see also Brosius, 1999).

Hydrological models at the heart of the Mekong River Commission's (MRC's) knowledge production

The core of knowledge production within the MRC Secretariat is still dominated to a great extent by technical and scientific expertise. Data-gathering is principally limited to data seen as relevant to the assessment and regulation of hydrological impacts of planned development projects. The backbone for all the assessments of the MRC consists of the Decision Support Framework (DSF) that has been developed over several years and is still being further improved. The DSF forms the foundation for the development scenario assessments of the MRC *Basin Development Plan* (BDP) and of the MRC's Water Utilization Programme (WUP), which are supposed to help implement key elements of the 1995 Mekong Agreement and inform and shape negotiations that address water-sharing issues between the member states.

The inherent simplifying aspect of models and other assessment tools, and the richness and diversity of living nature always create debates on how to interpret ecological and social complexities adequately. Interpretations of what is adequate and what is relevant to policy also vary quite differently in different knowledge domains. The examples below shed light on the MRC's assessment work and on the discussions that they have evoked. To date, the MRC's models have brought into focus rather limited aspects of the basin's ecological and social dynamics, which in turn results in overestimation of macro-level economic returns and underestimations of the risks that the tributary and mainstream dams, water diversions and other interventions are likely to bring about.

MRC's models: High investments, thin and controversial outputs

The original terms of reference of the MRC's Decision Support Framework were very ambitious, and the DSF was planned to consist of tools that would enable comprehensive basin-wide hydrological, environmental and socio-economic impact assessments in line with the principles of integrated water resources management (IWRM), which acknowledge the complexity of relations between water, environment and livelihoods. However, so far the DSF appears to remain a rather narrow platform, consisting mainly of hydrological components that, in turn, consist of models which simulate two-dimensionally the flow regimes and the main hydrological aspects of the river basin (Sarkkula et al, 2007). The hydrologists behind the models themselves have stated that the 'output of the models is quite narrowly hydrological – water utilized for irrigation and power generation; river flow and stage at key locations; volumes; inundated areas, depths and duration of inundation; and salinity levels' (World Bank, 2004, p1). But they also claim that 'these parameters can, in turn, provide insights into possible impacts on fisheries, flood management, saline intrusion, navigation and the environment' (World Bank, 2004). Because the models do not allow assessments on parameters such as water quality, and sediment transport in the river and the floodplain, the models' ability to give relevant insights on impacts upon the environment and fisheries has been questioned (Sarkkula et al, 2007).

The DSF has also received criticism on the basis that it has required high investments (the hydrological component has required approximately US$4.9 million), while outputs have been very limited. In particular, the outputs to the public domain have been almost non-existent. Lack of transparency with the findings is thus one of the core aspects questioning the legitimacy of MRC models (e.g. Affeltranger, 2008). When results have been published, they have been presented without giving information on the assumptions upon which they are based. One of the few outputs of the DSF process to the public domain has been the report *Modelled Observations on Development Scenarios in the Lower Mekong Basin* (World Bank, 2004), which emerged from a World Bank consultancy that drew on the DSF, but was not actually even an MRC output. The report was based on six scenarios:

1 baseline (representing the situation in 2000);
2 China dams (considers the Manwan and Dachaoshan dams) operating at the time of the report and the two largest proposed dams (Xiaowan and Nuozhadu dams);
3 low development (baseline + increase of water usage in line with the estimated population growth to 2020 + dams in China and most likely dams in Laos);
4 embankments (low development + increased number of built structures in Cambodian floodplains);
5 agriculture (low development + substantial increases in irrigation and inter- and intra-basin water transfers);
6 high development (includes all the previous ones + several tributary dams and a mainstream dam in Cambodia).

The purpose of this report was to inform the Mekong Water Resources Assistance Strategy then being shaped by the World Bank (which was later joined by ADB). The results of the modelling exercises were reported to show that there were few major risks related to the different scenarios, including the high development scenario (World Bank, 2004). The vulnerability of fisheries and other elements of the Mekong's ecology were acknowledged; but the report gave mixed messages. For example, although the tested scenarios were reported not to reveal significant negative impacts upon the fisheries except 'a small decline in fish feeding opportunities, the most pronounced reduction occurring in years of low flow' (World Bank, 2004),[5] the report also stated that:

> *Nevertheless, any development which directly impedes fish migration in the mid and lower reaches of the river will have significant negative impacts on fish production. Mainstream dams or weirs in the mid and lower Mekong are therefore most unlikely to be part of any balanced development scenario that complies with the objectives of the [Mekong] Agreement.* (World Bank, 2004)

The report was initially available to the public but was later (in 2006) withdrawn by the MRC Secretariat (MRCS) and made internal and inaccessible to the public. The reasons for this were not made public either. Interviewed experts and consultants of the MRC Secretariat have commented that after this, it was also difficult for them to make direct references to the report, and one could not talk openly about the different scenarios used. Baseline, low-development and high-development scenarios were within the MRCS renamed as flow regimes 1, 2 and 3. At the same time, the content and the assumptions of the scenarios were made inaccessible. The use of the modelled scenarios thus became much more technical, opaque and less informative for the public, and, hence, less open for public debate. Although the entire report became difficult to access, an excerpted hydrograph was widely used in World Bank, ADB and MRCS presentations to suggest that there was little change in the shape of the hydrograph under even the high-development scenario. This excerpt misrepresented the more nuanced messages that emerged from closer reading of the 2004 report.

A pretext sometimes used to restrict public release of such analyses is that they are developed by the consultants and await endorsement of the countries or of the MRC Secretariat. Yet, often there are also political reasons at stake: even though the report in question did not bring up very serious impacts of 'high development', it did, however, ask for a great degree of caution with the mainstream dams. This kind of statement would be perceived by at least some governments as a threat to their current development aspirations. A related explanation for the pressure to render the scenarios more technical might simply be that the countries do not want to discuss openly the different development plans. The limited information on the assumptions upon which the modelling exercises were based of course leaves little space for independent evaluation of the validity of the modelling results and their interpretations.

A second major occasion where results from MRC models were brought into the public domain was a working paper released by the World Bank and the ADB related to their Mekong Water Resources Assistance Strategy (MWRAS) for 2005 to 2010 (World Bank and ADB, 2006). The report strongly supports new large-scale water infrastructure projects in the Mekong Basin and states that the development has so far been too cautious. The MRC's models play a central justification role in the report. The assumptions behind the modelling results were not discussed in the report; but the modelling results were interpreted to show that there are no major risks related to new large-scale development plans:

> *The bottom line message of this Mekong Water Resources Assistance Strategy is that the analytical work on development scenarios has, for the first time, provided evidence that there remains considerable potential for development of the Mekong water resources.* (World Bank and ADB, 2006, p4)

And that:

> *The development scenarios modelling exercise demonstrated that the Mekong river system has significant tolerance for development, including of hydropower and water diversion for irrigation.* (World Bank and ADB, 2006, p31)

Such statements reflect the way in which an exercise whose design at best provides a hydrological building block to a basin-wide assessment of the complex linkages between hydrology, ecology and livelihood becomes a firm policy statement on the 'potential' and 'tolerance' of the river to accommodate large dams.

The report received criticism from several civil society groups, including, for example, the International Rivers Network and Towards Ecological Recovery and Regional Alliance (TERRA) (IRN, 2006; Middleton and Lee, 2007). It also received academic comment, including from the Australian Mekong Resource Centre (AMRC, 2007). The critiques included remarks on the use of the models. The way in which the report referred to the models was claimed to be over-simplifying and the capability of the models to assess environmental impacts was questioned because the models only simulate the water flow and do not address the ecology of the river. Central arguments were that bold claims on environmental and socio-economic impacts could be backed up with only narrowly hydrological results from MRC models (IRN, 2006; IUCN et al, 2006; AMRC, 2007), and that the macroscopic perspective could not address the likely localized negative impacts arising from infrastructure development (Middleton and Lee, 2007). Even the reliability of the models was questioned because the assumptions built into the model and the assessments of the robustness of the model have been unavailable to the public (IUCN et al, 2006). But even if the models were developed to a more reliable and credible standard, there is still always scope for social and political shaping of interpretation and presentation of the results.

The models have not only received criticism from civil society and university-based groups. Some of the interviewed experts in the MRC stated, as well, that the model base is not yet diverse enough, nor have there been adequate comparisons with other scientific models. Some hydrologists have stated that there are still major challenges for the MRC to build a scientifically validated and credible model platform (Adamson, 2007; Sarkkula et al, 2007). Some interviewed experts who have worked in the MRC criticized the consultants of the MRC's Water Utilization Project for working with too strong a private consultant mentality – keeping in mind the policy preferences of the institutions funding their consultancy – which was seen to downplay the limitations of the models and associated risks and uncertainties (Shackley and Wynne, 1996). Yet, studies in science and technology have found that while modellers themselves often do understand the uncertainties and limits of their models, the more distant users, such as experts closer to policy-making, become enchanted by the technologically sophisticated models and the unqualified modelling results, especially when these are self-serving (MacKenzie, 1990; Shackley and Wynne, 1996).[6]

Models such as the recently developed and more comprehensive WUP-FIN model (see Chapter 9) tend to be received within the leadership echelons of the MRC and by policy-makers with more scepticism and emphasis on the uncertainties than the previous model exercises because the results of these models have, for example, raised questions on the vulnerability of the Tonle Sap ecosystem and its aquatic productivity. As Shackley and Wynne (1996) have stated, the appreciation of uncertainty increases when there is a motivation to critically explore the basis of the knowledge claims. Thus, there will always be politics over uncertainties when modelling results are discussed in terms of policy implications.

The interwoven nature of facts and values: Models and their representation of the Mekong Basin

The social and political shaping of the interpretations of modelling results illustrates the blurred boundaries between science and politics. An even more profound issue is that definition of the scientific questions to be asked, and thus the relevant group of experts to answer them, is always also a value-laden and political act (e.g. Demeritt, 2006). Important questions here thus relate to why, despite the original ambitious plans of creating comprehensive and integrated assessments and the huge sums devoted to them, the *de facto* developed tools of knowledge production in the MRC have been reduced to relatively simple hydrodynamic models that do not allow ecosystem impact scenarios and, even less, impact assessments on natural resource-dependent livelihoods. And even more importantly: why in the first place do the models play such a crucial role in the MRC's knowledge production? A justification often given for the importance of the models is that there are such wide gaps in environmental information on the Mekong region that models are the only way to go forward with the assessment work. But the underplaying of the complexities of the ecology and livelihood interlinkages inherent in this sort of modelling also conveniently serves the economical rationalities and power relations shaping the MRC's approach to water issues.

The MRC's Decision Support Framework is, most importantly, built to support the objectives of the 1995 MRC Agreement. The relevant principles of the Mekong Agreement here include the 'reasonable and equitable utilization' (Article 5) of water by the riparian states and the maintenance of minimum flows on the mainstream (Article 6). These principles also guide the way in which the river is envisioned by the MRC. They also guide the knowledge production, as can be interpreted from the modelling report (World Bank, 2004, pv), which states that the rules required by the agreement 'define the key, monitorable indicators – in terms of times, locations, flow rates, levels, quantities, water quality or other variables – that are sufficient to define each country's opportunities and responsibilities in the Mekong Basin'.

Even though the MRC Agreement is often cited as a model for international river basins and as a promise of sustainable development in the developing world (e.g. Sonnenfeld and Mol, 2002), it has also been criticized as a manifestation of a state logic that violates the complex nature of the environment–livelihood linkages of large river basins (Fox and Sneddon, 2004; Sneddon and Fox, 2006). The principle of reasonable and equitable utilization is an important base for cooperation in international basins; but Fox and Sneddon (2004) argue that the interpretations of this principle should also be looked at critically. Because the main concern is the equitable allocation of water between the member states, the agreement establishes a vision of Mekong primarily as a watercourse, and not as a basin, with its complex socio-ecological dynamics. Following James Scott's (1998) idea on state simplifications, they interpret that this makes the river legible to state-centric reasoning. As a watercourse, the river can be reduced to its parts, divided between states, and rationally managed through the application of universal legal principles. Because of the allocation paradigm, the focus is mostly on the quantities of water and, especially, on the maintenance of the minimum flows in the dry season.[7] This leads to the neglect of the flood pulse-driven character of the Mekong's ecosystems for which the tempering of the peaks of flood and drought are damaging. The flood pulse system has been recognized by ecologists as crucial to aquatic productivity, and it sustains the rich fisheries of the basin and the livelihoods of millions of people who depend upon them (Lamberts, 2007; Sarkkula et al, 2007; Lamberts and Koponen, 2008; see also Chapter 9 in this volume).

Li (2006, 2007) has studied knowledge production in the development field, inspired by Nicolas Rose's concept of technical rendering. Technical rendering refers to processes in which the arena of intervention needs first to be represented 'as an intelligible field with specifiable limits and particular characteristics ... whose component parts are linked together in some more or less systematic manner by forces, attractions and coexistences' (Rose, 1999, p33, cited in Li, 2005). In his study of the 'anti-politics' of development, Ferguson (1994) has similarly described how knowledge production in the development field is used for planning purposes and the generated data is often *sui generis* because it identifies only those problems for which a technical remedy within the competence of the planners can be supplied. The knowledge production of the MRC seems to resonate with the rationale of technical rendering, which is an inherently depoliticizing process in its reduction of problems and their solutions as technical challenges and fixes, respectively. Through the use of the models, the river becomes dis-embedded from the ecosystem, livelihoods and meanings, and this facilitates the exercises of linking water back to society through economical reasoning and simplistic causal chains. The models generate powerful visualizations of the basin as a manageable system ripe for development interventions, which enables technocratic calculations and planned water allocations.

An example of this is a public statement by Jeremy Bird, the chief executive officer (CEO) of the MRC, that 'in the medium turn, we're going to see a situation where as a result of construction of dams upstream in China, there will be some significant increases in dry season flows in the Mekong which actually then might facilitate Thailand taking water from the river because then there'll be more water available during the dry season' (Radio Australia, 2008). The results of the modelling have thus produced, in MRC explanations, an image where hydropower dams are seen in a positive light because 'the excess water' they allow in the dry season can be effectively utilized for water diversions and irrigation in another place.

Knowledge production, including the MRC's models, is implicitly tied to the redistribution of rights to use the environment. The knowledge production and the production of political order (in terms of management and policies) should thus be seen as mutually constructing and reinforcing one another (Wynne, 1996, 2002; Forsyth and Walker, 2008). Like other technologies and innovations, however, models emerge in 'a garden of forking paths' (Williams and Edge, 1996), and the use or interpretation of the models is not predetermined. But the ways in which specific models developed by private consultants for the MRC have been used so far are clearly shaped predominantly by developmental values. This is not to say that different routes for their use are precluded. The questions raised by civil society on the models show that even though the model use was aimed at reaching closure over the discussion on the severity of impacts, this did not necessarily happen. And models are becoming more important as part of competing knowledge domains (e.g. advocacy coalitions, networks of action research or alternative discourses) shaped by different development aspirations.

An even more profound question, however, is whether the debate is limited to scientific facts. It is important to note that technical rendering refers, importantly, to a scientized rationality, which hides politics and depoliticizes development decisions. Because science is given such an instrumental role in legitimating policy, competing knowledge domains easily shift to questioning the science and presented facts, rather than questioning the reasons for policies or the specific ways in which the science is being framed and its results articulated. Critiques that restrict themselves to discussing the scientific validity of the knowledge production are thus problematic because they include a positivist expectation that political consensus about development plans will follow from scientific consensus of the impacts. If this paradigm is not opened, the politics behind seemingly science-based decisions are not really brought into the debate. Herein also lies the somewhat paradoxical nature of the new participatory approaches in knowledge production, which are discussed in the next section.

THE MRC, THE PARTICIPATORY TURN IN DEVELOPMENT AND NEW OPENINGS IN KNOWLEDGE PRODUCTION

The tools and technologies of knowledge production and the ways in which they have been used as presented above reflect the still relatively closed nature of the expertise within the MRC Secretariat. But the MRC has also faced pressure to open up the expertise and to enhance its participatory processes. Several scholars, Ulrich Beck and Mitchell Dean included, have emphasized the ways in which risks inherent in modernist control schemes have created a need to include the participation of groups such as civil society organizations and local communities who have previously been excluded by a scientific-technological rationality of risk assessment.

The current development orthodoxy recognizes public participation as necessary for achieving sustainable and socially just development. If earlier schemes were often doomed to fail because they were based on overly simplified representations, science now has to respond to the critique that it is missing contextualized and situated knowledge and practices. Participation has become something of a mantra even in the field of water management, which has for long been the exclusive preserve of technical experts.

In the Mekong Basin, voices have been raised with increasing intensity over the past decade about the undemocratic and unaccountable nature of the basin's water resources management. The MRC has typically responded to its critics by stating that as it is an intergovernmental organization, it principally serves its member states in ways that it is requested to. But recently, particularly in 2008, the MRC has begun to demonstrate a participatory turn. A first Public Participation Strategy was circulated in 2003 (MRC Secretariat, 2003) and a booklet about *Public Participation in the Lower Mekong Basin* was published in 2005 (MRC, 2005), but there has not been much follow-up action. The MRC's programme on fisheries has sought ways to create more space for local knowledge and spaces of inclusive decision-making at the community level. The MRC has also (somewhat reluctantly) participated in events such as Exploring Water Futures Together: Mekong Region Waters Dialogue (IUCN et al, 2006, 2007). More recently, two of its programmes in 2008 – Basin Development Planning (BDP) and Hydropower Programme – have hosted more extensive stakeholder consultations. These are not perfect; but for the MRC they are already significant steps on the road not travelled of more participatory engagement with non-state actors.

Many critical remarks about previous efforts have been raised. The participation strategy, for example, gives all power to the riparian states to decide who they consider relevant stakeholders and who they do not. This ensures that 'the invited spaces of participation are likely to reflect the dominant development aims of the riparian states' (Sneddon and Fox, 2007, p2175). Participation has been enhanced mostly within the MRC's own structures and member governments, and has not been extended to meaningful engagement with critical NGOs or local communities (Sneddon and Fox, 2006, 2007). It remains to be seen to what extent the expertise will be truly open to public review and whether there will be spaces where the technocratic risk assessment rationalities dominating the knowledge production can be meaningfully challenged. To take a step forward from 'tokenism', the debate needs to substantially expand beyond the received framings by experts on hydrology or policy-making.

The most relevant steps in relation to more participatory and inclusive knowledge production have been taken by the MRC Fisheries Programme (MRCFP). The methods developed in the Fisheries Programme hold the potential to produce knowledge where the complexities of the relations between water, ecology and livelihoods are not overlooked. In the programme's studies, local fishers have participated in identifying and tracking fish species and in evaluating the importance of the fisheries for local livelihoods. The findings of the studies have enhanced the knowledge base of the fish migration patterns and fish production, and also stressed the importance of fisheries for basin residents (Sneddon and Fox, 2007). There are, however, also signs that even within the Fisheries Programme, the appreciation of local knowledge still remains somewhat superficial. For example, the work of Thai Baan Research (see Box 13.1) has not received real acceptance from the MRC. Even though some space have been given, for example, in the MRC's stakeholder consultation for presentations on Thai Baan or Sala Phoum (the Cambodian version of Thai Baan) research findings, the MRC's own reports have not built on or even referenced the findings of Thai Baan Research. More fundamentally, local knowledge and participation of fishers in producing fisheries knowledge enrich a central knowledge base rather than, as is the principle of Thai Baan, knowledge from which fishers themselves manage their own livelihood interests. The knowledge base of the MRCFP, produced with participatory methods, has certainly been important for raising awareness of the importance of fisheries in a basin-wide framework; but it remains the experts and the technocrats who own and present this 'mined' knowledge.

The studies that acknowledge the different meanings which local communities attach to the fisheries hold the potential to produce knowledge that is less amenable to state-centric reasoning and technocratic trade-off paradigms. This, in turn, could enable the inclusion of alternative visions on the basin's development to the current debates. Ironically, the most crucial issue here is that while the MRC's Fisheries Programme has aimed to develop more appreciation for local knowledge, the whole programme has been increasingly sidelined from the MRC's core activities.

Box 13.1 Localized knowledge production through Thai Baan Research

Thai Baan Research – research undertaken by villagers – has recently emerged as a counter-hegemonic approach, aiming to reveal local knowledge about the environment and how villagers interact with it. It reveals their practical understanding of the complexity and dynamics of natural resources, the way in which resources have been used, and the moral economy of those who depend upon them for their livelihoods.

Thai Baan Research was established in 2000 when the Thai government agreed to open the Pak Mun Dam sluice gates to evaluate the social and environmental impacts from the dam's operation. In this case, for the Pak Mun villagers who tried many ways to voice their concern and register their grievances about the environmental, economic and social impacts of dam construction, Thai Baan Research was identified as a new way of influencing the contested dam project (see Chapter 3 in this volume). Many academic institutes were assigned by the Thai government to conduct various types of research; but the reports did not reflect the social and ecological realities as seen by affected villagers. Academics from Chiang Mai University, the Southeast Asia Rivers Network (a regional non-governmental organization) and villagers affected by the Pak Mun Dam developed the Thai Baan Research approach in order to collect data on issues such as local knowledge of fish, traditional fishing gear, natural plants and herbs, ecosystems and activities, which returned after the opening of the dam gates. The research conducted was published and submitted to the Thai government to coincide with findings submitted by academic institutes. This type of research was meaningful for the villagers because they were able to take control over the process and 'write' their own story on how they perceive and interact with their environment and how to live in harmony with it.

The findings of the Thai Baan Research at Pak Mun have gained acceptance by academics, the media and civil society groups. The methodology utilized has been adopted and replicated throughout other areas in Thailand, such as Rasi Salai, along the Upper Mekong in Chiang Khong, the Salween River along the Thai–Burmese border and the Songkhram River Basin in the northeast (see Chapter 7). This expansion has allowed for an informal network of researchers to develop, share information and learn from each other. In 2004, the Thai Baan approach was expanded to the Ramsar sites in the Lower Mekong in Cambodia and Vietnam. In Cambodia, Thai Baan Research (which is known as 'Sala Phoum') has been carried out by ethnic groups in Stung Treng, the NGO Culture and Environmental Preservation Association (CEPA), Health Unlimited (HU) and the World Conservation Union (IUCN)-led Mekong Wetlands Biodiversity Programme (MWBP). In Vietnam, Thai Baan Research is conducted by An Giang University, Tram Chim National Park, Lang Sen Preserve Forest, Care International and the MWBP.

The Thai Baan approach also provides a basis for more informed, balanced negotiations between local stakeholders and government. By working with local development institutions and gaining the support of provincial and national government agencies the Thai Baan is complementary to decentralization initiatives and national policy towards more integrated water resources planning and the establishment of river basin organizations.

Source: adapted from Sretthachau (2006)

Knowledge production and the entry of participation discourse in the riparian states

The MRC is not the only relevant level for observations on the changes of water-related knowledge production in the Lower Mekong Basin. The MRC Secretariat often reminds its critics that it is an intergovernmental organization. The MRC's ability to foster and implement participatory processes is, therefore, closely bound to the political cultures of its member governments. Here we briefly discuss the different country-level situations through selected illustrative examples.

The degrees of political freedom and spaces for civil society organizations (CSOs) differ significantly in the riparian countries, Thailand having the most active and effective advocacy groups and networks, and Laos and Vietnam having very limited space for CSOs. In all four countries, there have been struggles or contestations over large-scale water development projects (e.g. controversies over Pak Mun Dam in Thailand; dams on the Sekong, Sesan and Srepok rivers in Cambodia; Son La Dam in Vietnam; and Nam Theun 2 Dam in Laos PDR). The struggles have also, in some cases, resulted in attempts to create more participatory processes (see Chapter 3). A significant push for more participation has also come from aid agencies or influential international non-governmental organizations (INGOs) (Molle, 2005).

Nevertheless, the literal and societal translation of participation has varied from one national context to another. In Thailand, participation translated as *kaan mii suan ruam* – 'having a part in joining' – took hold as a discourse during the 1980s, first among NGOs, soon to be followed by government development programmes. Its interpretation, however, varied from one actor to another (Hirsch, 1990) along lines similar to those proposed by Arnstein (1969) in her ladder of citizen participation. Yet, with the democratization of Thai society and the promulgation of a progressive constitution in 1997, the notion of participation with empowering potential is well entrenched, even if bureaucrats often continue to interpret it as a willingness to engage with state-set agendas. In Laos, in contrast, a similar discourse of participation, translated directly from Thai (*kaan mii suan huam*), has a much less nuanced interpretation and remains at the level of preparedness to devote time, energy and resources to helping meet state and party-led development goals. Neither independent local NGOs nor a culture of challenge to policy articulated by the state are features of the Lao civil society landscape. In Vietnam, participation translated as *su tham gia* maintains a collectivist connotation of willingness to join the common cause (of development); but while the political structures remain largely state based, the culture of challenge and assertion of alternative ideas is quite vibrant at a local level. In Cambodia, the official discourse of participation as *kaa chaul ruam* translates as a compound 'enter-join', imparting a sense of joining a preset and usually state-sanctioned agenda. On the other hand, Cambodia has a vibrant NGO community with alternative perspectives, but which does not necessarily rely on participation as a key part of the alternative discourse.

The promises and limitations of participation in knowledge production

Even though participation has been very partial and rhetorical, and strategy papers face great challenges to translate into practices what would meaningfully open up expertise, the participatory turn is still on the upswing and, more importantly, is praised by many actors as a remedy to all ills, if properly implemented. It is therefore important to discuss more profoundly the promises and limitations of these attempts to shift the paradigm. Important questions include: what expectations exist related to more open use of expertise? In what ways could democratization of knowledge production matter? Could democratization of the assessment processes meaningfully influence development plans? In what ways is participation likely to fall short of promises?

The promises of participation relate to at least two different dimensions of knowledge production. One relates to the quality and validity of the assessments and the other to the power relations that shape the knowledge production, particularly where enhanced inclusiveness of different perspectives holds potential of democratization.

The examples given above of the model-based impact assessments demonstrate that the knowledge production of the MRC still fails to address the risks that the current development plans present to the basin's ecosystem and livelihoods. The exclusive nature of the expertise is one part of this story. External consultants or scientists are more likely to underestimate complex and interrelated ecology and livelihood systems as they often have no experience of the region they study prior to their assignment. In the Mekong region, there remain numerous unstudied areas where there is no accumulated 'scientific knowledge'. For example, the taxonomy, distribution, abundance and movement of fish is still far from sorted out. Often fishers themselves have the best available knowledge of the fishery and thus are more reliable experts than outside consultants. More inclusive approaches could enhance the actual 'scientific' quality of the assessments. As already discussed above, the MRC has taken this on board in a limited way in its fisheries assessment programmes. But this has not then reached the core of knowledge production, such as the modelling exercises that influence decisions by national governments on dams and diversions.

In principle, participatory approaches such as multi-stakeholder platforms and dialogues could contribute to more balanced framings of the knowledge production now driven by developmentalist objectives and values that mostly represent the interests and worldviews of powerful elite groups. But if one concentrates on participation only in knowledge production, the possibilities of contestations are narrowed and they may fail in contesting the technocratic core of the approaches that still reign at the MRC; in addition, they do not challenge the paradigm of looking for science to give indisputable answers in policy-making. As stated by Szerszynski (1996, p113), this kind of approach 'simply obscures the political nature of decisions about how we should live'. If the development plans can only be contested on scientific grounds, then the opponents are left to argue over scientific facts even though they might be actually opposing the plans for other reasons (Demeritt, 2006; see also Rayner, 2003). The 'fact'-driven debates also 'free' policy-makers from the inherently political nature of the decisions. This is why participation that is narrowed only to scientific knowledge production easily becomes a part of the 'anti-politics' machine of development.

Is the participatory turn necessarily counter-hegemonic?

At a more profound level, critical scholars have even questioned the very possibility of a participatory turn in development to form a counter-hegemonic force; indeed, they have shown how the participatory discourse may become an entrenched or even 'tyrannical' part of prevailing power structures and disempowering practices (Cooke and Kothari, 2001). This discussion does not refer specifically to knowledge production, and it also goes beyond analysing the failures of participation in specific programmes or organizations. Rather, it addresses the very rationale behind the whole paradigm shift. Henkel and Stirrat (2001), for example, have referred to the current emphasis on empowerment, the marginal distrust of the state, and celebration of 'local' knowledge as the 'new participatory orthodoxy'. This new orthodoxy has received different kinds of criticism, the conservatives claiming that the state and the experts actually do know the things better than the 'locals', and 'progressives' claiming that participatory policies do not lead to participation and empowerment because they are too naive and do not sway the dominant power structures. But as Henkel and Stirrat (2001) point out, most of these critiques fail to see that the participatory approach is actually a new form of governance, and not only a counter-hegemonic process. They state that:

> *Empowerment in this sense is not just a matter of 'giving power' to formerly disempowered people. The currency in which this power is given is that of the project of modernity. In other words, the attempt to empower people through the projects envisaged and implemented by the practitioners of the new orthodoxy is always an attempt, however benevolent, to reshape the personhood of the participants. It is in this sense that we argue that 'empowerment' is tantamount to what Foucault calls subjection.* (Henkel and Stirrat, 2001, p182)

The 'subjection' refers to the productive side of power relations, which in this case means that participatory processes shape the subjectivities of the participants and the ways in which the rationalities of the more powerful actors can be contested.[8] It is thus important to note that participation is not failing to fulfil its promises simply because it is trapped by the managerial discourse, but because it is actually an integral part of that discourse and of the present mentality of government. Dean (1999) has given several examples of how contemporary liberal rationalities of government endeavour to operationalize the self-governing capacities of the governed in the pursuit of governmental objectives, and he has also given many illustrations of how empowerment and self-government have become components of power relations.[9] These points are important to keep in mind as the MRC and the riparian states contemplate expanded participation. The main message of these discussions does not mean that the participatory techniques of governmental organizations form an end to political contestations; but they do not necessarily make contestation easier either.

CONCLUSIONS: GOVERNING WATER THROUGH DEPOLITICIZED KNOWLEDGE

The '*realpolitik*' in knowledge production about the Mekong should not be dismissed. Problems such as lack of transparency relate to the development interests of the MRC's member states, and secretive processes around knowledge production have been clearly intentional and guided by the interests of the basin's riparian populations. Yet, politics does not manifest in any simple way; the politics of knowledge production implies much more 'invisible' nuances of power relations than intentional misuses or falsifications of the assessment processes. Important questions relate to the values and aspirations that guide and shape knowledge production and the interpretations of the assessment results that are presented as neutral. In the case of hydrological models, the social shaping relates to the differing interests between model developers and users, and between different knowledge domains with different policy motivations, and importantly to power relations at and between different scales. Hydrological models form the current core of the MRC's knowledge production and they represent an area of expertise that is still relatively closed and exclusive. Motivated by state-centred reasoning, the models simplify the complex ecosystem and related socio-economic dynamics of the basin and back up visions of the Mekong as a watercourse and a system of water channels. This is why the results of the assessments tend to find the risks of the development interventions less critical than the more comprehensive studies that acknowledge the complexity of the Mekong's ecology. But even if there were more critical findings, it is not self-evident that they would actually influence decisions and policy-making.

In accordance with current international discourses of development and the requirements of both objective science-based risk assessments and participatory processes in the legitimizing of projects, participation in knowledge production is presented as key to redressing the power imbalances that shape the production of knowledge. But the participatory approaches of knowledge production do not necessarily challenge the overly scientized nature of politics. And better and more inclusive science and assessment procedures are still rarely a sufficient and appropriate remedy for the unequal power structures underlying policy decisions. Even in dialogue processes, the contested knowledge domains tend to be debated in terms that do not challenge the technocratic core of assessments and the interest of MRC member states, as well as their enduring developmentalist vision of the Mekong's future.

Water and river basin management are inherently political processes, involving decisions that affect different groups in different ways and negotiation of diverse interests and values. The MRC is an inherently political body, established to manage the interests of its member states regarding the use of water and related resources of the Mekong River. The role of the MRC as a knowledge production agency that influences policy and decisions on how the river is to be used, managed and developed should not be to pretend to take the politics out of decision-making, but rather to foster a political dialogue between and within riparian countries that is informed by a better understanding of the implications of particular decisions and policy approaches. The 'anti-politics' of knowledge production that we have described in this chapter obscure rather than enhance the embedding of river basin management within the realm of informed social and political negotiation.

NOTES

1 The inspiration for the chapter's title comes from the famous book by James Ferguson (1994) entitled *The Anti-Politics Machine: Development, Depoliticization and Bureaucratic Power in Lesotho*.

2 As Beck has stated, in premodern society, risks as such do not exist in the sense that there is no calculus of control or assessment. Rather, natural and unpredictable disasters remain in the domain of hazards to be coped with but not influenced or calculated.

3 Dean is a central scholar in the field of so-called governmentality studies that follow the Foucauldian approach in the studies of power and authority. The concept of 'governmentality' refers to ways of thinking about governing. It examines how we conduct ourselves and others in all our spheres of life. It links together elements of politics, institutions and subjectivities and examines how power relations shape collective and individual practices, subjectivities and identity formations. Dean (1999) has stated that the way in which Beck approaches risk within a narrative of the modernization process is based on over-totalizing assumptions about risk. According to Dean, risk should be analysed as part of governmentality and, thus, as a component of assemblage of practices, techniques and rationalities concerned with how we govern others and ourselves. So it should not be approached as a naturally occurring entity, but as a form of calculation about reality: a way of thinking about and representing events.

4 By *realpolitik* we mean the very visible power relations through which powerful actors, like the representatives of the riparian states in the case of the MRC, pursue their particular interests.

5 The data of water flow, height and area inundated used in the models were taken in the report as preliminary indicators of the fish production through estimations of changes in fish feeding opportunities. The report acknowledged that these should be seen as only preliminary indicators (World Bank, 2004).

6 Some, like Lahsen (2005), have criticized this interpretation as an oversimplification. Lahsen reminds us that modellers themselves also get seduced by the simulations they have produced because of the high stakes they have in the model development, which often requires several years, even decades, of dedication.

7 This does not mean that agreement on minimum flows would be easy. The equitable sharing and maintenance of minimum flows does not provide a straightforward set of targets and criteria of mutual acceptability to all the Lower Mekong states, as evidenced, for example, in the difficulty of agreeing on basic questions such as the definition of the length of the dry season. So far the differing national interests of the countries have been difficult to overcome (Hirsch and Mørck-Jensen, 2006). Vietnam and Thailand, in particular, have had a history of mistrust between them because of Vietnam's worries over Thailand's plans to implement large-scale water diversions. In this level of discussion, the Mekong Agreement and the DSF models can, of course, be seen as important elements in reaching balanced cooperation between the basin's governments. But the concept of national interest is tricky in the sense that it does not seem to mean that the diverse interests of all stakeholders within each country would be captured by it (Hirsch and Mørck-Jensen, 2006). Furthermore, such interests are embedded within wider sets of political relations, which in the case of Vietnam and Cambodia, for example, tend to supersede assertion of country interests around water.

8 This sheds light on why, for example, in the debates on the development plans in the Mekong Basin the focus often remains on what should be included in the impact assessment calculations (such as economic valuations of ecosystems), rather than on questioning the very rationales underlying the assessments of technocratic knowledge production.

9 The context that Dean mostly refers to is the failure of welfare government. More generally, it refers to the growing distrust of direct state interventions and how government now seeks to operate through free individuals. In development practice, corresponding processes include decentralization and participatory processes.

References

Adamson, P. (2007) 'Hydrological and water resources modelling in the Mekong region: A brief overview', in IUCN, TEI, IWMI and M-POWER (eds) *Exploring Water Futures Together: Mekong Region Waters Dialogue: Resource papers from Regional Dialogue*, Vientiane, Lao PDR, p132

Affeltranger, B. (2008) *Le contrôle de la vérité: (Géo)politique de l'information hydrologique. Le cas du basin du Mékong, Asie du Sud-Est*, PhD thesis, Laval University, Canada

AMRC (2007) *Unpacking the Mekong Water Resources Assistance Strategy*, Mekong Brief no 6, September 2007, www.mekong.es.usyd.edu.au/publications/briefs/mekong_brief6.pdf, accessed March 2008

Arnstein, S. R. (1969) 'A ladder of citizen participation', *Journal of the American Planning Association*, vol 35, no 4, pp216–224

Beck, U. (1992) *Risk Society: Towards a New Modernity*, Sage, London

Brosius, P. (1999) 'Green dots, pink hearts: Displacing politics from the Malaysian rain forest', *American Anthropologist*, vol 101, no 1, pp36–57

Cooke, B. and Kothari, U. (2001) 'The case for participation as tyranny', in Cooke, B. and Kothari, U. (eds) *Participation: The New Tyranny?*, Zed Books, New York, NY

Dean, M. (1999) *Governmentality: Power and Rule in Modern Society*, Sage Publications, London

Demeritt, D. (2006) 'Science studies, climate change and the prospects for constructivist critique', *Economy and Society*, vol 35, no 3, pp453–479

Ferguson, J. (1994) *The Anti-Politics Machine: 'Development', Depoliticization and Bureaucratic Power in Lesotho*, University of Minnesota Press, Minneapolis, MN

Foran, T. (2006) *Rivers of Contention: Pak Mun Dam, Electricity Planning, and State–Society Relations in Thailand, 1932–2004*, PhD thesis, University of Sydney, Australia

Forsyth, T. and Walker, A. (2008) *Forest Guardians, Forest Destroyers: The Politics of Environmental Knowledge in Northern Thailand*, University of Washington Press, Seattle, WA

Fox, C. and Sneddon, C. (2004) *Flood Pulses, International Watercourse Law, and Common Property Resources: A Case Study of the Mekong Lowlands*, Research Paper no 2005/20, UNU/WIDER, p20

Henkel, H. and Stirrat, R. (2001) 'Participation as spiritual duty; empowerment as secular subjection', in Cooke, B. and Kothari, U. (eds) *Participation: The New Tyranny?*, Zed Books, New York, NY

Hirsch, P. (1990) *Development Dilemmas in Rural Thailand*, Singapore, Oxford University Press, Oxford

Hirsch, P. and Mørck-Jensen, K. (2006) *National Interests and Transboundary Water Governance in the Mekong*, Australia Mekong Resource Centre, Sydney, Australia

IRN (International Rivers Network) (2006) *Mekong under Threat: New Strategy Promotes Dams and Diversions*, International Rivers Network Factsheet, http://internationalrivers.org/files/MekongUnderThreatEnglish.pdf, accessed March 2008

IUCN, TEI, IWMI and M-POWER (World Conservation Union, Thailand Environment Institute, International Water Management Institute and M-POWER Water Governance Network) (2006) *Feedback to the World Bank and Asian Development Bank from Participants and Conveners of the Mekong Region Waters Dialogue on World Bank/ADB Joint Working Paper on Future Directions for Water Resources Management in the Mekong River Basin*, Mekong Water Resources Assistance Strategy (MWRAS), www.sea-user.org/UserFiles/File/docs/ConvenorsMWRAS%20comments%2025%20Sep%202006u.pdf, accessed March 2008

IUCN, TEI, IWMI and M-POWER (2007) *Exploring Water Futures Together: Mekong Region Waters Dialogue*, Report from Regional Dialogue, Vientiane, Laos, World Conservation Union, Thailand Environment Institute, International Water Management Institute, M-POWER Water Governance Network, no 75, www.mpowernet.org/download_pubdoc.php?doc=3274

Lahsen, M. (2005) 'Seductive simulations? Uncertainty distribution around climate models', *Social Studies of Science*, vol 35, no 6, pp895–922

Lamberts, D. (2007) 'Little impact, much damage: The consequences of Mekong River alterations for the Tonle Sap ecosystem', in Kummu, M., Keskinen, M. and Varis, O. (eds) *Modern Myths of Mekong – A Critical Review of Water and Development Concepts, Principles and Policies*, Helsinki University of Technology, Finland

Lamberts, D. and Koponen, J. (2008) 'Flood pulse alterations and productivity of the Tonle Sap ecosystem: A model for impact assessment', *Ambio*, vol 37, no 3, pp178–184

Li, T. M. (2005) 'Beyond "the state" and failed schemes', *American Anthropologist*, vol 107, no 3, pp383–394

Li, T. M. (2006) *The Will To Improve: Governmentality, Developments and Politics in Indonesia*, Duke University Press, Durham, NC

Li, T. M. (2007) 'Practices of assemblage and community forest management', *Economy and Society*, vol 36, no 2, pp263–293

MacKenzie, D. (1990) *Inventing Accuracy: A Historical Sociology of Nuclear Missile Guidance*, MIT Press, Cambridge

Middleton, C. and Lee, G. (2007) 'Mekong Water Assistance Strategy: Justifying large water infrastructure with transboundary impacts', *Watershed*, vol 12, no 1, pp11–19

Molle, F. (2005) *Irrigation and Water Policies in the Mekong Region: Current Discourses and Practice*, IWMI Research Report no 95, International Water Management Institute, Colombo, Sri Lanka, www.iwmi.cgiar.org/pubs/pub095/RR95.pdf

MRC (Mekong River Commission) (2005) *Public Participation in the Lower Mekong Subregion*, MRC, Vientiane, Laos

MRC Secretariat (2003) *MRC Public Participation Strategy*, Draft Report, MRCS, Phnom Penh, Cambodia

Radio Australia (2008) 'Connect Asia Programme', 7 August

Rayner, S. (2003) 'Democracy in the age of assessment: Reflections on the roles of expertise and democracy in public-sector decision making', *Science and Public Policy*, vol 30, no 3, pp163–170

Rose, N. S. (1999) *Powers of Freedom: Reframing Political Thought*, Cambridge University Press, Cambridge

Sarkkula, J., Keskinen, M., Koponen, J., Kummu, M., Nikula, J., Varis, O. et al (2007) 'Mathematic modelling in integrated management of water resources: Magical tool, mathematical toy or something between?', in Lebel, L., Dore, J., Daniel, R. and Koma, Y. S. (eds) *Democratizing Water Governance in the Mekong*, Mekong Press, Chiang Mai, Thailand

Scott, J. C. (1998) *Seeing Like a State: How Certain Schemes to Improve the Human Condition Have Failed*, Yale University Press, New Haven, CT, and London

Shackley, S. and Wynne, B. (1996) 'Representing uncertainty in global climate change science and policy: Boundary-ordering devices and authority', *Science, Technology and Human Values*, vol 21, no 3, pp275–302

Sneddon, C. and Fox, C. (2006) 'Rethinking transboundary waters: A critical hydropolitics of the Mekong Basin', *Political Geography*, vol 25, no 2, pp181–202

Sneddon, C. and Fox, C. (2007) 'Power, development, and institutional change: Participatory governance in the Lower Mekong Basin', *World Development*, vol 35, no 12, pp2161–2181

Sonnenfeld, D. A. and Mol, A. P. J. (2002) 'Globalization and the transformation of environmental governance', *American Behavioral Scientist*, vol 45, no 9, pp1318–1339

Sretthachau, C. (2006) 'Localised knowledge production through Thai Baan Research', Paper presented at the Fourth World Water Forum, Mexico City, 16–22 March 2006

Sretthachau, C. (2007) 'Thai Baan Research (villagers' research): Local wisdom for resources management', in IUCN, TEI, IWMI and M-POWER (eds) *Exploring Water Futures Together: Mekong Region Waters Dialogue: Resource papers from Regional Dialogue*, Vientiane, Laos

Szerszynski, B. (1996) 'On knowing what to do: Environmentalism and the modern problematic', in Szerszynski, B., Lash, S. and Wynne, B. (eds) *Risk, Environment and Modernity: Towards a New Ecology*, Sage, London

Williams, R. and Edge, D. (1996) 'The social shaping of technology', *Research Policy*, vol 25, pp865–899

World Bank (2004) *Modelled Observations on Development Scenarios in the Lower Mekong Basin*, Prepared for the World Bank with Mekong River Commission, World Bank, Vientiane, Laos PDR

World Bank and ADB (Asian Development Bank) (2006) *Future Directions for Water Resources Management in the Mekong Basin: Mekong Water Resources Assistance Strategy*, World Bank/ADB Joint Working Paper, www.adb.org/water/operations/partnerships/mwras-June2006.pdf, accessed March 2008

Wynne, B. (1996) 'May the sheep safely graze? A reflexive view of the expert-lay knowledge divide', in Szerszynski, B., Lash, S. and Wynne, B. (eds) *Risk, Environment and Modernity: Towards a New Ecology*, Sage, London

Wynne, B. (2002) 'Risk and environment as legitimatory discourses of technology: Reflexivity inside out?', *Current Sociology*, vol 50 no 3, pp459–477

14

De-marginalizing the Mekong River Commission

John Dore and Kate Lazarus

Introduction

A new water governance paradigm is needed in the Mekong region to assist societies in making better choices about how to share and manage water for the production of food and energy. On mainstreams and tributaries, disputes exist, resulting from interventions to natural flow regimes and overt or default management decisions. These interventions are justified on the grounds of flood control, more irrigation for food or fibre production, urban or industrial supply, improving ease of navigation, or boosting energy production via hydropower. There are associated disputes about altered sediment and nutrient loads, groundwater use, water reuse and diversions (inter-state, intra-state, inter-basin and intra-basin). New regional water governance is vital because these issues have territorial, ecological and political dimensions that need to be managed via regional protocols, rules or benefit-sharing processes.

Numerous dams and water diversions are on the agendas of mobile private and quasi-public-sector developers, transnational capital providers, and the six governments of the region: Cambodia, China, Laos, Myanmar/Burma, Thailand and Vietnam. A recent count found 82 existing and 179 potential hydropower projects in the wider region (King et al, 2007) (see Chapters 1 and 2), many on Mekong River tributaries.[1] Planned dams and diversions will transform the waterscapes of the region.

Our vision is for a more deliberative water politics in the Mekong region. To be clear, when speaking of deliberation, we mean:

> *Deliberation is debate and discussion aimed at producing reasonable, well-informed opinions in which participants are willing to revise preferences in light of discussion, new information, and claims made by fellow participants. Although consensus need not be the ultimate aim of deliberation, and participants are expected to pursue their interests, an overarching interest in the legitimacy of outcomes (understood as justification to all affected) ideally characterizes deliberation.* (Chambers, 2003, p309)

Thus far, deliberation has been in short supply. This is partly because proponents of deliberation meet resistance from actors who prefer to reinforce contexts that are unfriendly to deliberation and favourable to pursuance of their vested interests. Many actors still believe, or at least rhetorically pretend or are instructed, that domestic criticism of public policy is unpatriotic. There is often an unhelpful conflation where dissent is mistakenly seen as synonymous with disloyalty. Enquiry or criticism of water resources development plans, which impact across state borders, is seen by many as encroachment on hard-won state sovereignty and legitimate national security concerns. Hence, the resistance to transnational deliberative politics should not be underestimated.

The Mekong River Commission (MRC) is mandated to engage in water resources development in the so-called 'Lower' Mekong part of the region – the Mekong River Basin in Cambodia, Laos, Thailand and Vietnam. Different people call on the MRC to be a social and environmental guardian of the basin; a platform for information exchange; a knowledge producer, synthesizer and broker; an investment facilitator; and convenor of multi-stakeholder processes demonstrating high-quality deliberative practice. Can it play all these roles simultaneously?

Since 1995, the MRC (and its predecessors since the 1950s) has been and remains the focus of substantial organization-building efforts. During recent years, the MRC has received much attention from people intent on using, improving, empowering or criticizing it. This chapter reflects on the practice and potential of the MRC at a time when all Mekong region governments need to make informed decisions about whether, or how, to proceed with major projects that will have dramatic, transformative, national and transboundary impacts.

UNDERUTILIZED

The Mekong River Commission has a contested governance mandate – embodied in the 1995 Mekong River Agreement – for the mainstream, tributaries and the lands of the basin within the territories of the Lower Mekong countries (Governments of Cambodia–Laos–Vietnam–Thailand, 1995; Browder, 2000; Öjendal, 2000). It is often referred to as a 'regional' initiative and endeavours to include China and Myanmar in some of its activities and outreach. This Mekong cooperation was originally catalysed via the United Nations and has a 50-year history (Bui Kim Chi, 1997; ESCAP, 1997).

Article 1 of the agreement commits the four member countries to cooperate in all fields of sustainable development, utilization, management and conservation of the water and water-related resources of the Mekong River Basin in fields such as irrigation, hydropower, navigation, flood control and fisheries.

The implementing organization for the agreement is the MRC, led by a governing Council at ministerial level, which meets once per year, and a Joint Committee (JC) of senior government officials, which meets formally twice per year, but increasingly now meets informally as the need arises (see Figure 14.1). The Council and JC are serviced by the *MRC Secretariat (MRCS)*, which is responsible for implementing Council and JC decisions, advising and providing technical and administrative support. The MRCS is currently located in Vientiane, Laos.

Although not specifically mentioned in the agreement, there are also National Mekong Committees (NMCs) established in each member country, set up differently in each country depending upon national government preferences. The heads of the NMCs represent their countries on the Joint Committee. NMCs are serviced by NMC Secretariats (NMCSs). It is important to note that there is a political dynamic between each of these five parts – that is, there is no homogeneous single 'MRC'. Any joint position needs to be collectively negotiated between the Council and JC members. Moreover, the MRCS must also manage its working relationships with the NMCSs, who are quick to object if they feel left out of MRCS activities, or if they perceive the MRCS to encroach into their national space. In turn, the NMCSs also have to establish their own role and working space within their national polities, with their functional power much less than key water-related ministries and agencies in each country.

The MRC also recognizes 'development partners' that include international lenders and donors – who at this stage still provide most of the finance for the MRC to function – international financial institutions (IFIs) such as the World Bank and the Asian Development Bank (ADB), and other 'internationals' such as the World Wide Fund for Nature (WWF), the International Union for Conservation of Nature (IUCN) and the International Water Management Institute (IWMI). More recently, knowledge networks involving various regional universities, policy research institutes and civil society organizations, such as the coalition implementing the Mekong Programme on Water, Environment and Resilience (M-POWER), are also increasing their engagement with the MRCS. At the national level, the NMCs and NMCSs have historically had less engagement with non-state actors or civil society organizations, particularly if these articulate alternative development narratives.

As in any large family, it is not possible for all the interaction to be smooth. The vaunted 'Mekong spirit' of cooperation often seems optimistically overstated;

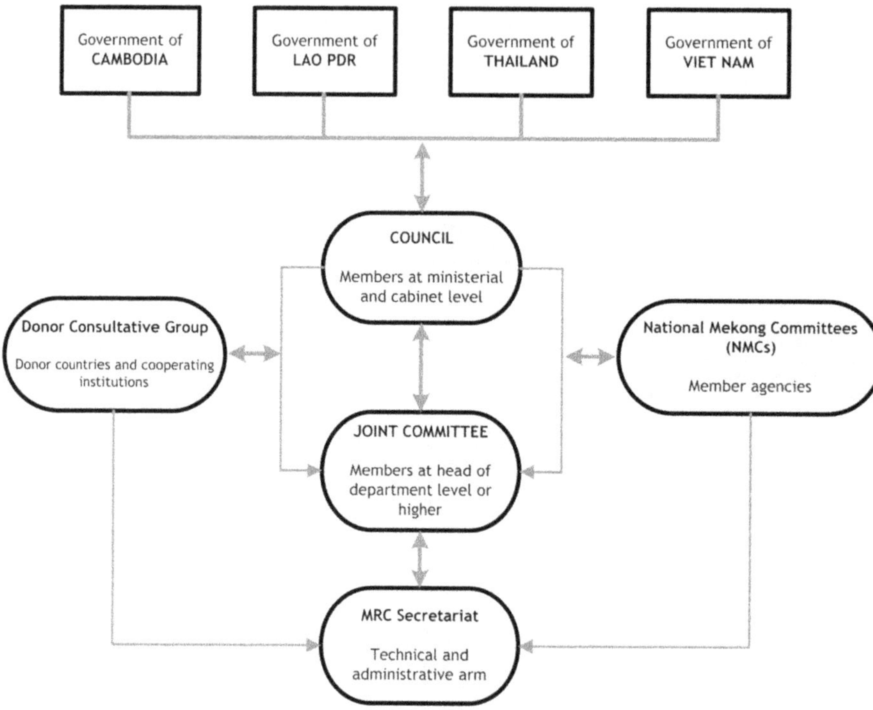

Figure 14.1 *Mekong River Commission structure*

Source: www.mrcmekong.org

but that is not to deny the importance of doing everything possible to encourage a constructive spirit between the countries.

It is apparent that for much of its brief history, the MRC has been underutilized. In reflecting on this, it is important to look at all members of 'the family', rather than just using the blanket term of MRC. Which parts of the MRC have been excluded or marginalized, why and by whom? Why have member governments chosen not to use their own river basin organization to engage in many of the major river basin development issues of this era? At present, individual national interests dominate over regional interests (Hirsch and Mørck-Jensen, 2006). Supporters of the MRC hope that it can become more a part of the solution to problems such as those illustrated in the following examples.

Sesan

On 4 March 2000, the water level in the Sesan River (a transboundary river flowing through Vietnam and Cambodia) rose suddenly, causing deaths and loss of livelihoods of fishers and farmers in north-eastern Cambodia's Ratanakiri Province. The unexpected surge was caused by a release of water from the Yali Falls Dam in Vietnam. Cambodian non-governmental organizations (NGOs) and local communities brought forward details of the damage and encouraged the national and international public to consider the implications of this transboundary incident. During the incident, the flow of information between Cambodian and Vietnamese officials was minimal, and there was virtually no communication between the provincial governments on either side of the border (Badenoch, 2001, p1). The MRC did not become involved before or in the tense immediate aftermath, despite having a mandate to do so.

Five years on, the topic of development on the Sesan, Sekong and Srepok rivers (often collectively referred to as the 3S) was still considered so sensitive that it was removed from the agenda of an MRC-convened conference on integrated water resources management (IWRM) that was held in tandem with the tenth anniversary of the signing of the 1995 Mekong Agreement. Eventually, the MRC has engaged in the process, principally via the NMCSs of Cambodia and Vietnam. More recently, the two countries are trying to better manage the hydropower operations on the Sesan in order to minimize the downstream impact upon local communities; but this has taken extensive advocacy by affected local communities and their supporters. There is now even an effort, facilitated by the ADB, to establish a new transboundary, sub-basin organization to 'manage development'. For a long time MRCS had been an onlooker, but it is now engaging more with ADB in this new initiative.

Commercial navigation

A commercial navigation agreement was signed by transport officials from China, Laos, Thailand and Myanmar in 2001 for the stretch of the Mekong River between Simao (China's Yunnan) and Luang Prabang (Laos). River trade between Thailand and China has since rapidly increased. Associated with the signing, a feasibility study was completed in late 2000, which supported, in principle, proposed alterations to the river, including rapids and reef removal. By September 2001, an environmental impact assessment (EIA), coordinated by the Government of China, had been prepared and sent to each of the other three governments. Thailand's government approved the EIA in January 2002 and it was subsequently approved in Laos in April 2002 (SEARIN et al, 2002). The MRCS was not used by either Thailand or Laos to inform or actively participate in the initial agreement negotiations. The secretariat became involved afterwards in offering to conduct an independent EIA of the project (an offer not taken up) and in commissioning evaluations that were extremely critical of the substandard EIA (Cocklin and Hain, 2001; Finlayson, 2002; McDowall, 2002). The intervention by the MRC was ineffective and extensive modification, via blasting, of the Mekong mainstream has since taken place in the northern reaches. The MRCS has played no substantive role in this 'Upper' Mekong navigation; however, at the time of writing, it is becoming more involved, and so this could change.

Thailand Water Grid

In 2003, then Prime Minister Thaksin Shinawatra relaunched the idea of a Thailand water grid, which would triple the area of irrigation in the country and require diversion of water from the Mekong River, and possibly other rivers in Laos and Cambodia, into northeast Thailand. For a time, the whole process was treated as a national secret by the Government of Thailand, with senior water academics fearful of the consequences to their funding and employment of criticizing or sharing any information about the scheme. There was no public deliberation within Thailand, and the MRCS was conspicuously silent, having been excluded from the process. Molle and Floch (2007) have observed that Thaksin's 'war on poverty' was presented as an unquestionable meta-justification used to silence opposition despite the fact that most water experts, commenting off the record, thought the rationale dubious and the scale of the scheme completely unrealistic. At the time of writing, the MRC (all parts) remains publicly silent about the merits or failings of touted water diversions by Thailand from the Mekong River, siphoning under the Mekong River from Laos, or, as some pundits remark, 'siphoning under the 1995 Mekong Agreement'. The grid, as well as every other Thailand water resources development scheme on the shelf, was relaunched in 2008 by the government of Prime Minister Samak Sundaravej (see Chapter 10 for more details). To our knowledge, no substantive information about this has been shared with the MRCS.

Mainstream dams

Construction since the mid 1990s of the Upper Mekong (Lancang) Dam cascade in Yunnan is so far the most significant human intervention ever made in the natural order of the Mekong River ecosystem, with substantial and undoubtedly complex transboundary ecological, social, cultural, economic and political impacts (Dore, 2003, p431). The regional/transborder nature of ecosystems requires regional/transborder political cooperation. China has plans to build up to 15 mainstream dams. Despite the MRC having an annual consultation with Chinese water officials, the MRCS has not noticeably affected China's construction agenda. Real-time flood data is now provided by China, and future consultations could fruitfully examine hydropower operation regimes in order to minimize negative downstream impacts – negotiations to do just that were propelled in 2008 by serious floods along the Mekong River, with the flood level reaching a height not seen since 1966. The impact of the Chinese dams is now included in MRC cumulative impact assessments and scenarios work; but dialogue by the MRC with Chinese officials has, to this point, been very limited. At least until 2008, exchanges have been more substantial beyond the MRC. Outside the MRC processes there has been increasingly substantive cooperation between Chinese colleagues and southern neighbours discussing many aspects of the Chinese mainstream development. This has included visits to Cambodia by Chinese water scientists, hosted by Cambodian non-state organizations. Increasing the depths of these types of dialogues and exchanges may be of critical importance in demonstrating the constructive possibilities of greater international understanding and perspective-sharing.

China's unwillingness to seriously engage with MRC has been problematic enough; but what is worse is when member country governments also choose not to use the MRC to share their own national water resources development intentions. More recently, there is renewed interest by all of the MRC member countries in building or investing in dams on, or diversions from, the Mekong River mainstream (see Chapter 2). At the time of writing, the only government to formally submit information to the MRCS about mainstream developments has been the Government of Laos, which in June 2008 advised that it is investigating eight dams on the mainstream. Despite the MRC Joint Committee having formally approved the Procedures for Notification, Prior Consultation and Agreement (PNPCA) in November 2003, there has thus far been only very modest compliance by member countries. The MRCS is hopeful that this action by Laos signals a new openness to sharing information about possible projects, and that the other member countries choose to take similar steps.

Tensions

The most recent strategic overview of the MRC that took place in 2006 involved much rewriting and negotiation before being finally endorsed by the JC and accepted by the Council. By this time there was something in the strategy for everyone, and the organization was assigning itself multiple, sometimes conflicting, roles. At the aspirational level of the text, there was little disagreement between stakeholders. It is hard to find anyone who disagrees with the stated goals and 'strategic IWRM' directions which frame the plan, although Molle would remind us to be wary of a 'nirvana concept', such as IWRM, which can 'obscure the political nature of natural resources management', and the fact that some of the goals may be 'frequently, if not always, antagonistic' (hence, the conflicts and the fact that 'trade-offs are necessary and hard to achieve') (Molle, 2008).

The differences that emerged were in the details and the intended emphases. The strategic plan preparatory process highlighted some of the tensions evident within the MRC and its wider constituencies, to which we now turn.

Box 14.1 Goals and strategic directions of the *Mekong River Commission Strategic Plan* (2006 to 2010)

The goals of the *Mekong River Commission Strategic Plan* are stated as follows:

- Promote and support coordinated, sustainable and pro-poor development.
- Enhance effective regional cooperation.
- Strengthen basin-wide environmental monitoring and impact assessment.
- Strengthen the integrated water resources management (IWRM) capacity and knowledge base of the Mekong River Commission (MRC) bodies, the National Mekong Committees (NMCs), line agencies and other stakeholders.

The 'strategic IWRM' directions of the plan are summarized as:

- *Economic development and poverty alleviation:* promote economic growth through the use and development of joint water resources in a manner that significantly alleviates poverty.
- *Integration through basin planning:* implement a participatory multi-sectoral basin planning process that integrates economic, social and environmental concerns across the Lower Mekong Basin (LMB).
- *Social development and equity:* ensure equity in the allocation of water resources and services across different economic and social groups; reduce conflict and promote socially sustainable development.
- *Regional cooperation:* integrate and coordinate water resource development and management between countries to optimize benefits from the joint resource and to minimize the risk of water-related conflicts.
- *Governance:* further and implement open, transparent and accountable institutions and regulatory frameworks that will promote IWRM at all levels.
- *Environmental protection:* protect the environment, natural resources, aquatic life and conditions, and the ecological balance of the Mekong River Basin from harmful effects of development.
- *Climate variability:* prevent, mitigate or minimize people's suffering and economic loss due to climate variability.
- *Information-based management:* ensure that water resource management decisions are based on best available information.

Source: MRC (2006)

Territorial domain: Mainstream only, or including the tributaries, basin wide?

Article 1 of the Mekong Agreement is clear that the territorial domain of the MRC is the entire Mekong River Basin. Acting in the China and Myanmar parts of the basin is difficult as these countries are not members, but multi-country overview of development in the LMB has also proved to be very difficult. At the time of the strategic planning process, many were disappointed that the MRCS had not been more involved in analysing and contributing to decision-making about development in the tributaries. The political role of the MRC had seemed reduced to mostly research and discussions about mainstream cooperation, with speculative emphasis on the impacts of Chinese mainstream developments, but with tributary development mysteriously scoped out of formal Council and JC discussions. Now mainstream projects are back on the agenda, and the MRC cannot again be silent. The MRCS recognizes this.

The Precautionary Principle: To apply or not?

Article 3 relates to the protection of the environment and ecological balance. It is of concern to many that the MRC has been too often subdued about the risks associated with many development projects – risks often borne involuntarily by those not clearly benefiting (or potentially benefiting) from project X, Y or Z. This silence has extended to the non-mention of the Precautionary Principle.[2] Instead, the mantra from the secretariat has been 'meeting the needs, keeping the balance' and acceptance of an ever-changing baseline. It is important to ask: whose needs, and what risks or trade-offs are considered acceptable in the quest for balance?

Constituency: Governments or wider society?

There was much discussion of the MRC mandate and expectations during the 2006 to 2010 strategic planning process. It was clear that the MRC did need to clarify its constituency and decide how much scope to give the MRCS to engage with a wider constituency than just the parts of the member state governments that have been tasked with MRC representation.

The final plan reflects the dominant attitude of the MRC towards engagement with non-state actors, suggesting that 'improved stakeholder participation can be accomplished by working through the NMCs who are best able to implement improved participation, including civil society and NGOs (MRC, 2006, p43). Many civil society organizations beg to differ as engagement between them and the state-centric NMCs has been at a very basic level, although this is now being stepped up. Many donors and consultants have had far easier access to the MRC than local civil society and Mekong academia.

Many people who have been involved in the MRC over the past decade have recognized that they need to bring other actors and subject matter into the mainstream of their processes and provide a mechanism for the expression and exchange of what may be widely and fundamentally differing views about upstream and tributary development, inter-basin diversions, etc. The 2000 annual report acknowledged that it is 'important that decisions on development include a "bottom-up" process and are not confined to a "top-down" approach. The voice of the people directly affected, and of other stakeholders such as community groups or NGOs, must be heard.' Moreover, it admitted that it 'has virtually no experience in this vital field' and that it must 'drastically accelerate activities to promote public participation' (MRC, 2001).

Soon afterwards, one of the authors of this chapter wrote that the MRC's lack of achievement thus far in genuine public participation is complex. The youth of the new version of the organization, the sustainability orientation and mindset of some of the agencies which dominate the National Mekong Committees, the politics between the member states, stinging criticisms by NGOs, realization of limited successes to this point, and operating rules that limit engagement with the wider basin community are all relevant. Collectively, this has resulted in the MRC lacking confidence and being constrained in the extent to which it has proactively engaged with the large range of Mekong region actors outside of the MRC family. In relation to hydropower and the Water Utilization Programme (WUP), there has been a hypersensitive wariness of member country intergovernmental politics. There is also some resistance to being 'lectured' at by NGOs and past and present Mekong country experiences of being 'directed by donors' (Dore, 2003, p424).

The drastic acceleration did not eventuate. At least until 2008, progress in this area has been slow. For example, the consultants who undertook an organizational review (discussed below), several years later, noted:

> *The Strategic Plan describes the importance of public involvement, public opinion, the civil society and NGOs in ensuring the success of integrated water resources management of the Mekong River Basin. However, it is the impression of the Review Team that the present attitudes and practices in MRC regard the member governments as the primary, if not the only, stakeholders that should be involved with MRC. A clear commitment and strategy for involving the civil society is lacking.* (Hawkesworth et al, 2007, p16)

Knowledge broker or investment promoter?

The approach to knowledge-sharing or knowledge-broking has varied during the first 13 years of the MRC. During this period the organization has had four chief executive officers (CEOs), punctuated by caretaker leadership.

During the Matoba-era of 1995 to 1998, the MRCS was a closed, state-centric organization, lacking in confidence and capacity, and with its potential constrained by the management style. It gave the impression of being a house for often independently operating donor projects.

Under the subsequent leadership of Joern Kristensen during 2000 to 2003, there was a clear shift towards being a 'knowledge broker', which implies enabling the constituency to both contribute to and receive knowledge. The new commitment was to being a 'learning organization' and a centre of knowledge and information exchange with a strong commitment to improving the livelihoods of the people of the basin. There continued an understandable privileging of state members – after all, it is an intergovernmental organization; but there was also a new openness to knowledge contributions from a wider set of actors beyond states. Kristensen restructured the operations of the secretariat into programmes and insisted that all those working in the secretariat building were accountable to him. These were all positive changes. Morale within the secretariat noticeably improved.

After a lengthy caretaker period during which the secretariat transferred from Phnom Penh to Vientiane, Olivier Cogels took up the CEO position for 2004 to 2007. The new leader was convinced that he would be the one to build the working relationship with China which had eluded his predecessors. Soon into his tenure he denounced any role of the MRCS being a 'watchdog' and launched a new push for the MRCS to be an investment promoter or facilitator. Both of these moves brought him into conflict with the knowledge-brokering role, as the promoter/facilitator was uninterested in any bad news about possible negative impacts of upstream, downstream or tributary development. Information exchanges, peer reviews and contestation, and characteristics of knowledge-building became more constrained. Morale within the MRCS staff dissolved as much analysis or commentary deemed counter-productive to the new mission – smooth sailing with the China relationship or investment promotion – was restricted. This tension was palpable during the strategic planning process. The authors' own observation of this situation was similarly detected by the organizational Review Team, who noted:

> *MRCS is starting to become known (among civil society organizations, scientific organizations) as an institution that will not release information that may illustrate negative environmental and social consequences of development projects. This is a threat to the credibility of the organization.* (Hawkesworth et al, 2007, p20)[3]

Preparing projects for investment or assisting societies to evaluate proposals?

MRCS engagement in project preparation was assumed during the drafting process for the strategic plan to be part of the new development promotion role. An alternative perspective was that a better role for the MRCS would be for it to support national actors (state and civil society) in order to examine development projects, their likely impacts, and their claimed merits and costs. It is this latter role that the MRCS has attempted to play with the Don Sahong Dam discussed below.

In any event, most perspectives about what the MRC should be doing ended up being included in the strategic plan, which in due course was adopted and quickly overtaken by the transformation in the region, part of which was the new avalanche of potential projects. A mid-term review of the MRC *Strategic Plan 2006–2010* is scheduled for late 2008. To the extent that it is possible, it is hoped this will remove some of the current ambiguities and, perhaps, make the roles of the various parts of the MRC a little clearer.

CASE STUDY: LAOS HYDROPOWER, DON SAHONG AND THE MEKONG RIVER COMMISSION (MRC)

No current development project better encapsulates the challenges facing the MRC than the present controversy over the Don Sahong Dam in southern Laos. If built, it would be the first dam on the mainstream in the LMB.

Hydropower

Laos is at the centre of the current hydropower surge in the Mekong region. According to the *Power Development Plan* in Laos (as of May 2008), there are 77 live hydropower projects: 10 are operational, 7 are under construction, 16 are under research and the remaining 44 have memoranda of understanding (MoUs) signed to move forward (see Chapter 2). There is a complex set of reasons driving the current surge. For the MRC, the explosion is a response to market demand: the increasing importance of regional trade and investment flows, rapidly growing energy demands (particularly in China, Thailand and Vietnam) and opportunities of an emerging regional power market have stimulated a new era of hydropower development in the basin, now mainly driven by private-sector actors (MRC, 2008, p37).

Soaring (albeit fluctuating) global energy prices and national commitments to energy security are also important drivers. Others include the ready availability of capital, at least until the advent of the global financial crisis; a new boldness by Mekong governments to move ahead; and very attractive concession terms for developers. Another driver that is now taking effect is the recognition of the changes that large new storage dams in China will have on the flow regime of the Mekong mainstream. When the Xiaowan and the Nuozhadu dams are completed in Yunnan, the dry season river flow will increase significantly and this will also make the LMB mainstream 'run of river' financially more attractive.

Fish

The Don Sahong story (see below) is not all about fish, but they are central; so before proceeding it is worth ensuring that the reader is familiar with the scale of the Mekong fishery (see also Chapters 9 and 12).

Recent MRCS research has estimated the LMB annual consumption of inland fish to be about 2 million tonnes by a population of 56 million people. About 90 per cent of the fish consumed in the LMB is from the wild-capture fishery. In addition, about 0.5 million tonnes of other aquatic animals (OAA) are consumed. Collectively, the inland fish and OAA are estimated to provide 47 to 80 per cent (country range) of the animal protein of the people of the basin (Hortle, 2007). This equates to about 17 per cent of the total global freshwater fishing catch and is worth in the order of US$2 billion. Other work by the MRC Fisheries Programme is showing that the bigger the flood (both in height and duration), the more fish you catch (in tonnes); and related to the previous point, the bigger the flood, the bigger the fish.[4]

These are extraordinary figures, showing massive reliance on a huge fishery. However, this data and information about threats to the fisheries seem to be having little impact upon river development policy-making (see Chapter 12). Bringing in fisheries is proving to be a challenge for local livelihood champions, economists, fisheries scientists and concerned political operators at all levels of decision-making. If it cannot be done at Don Sahong, it will be extraordinarily difficult anywhere else.

Don Sahong

In March 2006, the Government of Laos signed an MoU with a Malaysian engineering company, Mega First Corporation Berhad, to carry out a feasibility study for the run-of-river Don Sahong Hydro Energy Project (DSHEP) in the Khone Falls area, just north and upstream of the border between southern Laos and Cambodia.

In May 2007, a public letter from concerned scientists to governments and agencies responsible for managing and developing the Mekong River drew attention to and summarized 'grave environmental impacts, particularly on fish and fisheries but also on tourism and other significant aspects of economy and livelihood, causing damage that will far exceed the net returns from the project'. In their view:

> *While a degree of mitigation is sometimes feasible for some dams, the fisheries impacts of the Don Sahong Dam simply cannot be mitigated.... There is no prospect that a fish pass could make a significant difference to the blocking effects of this dam.* (Baird et al, 2007)

In June 2007, the concern about DHSEP was again summarized in a WorldFish Centre science briefing paper:

> *Khone Falls is a key site for all Mekong fish resources. At the falls, the Mekong drops some 20m to 30m from the Khorat Plateau to the Mekong Plain. Here the river forms a complex network of narrow braided channels, named hoo in Lao.... Of special significance are the 28 scientific studies that show how it serves as a bottleneck for fish migration in the basin. Hoo Sahong, the site of the proposed dam, is especially important as it plays a unique role in Mekong fish migration.... A dam on the Hoo Sahong would block the only deep channel that allows fish to migrate through the falls year round. This could effectively block dry season fish movements between the Lower Mekong plains and the Mekong basin upstream.... Data on the economic value of the Mekong fisheries, and on the impact of dams on fish migration, suggests that the economic costs from lost fisheries production could outweigh the expected economic benefits of the dam. This analysis suggests that if the proposed dam is to be considered further, a comprehensive scientific assessment would be required to evaluate the costs and benefits in the larger context of Mekong fisheries.* (Baran and Ratner, 2007)

In July 2007, a 'final draft' EIA report for DHSEP was completed and soon after submitted for evaluation by Lao authorities.[5]

The MRCS challenge

In September 2007, the MRCS was formally invited by the Government of Laos to contribute to its review of the EIA. This was a big step for Laos to include the MRCS in its internal processes. The staff within the MRCS supplied their best advice to Laos about the 'completeness, accuracy and adequacy' of the Mega First's consultants report, finding it deficient in many areas. Their report, prepared in November 2007, provides a clear critique of the EIA and offers objective advice to Laos. Included in their response, the MRCS pointed out the following:

- The geographic and economic extent of the impact on fisheries of the DHSEP has been underestimated.
- The proposed mitigation to allow upstream movement of fish cannot be proven to be effective prior to the DHSEP being built; and moreover the outflow from the turbines will attract fish to the blocked Hoo Sahong channel.
- The mortality of fish (all life history stages) that will be entrained through the turbines has been overlooked (MRC Secretariat, 2007, point 69).

The JC has encouraged the MRCS to be responsive to governmental requests for technical advice. In this Don Sahong case, the MRCS has responded to an in-confidence request from the Government of Laos. The MRCS analysis has not been released to the public, nor even shared with all member States. This is quite different to how the MRCS should be expected to act, as a transparent servant to all member countries.

In November 2007, the representatives from the Government of Cambodia pointed out their concerns about a Khone Falls Dam at the annual meetings of the full MRC, held that year in Siem Reap. Just prior to the MRC meetings, 201 citizens' groups and individuals from 30 countries wrote to the MRC demanding that it uphold the 1995 Mekong Agreement and that it protect the river and its people from the resurgent threats posed by the proposed mainstream dams. Also released at this time was a statement by MRC donors calling on the MRC to 'fully utilize its capacities, tools and mandate to assess hydropower development plans, with a view to transboundary environmental, economic and social impacts' (MRC Donors, 2007). The donors followed up with another letter in December, signed by the German ambassador to Cambodia, again asking for information about how the MRC procedures for 'timely notification, prior consultation and agreement' are being applied (Mann, 2007). The MRC Procedures for Notification, Prior Consultation and Agreement (PNPCA) were adopted by the MRC Council in November 2003 (and are discussed below).

In February 2008, Mega First signed a project development agreement with the Government of Laos and announced that its studies show the project to be viable. None of these studies have yet been publicly released.

In March 2008, 51 citizens' groups and individuals from the Mekong region wrote to the MRC asking it to engage more substantively and publicly in decision-making about development of the Mekong River. Premrudee Daoroung, director of the regional NGO Towards Ecological Recovery and Regional Alliance (TERRA), had this to say:

> *The new CEO must clearly state what steps the MRC will take in response to widespread concerns over the proposed mainstream dams. It can start by immediately releasing to the public all analyses relating to the Don Sahong Dam undertaken by the MRC.* (TERRA, 2008)

The new CEO responded in April 2008 that the MRCS would continue to work to develop a multifaceted understanding of the existing river system, prepare objective analysis of future development scenarios, provide advice on individual project proposals when requested by the member countries, and administer the procedures developed and negotiated (mostly during the Water Utilization Programme between 2000 and 2007) (Bird, 2008b).

In a recent interview, reflecting on the hydropower explosion, an analyst and campaigner for International Rivers acknowledged there is a 'catch-22' and that pleasing everyone is just not possible:

> *If the MRC provides advice to government agencies that is perceived as critical of proposed hydropower projects, this advice could be unwelcome, ignored, and then no longer sought, undermining the MRC's relevance in the eyes of the government agencies it considers itself primarily answerable to. Yet, by not providing this objective analysis and releasing it into the public domain, as it should do, the MRC faces a crisis of legitimacy in the eyes of the wider public that it is also intended to serve.* (Nette, 2008, interviewing Carl Middleton)

The analyst says 'as it should do', and we would agree; but under the current norms of MRC behaviour, without the permission of the Government of Laos, the MRCS could not publicly release its Don Sahong analysis and advice without being seen as having betrayed the trust of its member state. The MRCS technical staff would be delighted if their analysis and advice were put in the public domain, but would prefer that it was done by the Government of Laos. Many Lao officials would also be more comfortable if the Don Sahong decision-making process was more transparent and deliberative.

The Don Sahong is not yet built, and there may yet be more twists in the tale; but it is salutary to reflect on just how decisions actually get made about such projects. An actor in the Don Sahong case, who should not be identified because it is not possible to speak openly about matters like this, is concerned:

> *Development decisions in this region are almost entirely political. Technical matters play very little and sometimes no role in them. Water developments enable transfer of a dispersed, generalized wealth with no title – or, more correctly, traditional public title – into a focused economic resource with private title. This is a very attractive proposition for people in positions of power.*

A fisheries scientist searching to be effective suggested:

> *The real nature of politics and governance in the region is, indeed, one of the reasons why fisheries are not on the agenda, and that can be depressing to the citizens we are; however, that should not spare us from a critical analysis of our contribution, as scientists, to the development process.*

Another colleague very familiar with the interdependencies between ecosystems and local livelihoods reflected and recognized the need for more open deliberation:

> *Decision-making processes on dams are not based on rational assessments, and certainly not influenced by sound science regarding fisheries and their values. The irony is that we have had ten years of excellent research that has highlighted the importance of fisheries – and that this evidence (much of which comes from MRCS) is widely accepted. So we have had a great research success – but a failure in terms of influencing policy. I do not think science alone will have much influence – although good research, evidence and arguments are necessary. What is clearly lacking is an open discussion of the options and implications – and a process that draws on case study experience in this region (plenty to draw from) and opens up the debate to include people who are directly impacted.*

The Don Sahong example forces one to ask the question: how is it possible to have constructive, well-informed, deliberative processes before critical decisions are taken about water resources development? Thus far, the MRC has not been able to provide such a service to Mekong region societies. But things can change.

DE-MARGINALIZING

In early 2008, the MRC Joint Committee recommended, and the MRC Council subsequently appointed, new Chief Executive Officer Jeremy Bird, whose regional experience and existing working relationships ensure that he comes to the job with a solid grasp of the water politics of the place. He has previously worked in the Mekong region, including supervising cumulative impact assessment work in the Lao Nam Ngum River Basin (a sub-basin of the Mekong), and researching environmental considerations for sustainable hydropower development. The new CEO has laid out his vision for the period of 2008 to 2011 by proposing four areas of focus (Bird, 2008a) – regional and riparian; relevance; responsibility; risk reduction – which we use as departure points for the possible de-marginalization of the MRC.

Regional and riparian

The highest priority is for the MRC to become more regional, which is to us, in some ways, transnational. By this we mean addressing issues of joint concern to all the countries which share the land and waters of the basin, and to the extent possible, transcending solely national perspectives. Connected to regionalization is MRCS 'riparianization', which refers to the transition of the secretariat to an organization clearly led and directed by citizens of the MRC member countries. Given the commitment of the MRC to riparianization of all key MRCS positions by 2011, including the CEO position, it is understood by all that Bird will be a single-term CEO with only three years in which to make his contribution. All positions in the Council, JC and NMCs have always been taken by citizens of the member countries. However, the MRCS has been increasingly criticized for having too many 'international' (i.e. from beyond the Mekong region) staff in key positions such as the CEO, chief financial officer, chief of international cooperation, and programme managers.

Triggered by dissatisfaction with the overall performance, an independent review was commissioned in 2006 of the MRCS and the NMCSs. The consultants repeatedly encountered concerns about the staffing of the MRCS. They concluded that the overuse of internationals by the secretariat and the inadequate selection and retention procedures for riparians were preventing the MRCS from getting and keeping the best people from the member countries. Gate-keeping and control by the NMCs/NMCSs was identified as part of the problem. The recommendation was clear:

> *If there is going to be a successful professionalization and riparianization of MRCS, then it will be necessary to attract and secure the best qualified candidates, not just from government but from the civil society as a whole. The process should be managed on a strictly competitive basis and administered by MRCS itself.* (Hawkesworth et al, 2007, p37–38)

The key MRC donors agreed:

> *We strongly support the process towards riparianization. Riparian leadership, management and technical expertise in the MRC is critical to its long-term success and sustainability. To develop as a world class river basin management organization, the MRC employment procedures need to attract, appoint and retain the best and brightest from the Mekong member countries.* (MRC Donors, 2008)

MRC officials also agree with phasing down the role of internationals, but are finding it more problematic to make the riparian selection and retention systems more transparent and merit based. In early 2008, the JC rejected the recommendation, reducing the role of the NMCs. Donors are unlikely to accept anything less. So, it appears that all key positions in the secretariat will be 'riparianized' by 2011; but the processes for modernizing riparian recruitment and retention are still being negotiated.

Relevant and engaged

The MRC must demonstrate that it is relevant by actively engaging in development decisions taken in the basin. The MRC has too often been absent from, or silent about, substantial decisions being taken on water resources development in the basin. As pointed out earlier, the MRCS has had little involvement and usually very limited information about the hydropower development on the Mekong River mainstream in China, and on tributaries in Laos and Vietnam. It was excluded from the decision-making about 'channel improvement for navigation and trade' and the associated mainstream river blasting in the Upper Mekong above Chiang Saen in northern Thailand. In the past, it has also been excluded from speculations about possible Lao–Thai water transfers, and diversions from the Mekong to irrigate more of northeast Thailand. Moreover, in recent years, it does not seem to have been trying to engage in these important issues. In the absence of deliberative action by the MRC, other actors have sought to open up regional water resources development debates via multi-stakeholder dialogues (Dore, 2007; IUCN et al, 2007a, 2007b) and the establishment of transnational knowledge networks. That said, these previous exclusions or inactivity would look minor if the MRC cannot now contribute to decision-making about LMB mainstream dams and diversions, which is now publicly (since 2007) firmly back on the agendas of all four member countries. In the latter half of 2008, the MRC, via the MRCS, scaled up its engagement.

Responsible and accountable

More than ever before, the MRC is being called to account and to act on the mandate articulated in the 1995 Mekong Agreement: under the agreement, the MRC is to conduct 'assessment for the protection of the environment and the maintenance of the ecological balance of the Mekong River Basin' (Article 24) and should 'make every effort to avoid, minimize and mitigate harmful effects that might occur to the environment ... from the development and use of the Mekong River Basin water resources' (Article 7) (Rivers Coalition in Cambodia, 2007)

The MRC should clearly define its own responsibilities (i.e. roles, duties and obligations), and also understand those of other Mekong region water actors. In doing so, constituencies and accountabilities are clarified. Key questions include: what are the responsibilities of all stakeholders in a particular matter? Who is accountable to whom and for what? Are these responsibilities contested (Petkova and Veit, 2000; UN, 2006)?

Risk-reducing

The new CEO has expressed his desire for MRC to be risk-reducing, while the member countries are capitalizing on development opportunities. For these authors, risk assessment and risk management are an important element of water use and related development. In the past, most attention was usually given to investment risk by either public or private investors. There is now often a much stronger focus on the risks of all actors affected by a decision. Distinguishing between different types of risk is a good way to start.

Voluntary risk-taking includes risks taken in the normal course of business – for example, when a private company invests in a hydropower dam, or a public company invests in a water supply systems – or business partnerships between the public and private sectors. Involuntary risk-bearing is quite different. For example, people displaced by a new reservoir, or those whose water entitlement is reduced as the result of a reallocation, are involuntary risk bearers.

Risk analysis should not ignore voluntary risk-taking, but should also focus on involuntary risk-bearing (WCD, 2000, p207; Dore et al, 2004), whether it is fair and effective, and, if not, how can it be made so. Key questions include: for different options, what are the possible risks? Who are the voluntary risk takers? Who are the involuntary risk bearers? How might risk be equitably shared and, especially, how might involuntary risk be reduced?

The new CEO takes the view that:

> *The Secretariat has at least three roles in assessing and advising on opportunities and risks. One relates to the analysis of implications of projects, including the cumulative effects of national projects. This draws on work under a range of our programmes and, as I mentioned earlier, is being brought together by assessing various development scenarios under the Basin Development Plan. Another is to provide advice on specific projects where requested, including through our forthcoming Hydropower Programme. The third relates to administering the formal notification and consultation procedures under the 1995 Agreement, and, where required, providing technical advice under such procedures and facilitating negotiation of agreements.* (Bird, 2008a)

In comparison to his predecessors, these are extremely progressive statements embracing the tools of cumulative impact assessment and scenario-building, providing specific advice on projects and commitment to using the formal notification and consultation procedures.

Examining rewards and respecting rights

There are two other 'Rs' worthy of further attention by MRC. Thus far, there has been very little examination of rewards (winners and losers) and their distribution; and there has been an aversion to tread on the sensitive topic of often overlapping claims and rights.

The MRC could also emphasize the importance of identifying and unpacking rewards. This is not just the realm of economics, but rigorous economic assessment would be a good start. For different options, what are the possible multifaceted rewards or benefits (Sadoff and Grey, 2002, 2005)? Who stands to win? Who stands to lose? How might rewards be shared? Are there 'net' benefits? What is fair? What might be more ecologically, socially and economically sustainable?

The MRC could also display its concerns for the development of rights over and above territoriality and the sovereign rights of states. At various scales, water-sharing rights, or entitlements, may be assumed, negotiated, bestowed, contested, bought, sold, rented, traded, perhaps agreed upon, and sometimes ignored (UN, 2003; Scanlon et al, 2004). Rights analysis needs to be cognizant of a wide range of water-sharing regimes and the likely impacts of different options. An important departure point can be seeking answers to questions such as what is the history of water-sharing/management and use in a particular place or system? What are the entitlement claims of all stakeholders? Are these entitlements contested and, if so, on what grounds? Whose rights are affected by water resources development and allocation? How can these sometimes overlapping entitlement rights/claims be respected while searching for fair and effective workable agreements?

Conclusions

Governments need to make more informed decisions about whether to proceed with water resources development projects, taking into account comprehensive options assessment examining political, social, economic and ecological impacts – and drawing upon scientific evidence, situated local knowledge, and appreciating complexity and uncertainty. There has been an absence of informed discussion in the public space about the pros and cons of dams and diversions in Lower Mekong countries that have re-emerged on the agendas of national governments and transnational capital providers and developers. There is a need for transnational, transboundary public examination via high-quality, well-informed deliberative processes. This requires competent design, convening, facilitation, knowledge inputs and wise use of the media.

New flow regimes will have to be negotiated on Mekong River tributaries and, perhaps, the mainstream. Relatively little attention is being paid to how river flows will be 'managed' post-construction. There are many different possible scenarios. State and non-state actors need to become more familiar with flow negotiation tools and approaches that have the potential to ensure that all relevant issues and perspectives are taken into account in the inevitable negotiations ahead.

The MRC must increase its engagement in these issues. This will require applying existing and new research to discover methods appropriate for the Mekong region. Other essential ingredients are great diplomatic skill and social capital to allow equitable and informed negotiations to proceed. The MRC has deservedly received criticism for its performance thus far; but there remain many optimistic, latent supporters of the MRC initiative, hoping 'the family' will be enabled to capably respond to the current challenges. This will require the member governments, at the highest level, to 'de-marginalize' the MRC and its implementing parts, allowing them to make their best contributions.

A worthy goal is to make it normal practice in the Mekong region for important national and transboundary water-related options and decisions to be examined in the public sphere from a range of perspectives. Openness and deliberation are still far from being normal practice. The MRC, as mandated, has the opportunity and responsibility to play an important role in creating new, deliberative political space for learning and negotiating.

ACKNOWLEDGEMENTS

The authors wish to acknowledge the support of colleagues who read earlier drafts of this chapter: Louis Lebel, whose suggestions greatly improved the structure, François Molle and Andrew Noble. We also thank an anonymous reviewer and the Blue Moon Fund.

NOTES

1 King and his co-authors acknowledge that their data-compiling projects across the GMS is 'sufficient only for scoping purposes' as the data were 'compiled from a variety of sources and is unverified'. Existing projects are defined as: existing + those with financial closure + those under construction. Potential projects are defined as: committed + proposed + identified to any level of study. It is not implied that all 179 potential projects are necessarily going ahead. The figures used by King et al (2007) for Laos (11 existing, 32 potential) were assembled in 2006, and differ from the 2008 data quoted in the case study later in the chapter, which reported 77 projects at various stages from conceptualization/design through to operation.

2 The Precautionary Principle states that if a public action or policy may cause severe or irreversible harm, it should not be carried out despite the absence of full scientific certainty that harm would ensue. The burden of proof thus falls on those who would advocate taking the action.

3 The Review Team also noted that there was 'some concern among the staff about the consequences for themselves if they are too open with ideas and constructive criticism' (Hawkesworth et al, 2007, p17).

4 Presentation given by Chris Barlow, MRC Fisheries Programme coordinator, Vientiane, 20 June 2008.

5 The EIA was light in some technical areas (e.g. transboundary impacts), but spent considerable space exploring whether or not the development was a mainstream development (it is), pursuing a bizarre line that perhaps as the river is braided at this

point, the development could be seen as on a tributary; and, hence, whether, when and how it was compulsory, or not, for the Government of Laos to notify the MRC.

REFERENCES

Badenoch, N. (2001) *Transboundary Environmental Governance: Principles and Practice in Mainland Southeast Asia*, World Resources Institute, Washington DC

Baird, I. G., Barney, K., Bakker K., Beasley I., Blake D. J. H., Bush, S., Bourdier, F., Chapman, G., Cunningham, P., Davidson, P., Dearden, P., Fisher, R., Ganjanapan, S., Hirsch, P., Hogan, Z., Hvenegaard, G., Jensen, J., Jonsson, H., Khim, K., Kingsford, R., Le Billon, P., Lopez, A., Manorom, K., Maxwell, A., Meynell, P.-J., Neave, M., Poulsen, M. K., Springer, S., Stacey, P., Steinmetz, R., Sverdrup-Jensen, S., Townsend-Gault, I., Welcomme, R. and Williams, F. (2007) 'Letter from scientists concerned for the sustainable development of the Mekong River', Australian Mekong Resource Centre, Sydney, Australia

Baran, E. and Ratner, B. D. (2007) *The Don Sahong Dam and Mekong Fisheries*, WorldFish Centre, Phnom Penh, p4

Bird, J. (2008a) 'Introductory speech by new CEO', Mekong River Commission, 7 April

Bird, J. (2008b) 'Requests from Mekong Civil Society', Mekong River Commission, 11 April

Browder, G. (2000) 'An Analysis of the Negotiations for the 1995 Mekong Agreement', *International Negotiation*, vol 5, no 2, pp237–261

Bui Kim Chi (1997) *From Committee to Commission? The Evolution of the Mekong River Agreements with Reference to the Murray-Darling Basin Agreements*, PhD thesis, University of Melbourne, Australia

Chambers, S. (2003) 'Deliberative democratic theory', *Annual Review of Political Science*, vol 6, pp307–326

Cocklin, C. and Hain, M. (2001) *Evaluation of the EIA for the Proposed Upper Mekong Navigation Improvement Project*, Prepared for the Mekong River Commission, Environment Programme, Monash Environment Institute, Melbourne, Australia

Dore, J. (2003) 'The governance of increasing Mekong regionalism', in Kaosa-ard, M. and Dore, J. (eds) *Social Challenges for the Mekong Region*, White Lotus, Bangkok, pp405–440

Dore, J. (2007) 'Mekong Region water-related MSPs: Unfulfilled potential', in Warner, J. (ed) *Multi-Stakeholder Platforms for Integrated Water Management*, Ashgate, Aldershot, pp205–234

Dore, J., Lebel, L. and Manuta, J. (2004) *Gaining Public Acceptance: A Strategic Priority of the World Commission on Dams*, M-POWER Working Paper (MP-2004-02) provided as a commissioned report to the UNEP Dams and Development Project, Chiang Mai University's Unit for Social and Environmental Research, p40, www.unep-dams.org

ESCAP (1997) 'ESCAP and the Mekong cooperation', *Water Resources Journal*, September, pp1–6

Finlayson, B. (2002) *Evaluation of Hydrological and Other Aspects of the EIA of the Navigation Channel Improvement Project of the Lancang-Mekong River from China–*

Myanmar Boundary Marker 243 to Ban Houei Sai of Laos, Centre for Environmental Applied Hydrology, University of Melbourne, Australia

Governments of Cambodia–Laos–Vietnam–Thailand (1995) *Agreement on the Cooperation for the Sustainable Development of the Mekong River Basin*, Signed in Chiang Rai, Thailand, 5 April 1995, www.mrcmekong.org

Hawkesworth, N., Fox, J., Sokhem, P., Broderick, J., Hoanh, C. T., O'Keefe, V., Phommahack, S. and Viriyasakultorn, V. (2007) *Independent Organisational, Financial and Institutional Review of the MRC Secretariat and the National Mekong Committees*, Mekong River Commission, Vientiane, www.mrcmekong.org/free_download/report.htm

Hirsch, P. and Mørck-Jensen, K. (2006) *National Interests and Transboundary Water Governance in the Mekong*, Australian Mekong Resource Centre, University of Sydney, Sydney, Australia

Hortle, K. G. (2007) *Consumption and the Yield of Fish and Other Aquatic Animals from the Lower Mekong Basin*, MRC Technical Paper no 16, Mekong River Commission, Vientiane, p87, www.mrcmekong.org/download/free_download/Technical_paper16.pdf

IUCN, TEI, IWMI and M-POWER (2007a) *Exploring Water Futures Together: Mekong Region Waters Dialogue*, Report from Regional Dialogue, Vientiane, Laos, World Conservation Union, Thailand Environment Institute, International Water Management Institute and M-POWER, p75, www.mpowernet.org/download_pubdoc.php?doc=3274

IUCN, TEI, IWMI and M-POWER (2007b) *Exploring Water Futures Together: Mekong Region Waters Dialogue: Resource papers from Regional Dialogue*, Report from Regional Dialogue, Vientiane, Laos

King, P., Bird, J. and Haas, L. (2007) *Environmental Criteria for Hydropower Development in the Mekong Region*, Technical report from joint initiative of Asian Development Bank (ADB), Mekong River Commission (MRC) and WWF, Vientiane

Mann, F. M. (2007) 'Follow-up by MRC donors to the 12th meeting of the Council of Mekong River Commission of 15 November 2007', Letter from the Ambassador to Cambodia of the Federal Republic of Germany, 20 December

McDowall, R. (2002) *Evaluation of Fisheries Aspects of the EIA of the Navigation Channel Improvement Project of the Lancang–Mekong River from China–Myanmar Boundary Marker 243 to Ban Houei Sai of Laos*, National Institute of Water and Atmospheric Research, New Zealand

Molle, F. (2008) 'Nirvana concepts, narratives and policy models: Insights from the water sector', *Water Alternatives*, vol 1, no 1, pp131–156

Molle, F. and Floch, P. (2007) *Water, Poverty and the Governance of Mega-Projects*, M-POWER Working Paper MP-2007-01, Mekong Programme on Water Environment and Resilience, www.mpowernet.org/mweb.php?pg=92

MRC (Mekong River Commission) (2001) *Annual Report 2000*, MRC, Phnom Penh

MRC (2006) *Strategic Plan 2006–2010: Meeting the Needs, Keeping the Balance*, MRC, Vientiane, Laos

MRC (2008) *Documents for Informal Donor Meeting*, 19–20 June 2008

MRC Donors (2007) 'Joint statement by donors at 12th council meeting of the Mekong River Commission, 15 November 2007', Siem Reap

MRC Donors (2008) 'Joint statement by donors at informal meeting with Mekong River Commission, 20 June 2008', Vientiane, Laos

MRC Secretariat (2007) *Review Report of the Environmental Impact Assessment Report Prepared for Don Sahong Hydropower Project, Lao PDR*, Draft 19, November, Mekong River Commission Secretariat, Vientiane, Laos, p21

Nette, A. (2008) 'Concerns rise with planned Lao dam on Mekong', *Inter Press Services*, 28 March

Öjendal, J. (2000) *Sharing the Good: Modes of Managing Water Resources in the Lower Mekong River Basin*, Department of Peace and Development Research (PADRIGU), Goteborg University, Goteborg, Sweden

Petkova, E. and Veit, P. (2000) *Environmental Accountability beyond the Nation State: The Implications of the Aarhus Convention*, World Resources Institute, Washington DC, www.wri.org/wri

Rivers Coalition in Cambodia (2007) *Cambodian Civil Society Calls for the MRC to Address Its Shortcomings*, Press release, 16 November

Sadoff, C. and Grey, D. (2002) 'Beyond the river: The benefits of cooperation on international rivers', *Water Policy*, vol 4, pp389–403

Sadoff, C. and Grey, D. (2005) 'A cooperation on international rivers: A continuum for securing and sharing benefits', *Water International*, vol 30, no 4, pp1–7

Scanlon, J., Cassar, A. and Nemes, N. (2004) *Water as a Human Right?* World Conservation Union (IUCN), Cambridge

SEARIN, Project for Rivers and Communities and Love Chiang Khong Group (2002) *Mekong Rapids under Fire*, Southeast Asia Rivers Network, Project for Rivers and Communities, Love Chiang Khong Group, Chiang Mai

TERRA (Towards Ecological Recovery and Regional Alliance) (2008) *MRC's Looming Crisis of Legitimacy and Relevancy Challenges New CEO*, Press release, 27 March

UN (United Nations) (2003) *Water for People: Water for Life*, World Water Development Report 1, United Nations, www.unesco.org/water/wwap/wwdr1/table_contents/index.shtml

UN (2006) *Water: A Shared Responsibility*, World Water Development Report 2, United Nations, New York, www.unesco.org/water/wwap/wwdr2/

WCD (World Commission on Dams) (2000) *Dams and Development: A New Framework for Decision Making*, World Commission on Dams, Cape Town, www.wcd.org

15

Contested Mekong Waterscapes: Where to Next?

François Molle, Louis Lebel and Tira Foran

Introduction

The Mekong region has gone through massive human and material transformations (Rigg, 1997; de Koninck, 2005). Even as wars, expanding land frontiers, urbanization and industrialization have profoundly remodelled landscapes and societies, rivers and wetland ecosystems have remained persistent defining elements of rural livelihoods and agricultural waterscapes. Large-scale water resources development, although locally significant, has long remained short of the grand projects of 'harnessing' and 'taming' the Mekong River and its tributaries pushed forward by various regional organizations, governments and investors during the second half of last century. Regional conflicts and an obvious poor fit of many grand projects to local conditions and actual water/energy needs have thwarted large-scale investments (Kirmani, 1990).

Rising demands for energy, recent soaring fossil fuel and agricultural prices, and improved relations among China and other countries in the region have contributed to a renewed groundswell of interest in hydropower and irrigation projects. As earlier chapters show, many old projects are being dusted off; earlier concerns with environmental and social impacts are being addressed, or dismissed, with a fresh rhetoric of mitigation, trade-offs and best practices. Development banks and governments liken poverty alleviation to investments in infrastructure, while powerful new actors – private companies and banks from the region – have entered the scene and are reshaping patterns of water governance. The long imagined grand waterscapes of the Mekong region are once again being promoted, pursued and contested.

Powerful coalitions are bent on instilling a sense of inevitability well incarnated in the statement that 'development cannot wait'.[1] Investments in health and education, as well as water and electricity infrastructure, are desirable and necessary to improve the lives and living conditions of people in the Mekong region. However, experience illustrates that 'the subsidized construction of massive infrastructure is most unlikely to provide the optimal result in this respect for the poorer sections of the populations' (Phillips et al, 2006).

To different degrees according to place and time, the riparian countries of the Mekong all present tales of land, water and natural resource concentration: dam construction and reservoir water bodies displace residents; exclusionary forest zoning and watershed classifications dispossess ethnic minorities; dams that impact upon fish migration and natural flood regimes disrupt fisheries in rivers and wetlands; more powerful and organized groups redirect and seize opportunities and benefits derived from flood protection measures and irrigation schemes.

Unchecked states, bureaucracies and attendant private or political interests almost invariably fail to achieve a balance between economic, social and environmental dimensions of development; and it is hard to avoid the conclusion that more balanced outcomes have largely resulted from various forms of contestation.

Sewell and White (1966) once pondered why the human dimensions of water management were seldom considered or studied 'before the bulldozer moves in'. They found reasons 'rooted partly in engineering practice, partly in lack of funds and trained personnel, and partly in lack of analytical techniques'; in other words, regrettable (but hopefully transient) deficiencies in scientific tools and the understandable eagerness of the engineering profession to get the job done led to a predominance of infrastructural considerations over human concerns. This final chapter, largely drawing on the preceding chapters, broadens Sewell and White's question and answers by reflecting on patterns of water governance in the past half century. We identify a much wider set of interests in water resources development and management – from financiers and politicians, owners and operators of new infrastructure, to consumers, academics and organized community and civil groups or international non-governmental organizations (NGOs) of various stripes – and distil some of the key features of water politics in the Mekong region. We highlight how the governance and transformation of waterscapes in the Mekong region could move along a more fair and sustainable direction.

Water governance in the Mekong region

Understanding impacts

The issue of social and environmental impacts has bedevilled many projects; a lot of politics revolves around how these impacts are identified, framed, assessed, valued, mitigated and compensated. Social and environmental impacts are usually identified by (often mandatory) impact assessments. These studies are often not made available to the public, sometimes undertaken after construction has started, and premised upon an approach of mitigating impacts. Impact assessments tend to be seen as a 'bureaucratic "hoop" to be jumped through in order to start construction, not as an authentic mechanism to decide whether or not the dam should be built' (Friesen, 1999). States are often content to take impact assessments as just another perfunctory step towards project approval or completion.

Salinization problems and conflicts over floodplain management around the Rasi Salai and Hua Na dams on the Mun River in Thailand have generated debates about environmental impact assessments (EIAs) and critiques about the ways in which substandard assessments are used to 'green wash' projects (see Chapter 10). For example, the first EIA performed for the Songkhram Irrigation Project in 1992 by consultants was rejected by the National Environmental Board (NEB), after finding that these were exact copies of EIAs that had been done earlier for another large-scale water diversion project, the Khong-Chi-Mun (Breukers, 1999; see Chapter 7). Likewise, the 1993 preliminary EIA of the Theun-Hinboun Project in Laos by Norconsult was rejected by the Asian Development Bank (ADB) because of its poor quality (Barney, 2007), while the independent assessment of the Chinese navigation project commissioned by the MRC in 2001 found that the EIA was 'substantively inadequate and in many places fundamentally flawed' (Hirsch and Mørck-Jensen, 2006). Problems like these have led local groups to engage in grassroots research in order to mobilize local knowledge and to produce 'people's EIAs' (Manorom, 2007; see Chapter 13).

In carrying out assessments, pre-existing benefits are frequently downplayed. In the Nam Songkhram wetlands (Chapter 7), people rely on diverse natural resources that provide both food and income. Many of these benefits are not evident in enumerations of cash incomes or macro-economic analyses measuring poverty levels.

Likewise, social and economic impacts are frequently glossed over. Differences between the number of people expected to suffer impacts both at the stage of the feasibility study and subsequently are often very large. In the case of the Pak Mun Dam, for example, the first studies had identified 243 households, while concerned people inventoried 1649 and the final compensation was extended to at least 1821 households, or even 6200 households if compensations for lost fisheries are included (see Chapter 3). In many cases, people receive 'too little too late', if anything. Compensations come under two guises: direct financial compensations, or indirect economic opportunities or subsidies to develop other activities.

Early water projects in the region, many of them carried out in a context of war or political tension, have been planned and implemented as indisputable acts of national security. Inaugurated in 1971, the Nam Ngum 1 Dam in Laos displaced 800 families who did not receive compensation. In northern Vietnam, between 50,000 and 60,000 mainly ethnic minority people, the majority of whom continue to suffer impoverishment, were removed to make place for the Hoa Binh

Dam, initiated in 1971 at an estimated cost of US$1.5 billion (Hirsch, 1998). In northeast Thailand, populations resettled during the construction of the Ubol Rat, Lam Pao or Nam Oon dams suffered many hardships.[2] Hori (2000) reckons that dam projects were saddled with many 'severe problems with compensation for land' that were based on low market values, with delayed payment not adjusted to account for inflation, and moneylenders and middlemen taking advantage of the situation to lend money at high rates.

Financial compensations are often minimal and quickly absorbed in the purchase of goods. But if they are quite generous – as in the case of the Pak Mun Dam, where high sums were eventually proposed to affected people – they may buy people's support for the project. This, in turn, can result in splitting protesters, and even in a desire by others to also be 'affected' and receive payments for land that exceed market values (see Chapter 3).

Other forms of compensation include financial subsidies or support for economic activities such as irrigation for intensification or diversification of agricultural production. Irrigation, however, does not necessarily benefit those who have lost their land. Promises of profitable cash crop production (which are also heavily resorted to at the time of the feasibility study and seldom materialize as planned) often amount to wishful thinking. The Theun-Hinboun Hydropower Project in Laos initially lauded by the Asian Development Bank (ADB) as a project with 'little for the environmental lobby to criticize' eventually resulted in widespread impacts upon fisheries, river embankments and riparian agriculture. Activities launched as compensation and mitigation measures have largely failed despite hikes in the project budget (Blake et al, 2005; FIVAS, 2007). Nearby at the Nam Theun 2 Dam (NT2) in Laos, the viability of a proposed cash-crop agriculture programme for resettled villagers has been questioned due to the poor quality of soils, experimental cropping methods and hypothetical markets. Another promise is that of aquaculture within reservoirs. In the Water Grid Project in northeast Thailand, promises were clearly self-serving desktop exercises. Consultants in charge of project design simply took for granted that 'farmers would have to greatly change their farming practices in order to shoulder water fees ... [and] switch from rice cultivation to other cash crops, which consume less amounts of water than rice' (*Bangkok Post*, 2004b). The problem of marketing was solved by merely 'recommending' a contract farming system with agribusiness companies 'to ensure that farmers can sell their produce at reasonable prices'.

Such assumptions typically make light of social and ecological complexity and production-cum-marketing risks that characterize much of smallholder production. In practice, many farmers do not have the skills, the knowledge, the capital or the labour force to engage in new activities with strong links to unstable markets (Cornford and Matthews, 2007), let alone the frequent cases where market opportunities remain elusive. Construction companies are usually well equipped to face technical challenges; but transforming 'aggregations of houses, community facilities, cleared patches of still-smoking vegetation and disrupted families into self-managed and self-sustaining communities, viable in all dimensions' (McDowell et al, 2008) is another story altogether (see Chapter 4). Without ongoing pressure and scrutiny, such immense challenges frequently lead to failure; meanwhile, the project proponents and their consultants pack up their bags to move on to the next project.

International development banks or bilateral cooperation organizations sometimes have to withdraw or refrain from associating with sensitive projects. The Tasang Dam on the Salween River in Myanmar/Burma has been part of a master plan for a Mekong region power grid; but the ADB backed away after 'serious socio-environmental concerns' were identified and the project did not pass the ADB's filters (see Chapter 5). Despite occasional complacency with shoddy EIAs, development banks foster a culture of impact assessment[3] that, however, find limited echo with governments in the region. While governments have passed *ad hoc* legislations that reflect changing societal values, these are often to fulfil lending and other requirements of banks and donors and 'do not generally measure up to contemporary international standards' (Hirsch and Mørck-Jensen, 2006).

Modes of engagement

How governments engage with the wider public in planning and implementing water projects varies tremendously from country to country in the Mekong region. Degrees of openness and transparency and the availability of channels for public participation and dissent are an initial measure of the quality of water governance. Practice differs from one case to the other, ranging from secrecy, forced displacement and overt suppression of dissent, buying-out or co-opting by representation in *ad hoc* committees or rewarding with well-paid consultancies, to genuine attempts to accommodate and incorporate diverse inputs from society.

As an example from one extreme, the construction of the Tasang Dam carried out under the military regime of Myanmar has constantly faced accusations of human rights abuses and potential widespread environmental damage (see Chapter 5). Over the past ten years, the Myanmar army is believed to have relocated more than 60,000 villagers from areas adjoining the dam and the inundation zone (SSEO, 2006); forced labour, rapes and killings are being linked to the regime's intimidating preparations to build the dam (EarthRights International, 2005).

Planning processes often unfold behind closed doors. The memorandum of understanding (MoU) signed by the Electricity Generating Authority of Thailand (EGAT) regarding the build–operate–own Hutgyi Dam in Myanmar states that 'each party shall strictly keep confidential any and all technical, legal and commercial data and information' (see Chapter 5). Likewise, in China, despite new regulations in 2003 providing for public input on the EIAs for large projects such as the Nu-Salween River dams, authorities in the development companies and the National Development and Reform Commission insist that the Nu is a transboundary river and therefore detailed hydrological data on the river are of national security concern and the EIA cannot be made public (see Chapter 5).

In Thailand, the planning of large-scale irrigation schemes in the northeast ('Green *Isaan*', 'Khong-Chi-Mun', or 'Water Grid' projects) have been shrouded in secrecy with occasional media releases and official declarations creating confusion rather than clarity around implausible targets and dubious assumptions (see Chapter 10). Water planning in the Songkhram Basin (see Chapter 7) and several weirs constructed on the Chi-Mun mainstream and tributaries have also been imposed without space for deliberation. The Pak Mun Dam sticks out as an exception (see Chapter 3), where contestation – after construction – compelled the scrutiny of the dam by outsiders, including a team from the World Commission on Dams (WCD) and Thai academics. But political space itself may be a temporary phenomenon. Efforts in 2004 by Thailand's minister of natural resources and environment to reform the EIA procedures in the face of rapid economic growth and to promote participation from the public were not rewarded, and perhaps resulted in his subsequent removal and replacement by a politician supportive of the Water Grid Project (see Chapter 10).

Other multilateral actors involved in water resources development have tried to take the 'participatory imperative' more seriously. The 2000 annual report of the Mekong River Commission (MRC) acknowledges that it is 'important that decisions on development include a "bottom-up" process and are not confined to a "top-down" approach. The voice of the people directly affected, and of other stakeholders such as community groups or NGOs, must be heard.' In 2005 the MRC (2005) issued a strategy document on stakeholder consultation and public participation strategy. The 2006–2010 plan, however, partly reflects the efforts of several donor states rather than the conviction of riparian governments, which 'see participation as, at best, a tool of antidevelopment northern environmental groups or troublesome local NGOs and, at worst, as worthless' (Sneddon and Fox, 2007). The MRC's uneasiness at engaging with non-state actors is perceptible in that it refers stakeholder participation back to the National Mekong Committees (NMCs) arguing that they can best implement it. But many civil society organizations have been unable to engage with the state-centric NMCs beyond a very basic level (see Chapter 14).

The public does not always wait to be consulted. If sufficient safe spaces are available, the actions of advocacy groups in civil society can be an important driver of decision-making processes. Moreover, public mobilization may be a prerequisite to compensation. After construction of the Theun-Hinboun Hydropower Project, for example, it gradually emerged that the project reduced fishery catches by between 30 and 90 per cent along the three rivers it affected, and impacted upon the livelihoods of 30,000 people living downstream and upstream of the dam (see Chapter 2). It is only after independent external investigations and pressure from the International Rivers Network and the Association for International Water and Forest Studies (FIVAS) that the ADB acknowledged that the project impact area should be expanded, and established a ten-year US$4.7 million mitigation and compensation plan (Barney, 2007). In the case of the Pak Mun Dam (see Chapter 3), villagers had to fight for each compensation (houses, land and lost fisheries) and their 2000 blockade of the dam appears to have done what many months of sit-in demonstrations outside Government House since 1994 could not: it conveyed to EGAT senior management that they needed to take much more active measures to address local concerns and resolve the conflict.

The lack of transparency in the planning process and in the operation of infrastructures nurtures *ad hoc*, emotional and often self-serving interpretations of events. A good illustration of this was the higher than usual flooding that occurred in the Mekong River in August 2008, which triggered a rapid response from a coalition of local and international organizations typically opposed to mainstream dams (see Chapter 11). These organizations were quick to assert that the serious flood conditions were, in part, a result of operations of dams in Yunnan Province of China, while the MRC stood in defence of China, saying there was no evidence that upstream dam operations had any impact upon the severity of the flood (Wipatayotin, 2008). It followed up quickly with more detailed analysis (MRC, 2008). Poor analysis of, and lack of access to, credible information tend to beget suspicion and conflict.

The NT2 dam in Laos, although not yet completed, has been heralded as a success story and an example of 'doing dams right' by the World Bank (Porter and Shivakumar, 2008). The World Bank and other NT2 proponents have claimed that the project achieved public acceptability in Laos through consultation processes that occurred throughout the project development period and that social and environmental impacts have been adequately addressed through 'skilful management, effective communications and technical expertise' (Porter and Shivakumar, 2008). Despite unprecedented attention and funding devoted to mitigating socio-environmental impacts, it is apparent that constant scrutiny by NGOs and other outsiders has led donors and developers to improve standards. As a senior staff from the Italian–Thai company involved in NT2 admitted,[4] the technical alternative with regard to the tailrace channel, which greatly lessens the impact of the dam's releases upon the Xe Bang Fay Basin residents, would not have been implemented had NGOs not provided pressing advocacy on the negative impacts of the project.

The World Bank and others touted the transparency and participation of the process, and pointed to the stacks of studies assessing NT2's environmental and social impacts (see Chapter 4), while *The Economist* (2005) claimed that local people had been consulted until 'they were blue in the face'. While the NT2 project can be credited with several innovative aspects, such as the presence of independent monitors, a revenue management framework and a commitment to public reporting largely adhered to, shortcomings are also apparent as several social and environmental commitments are either loosely or not completely adhered to (see Chapter 4).

More crucially, such efforts at improving governance, meant to pave the way for improvements in the planning of subsequent dams,[5] may have brought about the opposite result: since NT2, the governments of Laos, Cambodia and Vietnam have started bypassing international development banks in favour of private operators and bilateral agreements. Thus, rather than conduct a transboundary study required by the ADB that ultimately could have required them to pay compensation to the affected villagers, Electricity of Vietnam (EVN) ultimately acquired funding from Russian sources to go along with the Sesan 3 Dam (Hirsch and Wyatt, 2004). EVN has also welcomed other foreign assistance and funding sources with limited social and environmental conditionalities, including Chinese companies (for the Lao Cai Hydropower Station), the Indian Export-Import Bank (which provided a loan for the Nam Chien Hydropower Plant), and Russian financial and technical support (e.g. Son La and Sesan 3 Dam projects) (see Chapter 2).

Finally, not all public responses to interventions in rivers for hydropower, irrigation and flood management are organized and explicit. Perhaps more often than is realized, individual local water users and people at risk adapt to changed flows, burdens, ecosystem conditions and opportunities. The aggregate response of many farmers, irrigation districts or city wards can significantly change the effective way in which water is governed. Benefits, burdens and risks can be redistributed without a word being said or placard hoisted. The significance of individual agency can be illustrated with responses to flood interventions. Bangkok, more than any other city, has acquired its flood protection system largely by fragmented accumulation (see Chapter 11). After major events, different parts of the city take action, acquiring pumps and building canals, river walls or dikes. Within the larger metropolitan bureaucracy, different districts 'game' each other's flood protection operations. The result in both cases is that actual movement of flood waters is often complex. Adding new measures in such a complex system invariably creates side-effects for others on the wrong side of the wall, end of the tunnel or receiving end of a drain (see Chapter 11). Each new intervention triggers a series of compensatory responses, both operational and infrastructural.

Interests and ideologies

Large-scale public investments provide opportunities for private gains to powerful players that typically include local/national politicians, bureaucrats, firms and funding partners. These groups are often associated in 'iron triangles' (Woodall, 1993) or 'iron rectangles' (Molle, 2008b) – systems of vested interests that encourage bribery, bid-rigging, the exchange of favours, or simply overestimation of benefits and neglect of costs in order to secure a steady flow of projects. Collusion between business, politics and bureaucrats in the water sector is a commonality shared by virtually all countries (Repetto, 1986), is thus not specific to the Mekong region, and has been well documented in countries such as Japan (Feldhoff, 2002) and the US (Reisner, 1986; McCool, 1987).

In the Mekong region, associations of groups with vested private interests are best documented in the case of Thailand (see Chapter 10). Chai-Anan Samudavanija (1995) underlines how 'in the name of "economic development" the military and bureaucratic complex acquired additional financial sustenance through sponsoring infrastructure construction in rural areas' and points to the corruption associated with these projects that has helped the various patron–client networks maintain their political authority. Bruns's (1991) study of water resource development in northeast Thailand shows evidence of how 'irrigation projects are large and visible rewards that politicians can offer in exchange for support'. Members of parliament are active in lobbying the Royal Irrigation Department (RID) for projects either at the request of their constituencies or in self-interest.

Pondering over the announced megaprojects, Ekachai (2008) concludes that 'the construction business, the local godfathers-cum-politicians and the bureaucracy will get richer from these mega-projects. Not the villagers. Not Mother Nature. But that is not the government's concern.' At its worst, such collusion in schemes involves notorious godfathers, as shown by the relationships between high-ranking officers and mafia leaders from Khon Kaen who assisted in the Green *Isaan* Project (Phongpaichit and Piriyarangsan, 1996). Myanmar's largest construction company, Asia World Co, was founded in 1992 by Lo Hsing Han who also controls one of the largest armed drug trafficking gangs in Southeast Asia (see Chapter 5).

Powerful and well-connected politicians or companies easily capitalize on large-scale water projects. In the lower Songkhram Basin, SunTech Group Ltd, which had acquired close to 10,000ha of floodplain land at very low prices (Blake, 2008) and used state subsidies for eucalyptus plantations and for establishing a modern vegetable canning factory, saw possible projects in the lower basin as an unexpected opportunity to receive compensation for land after its undertaking completely failed (see Chapter 7). Flood protection schemes also allow officials or wealthy individuals to influence plans so that their land ends up protected by dikes; they can also buy land targeted for flood security and resell it for profit (see Chapter 11). In eastern Bangkok, public pressure from groups affected by floods has led to river walls along 80km of the river at a high cost of 0.1 million to 0.3 million baht (US$6000) per metre, a type of investment that provides benefits to many private interests.

Overlap between private interests and political functions is also apparent in the case of the Thai MDX company, which is steered by a former minister of foreign affairs and a minister of commerce once bent on opening neighbouring markets to Thai companies. Convergence of bureaucratic and private interests is not new. During the Vietnam War, the endless and costly dredging works undertaken in the Mekong Delta at public expense were frequently met with scepticism about whether aid was merely enriching French and American interests (see Chapter 8), including American construction firms such as RMK-BRJ[6] (now part of Halliburton). In the present era, concerns have surfaced about the influence of politically connected contractors, as well as planning and engineering departments in Hanoi and Ho Chi Minh City, in infrastructure development plans in the Ca Mau Peninsula (see Chapter 8). In some cases, private interests may emancipate themselves from bureaucratic control to the point that in late 2006 the Chinese Minister of Water Resources referred to the 13-dam cascade planned on the Nu-Salween River as a case of 'predatory development'.

Chomchai (1994) notes the pressure exerted, as early as the 1950s, by international aid organizations to formulate development plans and 'mobilize maximum foreign assistance', while Kirmani (1990) sees the Mekong Project as 'a classic example of external effort, external management and external planning with little involvement of the beneficiaries'. Aid is sometimes part of fishy deals, as illustrated by the Green *Isaan* Project, for which Margaret Thatcher's government was ready to grant US$100 million and loan US$500 million if agreement was found on a planned major package of military equipment purchase (see Chapter 10). The interests of development banks and their pervasive 'lending culture', where staff incentives are aligned with the maximization of loans, are also influential in the decisions to invest. While banks pretend to have policies driven by borrowers' demands, it is readily apparent that their policies are often supply driven and internally defined. This is illustrated by an ADB official who stated that 'ADB has decided to prioritize water investments ... the President has decided to double – up to $2 billion a year – investments in the water sector' (Drooj, 2006).

In sum, the pivotal drivers of large-scale water resources development lie within webs of interests that associate the most powerful political, bureaucratic or business groups or sectors of society together with foreign companies or international organizations. Decision-making appears to be highly political, in the broad sense of the term, and only marginally based on technical or economic fundamentals.

But water resources development is also predicated upon viewpoints, values and ideologies (Molle, 2006; see Chapter 10). The vision of nature as a threatening environment that must be 'harnessed' or 'tamed' through massive injection of capital, technology and concrete has fuelled much of the 20th-century 'hydraulic mission' and is still a very pervasive mental framework (see Chapter 1). This is apparent in grandiloquent language, such as the promotion of 'megaprojects' expected to 'eradicate poverty'; the ideology of 'big is beautiful' is also perceptible in minor details, such as the names of some construction companies in the region – for example, the Malaysian Mega First Corporation Berhad, which is involved in the Don Sahong Hydro Energy Project in Laos (see Chapter 14).

Extreme flood events provide opportunities to call for and strengthen control strategies. After the 1966 floods in the Mekong Basin, the executive secretary of the United Nations Economic Commission for Asia and the Far East (ECAFE) declared that the flood had 'deepened the determination of all of us engaged in the Mekong effort to convert the wasted and destructive powers of the Mekong untamed into a giant tamed and harnessed to the uses of mankind' (Jenkins, 1968). Likewise, the study by NEDECO/TEAM (1983) on the Songkhram Basin came up with a plan to tame the 'unruly' Nam Songkhram River. Another central argument (see Chapter 7) – albeit ubiquitous and not specific to the region – is that 'water flows to the Mekong unused' (Roongrueng, 1999), a typical statement insensitive to wider ecosystemic functions of the water regime, as well as to pre-existing people's livelihoods, echoed in 1995 by the foreign minister of Thailand, who found it 'a pity to let the Mekong River, with its abundance of water resources, just flow to the sea' (cited in Friesen, 1999; see Chapter 10).

Such approaches and views of nature have their root in colonial practice and in the iconic model of basin-wide 'comprehensive development' of the Tennessee Valley Authority (TVA), applied to northeast Thailand by the US Bureau of Reclamation in 1965, where almost every single tributary to the Chi and the Mun rivers was planned to be dammed in its upper course (Floch et al, 2007); the same model has been projected at the Mekong Basin scale. The culture of full control was strengthened by many visits from the Mekong Committee and Thai technical departments to the TVA, the Bureau of Reclamation and the Columbia River Basin Development Headquarters (Darling, 1962; Hori, 2000; Biggs, 2006).

The 'great potential' of this 'majestic river' was praised early on by Wheeler's (1958) study and engineers would marvel at the 'potential' of all the 'promising dam project' sites they would identify and at the 'tremendous potentialities for power production, irrigation, navigation and flood control' that C. H. Schaaf, the first executive agent of the Mekong Committee, saw lying in this 'sleeping giant'. The heyday of heroic and enthusiastic engineering is epitomized in Hori's (2000) account of the early Mekong development plans when 'the Japanese team's grand vision of development in Cambodia' included the Stung Sen Dam, whose 'grand scale … amazed ECAFE'.

These dreams did not remain unchallenged. US geographer Gilbert White and his colleagues (White et al, 1962) warned that the Lower Mekong countries could not 'stand the luxury of monolithic concrete structures whose immediate return is inflation of national ego'. When concerns related to the social and environmental impacts of the proposed Mekong Development Scheme emerged, the US Agency for International Development (USAID) also commissioned a study on the 'social feasibility' of the Pa Mong Dam (Ingersoll, 1969). But the report did not receive much attention from the Mekong Committee. As reported by Ingersoll (1969), C. H. Schaaf responded that 'he had wanted no criticism of the Mekong River project: it was good, all good, nothing but good'. Despite greater emphasis on social and environmental issues, it is apparent to many observers that the developmentalist vision of resources use in the Mekong is well and alive, entrenched in narrow conceptions of sovereignty, and has been only marginally swayed by contestation (Friesen, 1999; Fox, 2000; Hudson-Rodd and Shaw, 2003; Goh, 2004; Hirsch and Mørck-Jensen, 2006).

Discursive practices

A good deal of the debates and conflicts around water development or management decisions are linked to various, often antagonistic, discourses. These discourses (and associated options, ideas, values and narratives) can be observed in confrontation at meetings, public hearings and multi-stakeholder platforms, as well as in written texts and the media. The discursive dimension of power, although often ignored, is a key element of governance. Several chapters in this volume have evidenced different components of discursive power – from weaving narratives, labelling peoples and conjuring up meta-justifications – in debates over water resources development and management in the Mekong region.

Narratives are 'a story with a beginning, middle and an end' (Roe, 1991). They define a problem, explain how it comes about, and frame it in a way that suggests particular courses of action while ignoring others (Keeley and Scoones, 1999). Likewise, positive narratives associate a desirable outcome, often reduced to an alluring rosy picture or a catchy motto, with obvious solutions, generally provided by a benevolent state bent on distributing the fruits of growth and development. Narratives and the visions and solutions they promote are frequently legitimized by association with powerful 'nirvana concepts' (e.g. good governance and integrated water resources management) that are by nature consensual and serve as a means of closing debates (Molle, 2008a). Projects such as NT2 are being repackaged as environmental management projects or, rather, 'not as a project *per se*, but as a vehicle through which to make a considerable progress in the effort of poverty reduction' according to Shengman Zhang, the World Bank's managing director in 2003.

Regional politics have promoted the 'Mekong spirit', described by U. Nyun, executive secretary of ECAFE during the 1960s, as 'the great goodwill, the friendly spirit of collaboration, the abundant enthusiasm which animates Mekong work', and conveyed an ideal of solidarity, cooperation and mutual help expected to keep the committee members 'above ideological and political disputes' (Menon, 1972). According to the all-purpose phrase of the committee founding document, activities were carried out 'for the benefit of all the people of the basin, without distinction as to nationality, religion or politics'. This 'Mekong spirit' rhetoric has endured over time and has helped to shape the eventful history of regional politics and development of the Mekong River as a success story, and fuelled a powerful narrative of converging goodwill and cooperation (Goh, 2004; Cornford and Matthews, 2007).[7]

The desirability – and inevitability – of developing the Mekong are the obvious feelings conveyed by most of the literature: 'A simple enumeration of the needs that could be satisfied by harnessing the water of the river gives an idea of the necessity to develop the Mekong Basin.' Hydropower generation, irrigation, flood protection, navigation and even fisheries (which need to be 'increased and diversified to produce the proteins needed by the population') are within reach to improve incomes and 'ensure a political environment that offers better hopes and a larger stability' (CCILMB, 1970). After recalling Li Ping's irrigation development of the Chengdu plains in the third century BC, which transformed '5000km^2 of semi-desert into one vast market-garden', another CCILMB (1972) report describes the Mekong development plan as 'several thousands times more ambitious than Li Ping's and, in fact, one of the largest water resources development schemes ever devised'. Visions of wealth and plenty associated with development, capital investments and water resources development have been distilled by leaders such as Subin Pinkayan, the former Thai minister of foreign affairs and minister of commerce, who once announced he wanted to turn the Southeast Asian mainland into Suwarnabhumi, or a 'golden land' (see Chapter 5). In the 1960s, then Prime Minister of Laos Prince Souvanna Phouma announced that the Nam Ngum Dam would irrigate 100,000 acres (40,468ha), transforming them into 'orchards and gardens' (Jenkins, 1968). Laos is now poised to become 'the battery of Asia', or even 'another Switzerland crossed by roads and railways, a country of services and hydropower'.[8] No doubt, all of these grand prospects and plans aiming to exploit water and other resources are legitimate; but casting expected benefits in such a glaring light often serves to justify and impose projects indiscriminately (as shown by experience worldwide), rather than establishing improved decision-making processes.

In the eyes of water engineers and power planners, the limited exploitation of the Mekong River system's hydropower potential – in a region undergoing rapid economic growth – is a global rarity (Ratner, 2003; see Chapter 2). When negative impacts are acknowledged, they are generally framed in a discourse of trade-offs and mitigation. Already in 1972 the Mekong Committee boasted 'the methods it employs to make sure that the benefits of development will be maximized and the costs – including ecological costs – are minimized' (CCILMB, 1972). Three and a half decades later, the Mekong Water Resources Assistance Strategy (MWRAS) strategy claims that livelihood restoration programmes for affected communities can mitigate negative impacts from the projects, and the compensation schemes or alternative opportunities offered to these communities might even result in 'win–win' situations (see Chapter 2).

If losses are unavoidable, these are, nevertheless, framed as an inevitable 'sacrifice' for the common good of the nation and undervalued. For example, drawing attention to the 'almost cataclysmic changes in the ecology' that would result from basin development plans, Tubb (1966), a United Nations Food and Agriculture Organization (FAO) fisheries official concluded, however, that such development could and should not be avoided because of the 'greater economic value' of planned water uses. The importance of capture fisheries is constantly diminished by an enduring narrative of doom (see Chapter 12). The narrative suggests that poor people fish, and that people are poor because they fish; that resources are declining and facing 'the tragedy of the commons'; and that natural fisheries can be aptly replaced by modern techniques of aquaculture and 'alternative sources of income (as provided by irrigated agriculture) and development generally (as facilitated by the availability of power)' (World Bank/ADB, 2006). The possibility of potential negative impact upon the capture fisheries is not necessarily denied, nor even downplayed; but the inevitability of trade-offs, with their connotations of 'balance', is reaffirmed (see Chapter 12). Another pervasive framing is that of floods as a threat and catastrophe (see Chapters 7 and 11).

Where the inevitability of negative impacts is not easily accepted, problems and solutions can be framed with visions of threats and doom. 'Water crisis looms', says a study on the Water Grid Project (*Bangkok Post*, 2004b), while the *Bangkok Post* (2008) discusses the hypothesis that at some time in the future Thailand would not be able to feed its own people, and a senior official justifies water transfer to the Phetchaburi Province that runs the risk of 'becoming a "desert" because the province received less rainfall than the amount of water evaporating from its soil' (*Bangkok Post*, 2004a). Promotional material printed for the Khong-Chi-Mun Project included drawings of *Isaan* as a piece of cracked soil traversed by unused rivers (see Figure 10.4 in Chapter 10).

Such framing of development issues in the Mekong region generates counter-framing. For example, researchers involved in the Nam Songkhram Basin (see Chapter 7) try to undermine the negative framing of flooding by stating that 'the local people consider it a disaster when there is no flooding'. The negative vision of wetlands as 'swamps' is likewise opposed by labelling wetlands as 'nature's supermarket' where you need no money to 'shop' for the large variety of different resources they provide (MRCS/WUP-FIN, 2007). Critics of the Pak Mun Dam countered narratives of doomed capture fisheries by demanding that the government open the gates of Pak Mun to restore fisheries and livelihoods; they subsequently produced their own study showing positive restoration effects (see Chapter 3).

Another common discursive practice is labelling, which consists of simplifying the complexity of some particular categories of people, the range of interests they represent, and the diversity of both their experience and their resource endowment (especially the environmental constraints that they may face) (Sutton, 1999). In particular, some groups are frequently associated with labels that bear a strong positive or negative undertone. 'Farmer' is usually used as a positive label when mobilized to justify new water projects. 'Farmers' associated with an image of *Isaan* that emphasizes drought, parched soils and migrating rural population are a handy way to justify bringing more water to the region, irrespective of the fact that, on average, farming now only represents a portion of rural household incomes (and often a minor one). The labelling of *Isaan* as a poor and drought-prone region (see Chapter 10 and Bell, 1969) has featured prominently in all projects to divert the Mekong River.

Right until the last decade of the 20th century, the emotive term 'communist' has been used to demonize and disqualify protest or dissent (Sretthachau, 1999). In the Pak Mun Dam controversy (see Chapter 3), during the early 1990s the police described people who distributed leaflets, wrote letters and attended demonstrations as 'communists', or more commonly categorized them as a 'minority' or as 'paid' agents (*Bangkok Post*, 1991). Likewise, in certain Thai official discourse, NGO has become a dismissive term: some officials distinguish 'bad' NGOs (those who engage in pressure politics and civil disobedience) from their 'good' (non-political) counterparts. Bad NGOs are sinister: they incite villagers (who are normally placid) to engage in sophisticated and disruptive 'mob' protests, so goes the dominant framing (Missingham, 2003; Foran, 2006).

A third form of discursive practice is to make appeals to justifications and goals with which almost everybody agrees. Meta-justifications are frequently mobilized to justify a particular project: they usually associate a sense of urgency with a general objective that can hardly be challenged, such as 'development can't wait!' and 'poverty eradication'. Invoking higher-level overriding benefits tends to make local counterclaims parochial, 'selfish', non-legitimate or 'backward'. Further misrepresentations of the debate consist in overemphasizing expected benefits (rural income will increase, farmers will grow two crops, etc.) without consideration of costs, thus avoiding discussions about alternative investments either in the water/agricultural sector or in the wider economy.

A particular strand of meta-justifications includes arguments that stress national security, or food self-sufficiency objectives, that inherently refer to the state's prerogatives and core duty. 'Securitization' of development objectives has been particularly prominent during the Cold War (see Chapter 10). Again, while such objectives may be desirable, they are frequently mobilized to justify both sound and poor projects indiscriminately. In Laos, internal resettlement is a key policy: it is justified by the government's expressed goals of 'poverty alleviation', 'rural development' and 'nation-building'. Ethnic minority populations living in mountainous areas are frequently seen as 'holding the country back' from achieving 'development' (Baird and Shoemaker, 2007). The government's alleged goals of opium eradication, swidden agriculture reduction, and improvement of accessibility to government services, tinged by security and 'nation-building' concerns, eventually translate into forced cultural integration and massive internal displacement with severe social impacts.

People have learned – in the case of Thailand, after several decades of democratizing struggles – to challenge such sweeping development or security narratives. The Assembly of the Poor (AOP, 2000), for example, underlined that the 'sacrifice for the country's development', explicitly requested by the government, 'involved destruction of our lives and communities' and was unimpressed by promises of a brighter future, adding that 'we were never poor until the day that you appeared in the name of "development"'. As for the Pak Mun Dam, while the project was predicated upon the need to electrify northeast Thailand for development, its contribution in 2008 – if working at its design capacity, which it was far from achieving – would have lowered instantaneous electricity peak demand by a mere 0.6 per cent, equivalent to providing electricity for two large shopping malls in Bangkok (see Chapter 3).

SHIFTING WATER GOVERNANCE

The preceding sections convey a rather bleak picture of the governance of water resources development in the Mekong region. We have identified a whole gamut of politics, attitudes and discourses, ranging from 'hardcore' developers insensitive to social-environmental issues, on one extreme, to, on the other extreme, activists opposed to any infrastructure, often seen as emblems of the transformation of nature into capital to benefit an elite (Parnwell and Bryant, 1996). Avoiding either extreme, this book has reflected on why current water governance is lopsided, and how debates and decision-making processes could be improved so as to ensure economically, socially and environmentally sound outcomes.

Five complementary and interdependent paths to improved water governance can be found in this volume. Each path attracts a different political traveller according to varying inclinations and professional backgrounds. The first path is that of knowledge production: that of conventional science, but also of alternative knowledge registers and narratives. The second path is centred on concepts of negotiation and deliberative democracy, and seeks to bridge antagonistic viewpoints, foster social learning, and reach agreements or build consensus. The third path focuses on establishing rules, standards and norms in order to frame and constrain behaviours and to limit externalities. The fourth path is that of advocacy, where a more direct political struggle is seen as the most effective way of empowering marginalized groups, voicing their concerns and tilting the balance of power. Last, in the particular case of the international rivers of the Mekong region, efforts at improving transboundary management of resources may also shape and improve the evolution of waterscapes in the region. These paths are reviewed here in more detail.

Co-producing knowledge

A first aspect of knowledge production is the generation of conventional science. Experts who are called to assess a particular project or to recommend adequate policies draw on a body of knowledge that is perpetually in the making. For example, whereas few Mekong river fish were regarded as migratory during the 1960s (Hori, 2000), specialists now estimate 'that over 70 per cent of the total fish catch in the Lower Mekong Basin is dependent on long-distance migrant species' (Dugan, 2008). The ecological impacts of dam development in the basin are not well captured by conventional crude hydrologic models; investigating impacts on the Tonle Sap ecology (Chapter 9) or coupling ecological models of primary productivity with the 'flood pulse' (Lamberts and Koponen, 2008) provides further and badly needed insight on expected changes.

But many times, as shown earlier in the discussion of the politics of knowledge (see also Chapters 12 and 13), distorted common wisdoms do not only reflect a possible imperfect knowledge, but also the particular narratives that are propelled by interest groups and that mirror power structures. It is therefore necessary to work on counter-narratives that re-establish a better balance in perceptions and understanding of reality. The myths and misperceptions about the projection of floods as disasters, or of fisheries as a doomed resource, must be combated with new knowledge that sometimes has to emancipate itself from the usual channels of scientific production, as shown by the example of the Tai Baan Research (see Chapters 7, 11 and 12). In some cases, it is the very existence of a denied fact that must be established through investigation, as in the case of NGOs documenting the impact of the Yali Falls Dam in Cambodia (Öjendal et al, 2002; Hirsch and Mørck-Jensen, 2006).

In other cases, the very scientific narratives called in support of a project are partially or flatly erroneous. Thailand's Prime Minister Samak Sundaravej, for example, recently refloated the idea of building the controversial Kaeng Sua Ten Dam on the Yom River in upper northern Thailand 'to protect Bangkok from flooding' (*Bangkok Post*, 2008), although consultants already demonstrated in the 1980s that the dam would have a completely marginal impact upon flooding in the lower part of the Chao Phraya Basin. Myths regarding the relationships between upland forest uses and downstream floods and droughts have also justified the expansion of state enclosures (national parks, forest or wildlife reserves, etc.), afforestation schemes, and the removal of ethnic communities in the uplands of northern Thailand; these myths have now come under greater scrutiny and been increasingly challenged by scientific research (CIFOR, 2004; Forsyth and Walker, 2008). The link between large hydropower development and poverty alleviation has also been cemented in development discourse in the region and notably in China (see Chapter 5), and alternative proposals have to deal with demonstrating the frailty of that link.

The 'risk society' described by Ulrich Beck (1992) associates the emergence of multiple ecological crises with the contestations of formal authority by social movements. The status of knowledge is now contested and risks that were perceived to be safely managed by experts have become subject to public debate. The Mekong River Commission (and, to some extent, development banks) have responded to these trends by repositioning themselves as knowledge brokers; but their scenarios, impact assessments and other cost-benefit analyses have also generated intense debates. Hence, those who travel the path of knowledge production invariably face self-serving arguments, narratives rooted in bogus science and tunnel visions. While exposing harmful untruths is a matter of urgent necessity, all knowledge brokers work in highly politicized contexts that necessarily have a bearing on the knowledge that they produce.

Debating alternatives

The negotiation path is predicated upon the observed possibility that actors engaged in controversies might not just negotiate based on predefined positions, but may actually somehow learn from each other, accept trade-offs and losses, gradually change their positions and viewpoints, and arrive at shared decisions or agreements, if not consensus (Dryzek, 2000; Warner, 2006; Karl et al, 2007).

At both local and basin scales, states' views are often enforced with little discussion; few opportunities exist for defining modes of co-management. In the coastal part of the Mekong Delta, for example, the authorities have gated the outlets to the sea in order to conserve fresh water inland and to foster rice multiple-cropping. But this decision undermined brackish water shrimp farming in the area and led to protests and structures being destroyed (Hoanh, 2003; see Chapter 8). This, in turn, forced the authorities to discuss with local villagers and enabled the definition of an agreement that allowed both rice and shrimp farming through adequate operation of gates. In the case of the NT2 Dam in Laos, as mentioned earlier, discussions with affected populations and NGOs allowed a technical alternative that minimized impact upon villagers in the Xe Bang Fai Valley to be found at an equivalent cost.

Decision-making is thus an (often long) process of social learning where the room for manoeuvre of actors and interest groups becomes constrained by public exposure of their interests, strategies and discourses, which may then be contested and scrutinized. Accessible policy conferences, public hearings and multi-stakeholder platforms all provide opportunities to advance in this direction (Warner, 2006; Dore, 2007).

Forums such as the Exploring Water Futures Together dialogue held in Vientiane in 2006 (IUCN et al, 2007) and the MRC's Hydropower Forum in 2008 brought together a diverse group of stakeholders, including elite policy-makers, developers, development bankers and advocates on behalf of vulnerable people. At their best, such events give space and legitimacy to suppressed narratives, such as the narrative of how the impacts upon fisheries from the planned dams to be sited on the mainstream Mekong River cannot be mitigated, as recent scientific research shows that the diversity of Mekong fish species and their migrations makes it impossible to mitigate impacts using fish passes or aquaculture.

It can be extremely difficult to persuade elite actors to participate in such multi-stakeholder events. But positive interactions build rapport and trust that could catalyse more sustained interaction.

Promoting standards

One particular pathway that potentially helps to charter the boundaries within which the different parties may act is to seek agreement on codes of conduct, or 'standards'. The World Bank and the ADB have, for example, defined guidelines for project planning and implementation that standardize procedures and establish policies on issues such as data disclosure, social and environmental impact assessments, resettlement and compensation of people displaced by dams.

In the energy sector, international standards in electricity planning, such as integrated resources planning (IRP), are now common in many developed countries (see Chapter 2). Energy planning in the past has been biased towards overestimating future demand in energy projections, leading to energy surpluses and over-investments in new capacity that are socially and economically wasteful (but lucrative to developers). The planning processes currently in place, both at the national and regional levels, fall well short of these standards. In Thailand, IRP is well known; but incentives to adopt it are not compelling. Plans conducted according to IRP principles would include more demand-side energy-efficiency measures. These would lower energy sales and construction of capacity, but are not 'attractive' in a context where utilities are allowed to recover their costs plus fixed rates of return on their investments. Not surprisingly, IRP in North America is typically a requirement imposed upon utilities by regulatory bodies.

But in contexts where regulatory regimes are still weak, self-regulation frameworks deserve mention. The Equator Principles, for example, are a set of guidelines aimed at private financiers of large infrastructure projects, particularly projects over US$10 million that are 'project financed'. Project finance is a method of raising large amounts of capital from both equity investors and lenders, including both commercial and development banks. Loans are typically secured by cash flows from a project company (a new organization that is legally separate from the investing parent firms). In the event of financial distress, lenders have limited recourse to the assets of the project company, but no recourse to the assets of the parent firms (Vaaler et al, 2008). This feature means that investors and sponsors have strong incentives to get projects built and operating on time, and weaker incentives to consider negative external effects (see Chapter 4).

The Equator Principles provide general guidance to investors for project evaluation, including initial risk screening; whether impact assessment is required and, if so, what standards to use; public disclosure; independent review; and compliance monitoring. The principles are based on existing guidelines and safeguards of the International Finance Corporation, in turn modelled after those of the World Bank. While these principles have been criticized as green 'window dressing', it is also apparent that many banks engaged in project finance have not signed up to the Principles, suggesting that complying with them imposes additional costs (Scholtens and Dam, 2007). Instead, signatories to the Principles tend to be larger banks with active Corporate Social Responsibility programmes. As of 2008, the Equator Principles were still new to the Mekong region. No regional banks had signed up. Further analysis of projects funded by Equator signatories (such as the Theun-Hinboun Expansion Project led by ANZ Bank) is necessary to tell if the Principles produce better projects.

These standards, predictably, are neither used nor accepted by all parties. Private- sector hydropower developers from Thailand, Vietnam, China, Malaysia and Russia, often backed by influential political players, government bureaucracies and financiers from their own countries, which have recently entered the dam building arena, are determined to build hydropower dams or irrigation schemes without becoming entangled within burdensome environmental and social dilemmas that have often dogged projects in the past. As a result, Mekong country governments may be tempted to make deals with such developers,[9] while development banks may be tempted to weaken their own standards to stay the course and to remain 'competitive' (see discussion on the ADB in Chapter 2).

Even actors attempting to adhere to standards often end up retaining information or ignoring events. For example, as part of its decision to proceed with NT2, the World Bank commissioned a study based on IRP principles (Greacen and Palettu, 2007). The study (du Pont, 2005) showed that feasible demand-side management, energy conservation measures and renewable energy generation in Thailand would exceed the output of NT2 and would provide energy to the customer at a cost approximately 25 per cent less than NT2 (see Chapter 4). However, the bank did not publish du Pont's study until after its board had approved NT2.

The low attractiveness of these standards is linked to the additional costs and time delays that they impose on planners and project developers. The maximization of profit works to edit out of the picture these nagging social or environmental impacts that will come in the way of bulldozers and building concrete infrastructures. What are the incentives for operators to adhere to constraining standards, especially in a context where those who adhere lose a competitive edge with regards to those who don't? Just as in the case of polluting industrial activities, profit and competitiveness are tightly linked to the non-consideration of the externalities generated.

In general, 'best practices' or standards address issues of concern to wider society through eliminating or minimizing externalities and sharing project benefits (see Chapter 2). Such practices may thus reduce political risk – for example, from protests or legal measures that could delay project construction or add unforeseen additional costs. Governments may also have an interest in selecting developers with a sound reputation to avoid political turmoil or social protests that could tarnish their reputation. However, commercial or strategic short-term interests often override the consideration of precautionary measures. Where corruption is high or local protest stifled, project developers perceive low political risk and feel less inclined to implement best practices. Past dam projects, unfortunately, confirm that compensation schemes and other concessions from dam builders and governments have generally been secured only after substantial mobilization or protest (see Chapters 3 and 4). Overall, weak regulatory regimes seem to require more direct political action in order to improve governance outcomes.

Contesting decision-making

Contexts with prevailing top-down, state-centred decision-making and weak political representation of marginal categories of population – not to mention cases where political freedom is restricted – are prone to producing projects where social and environmental impacts are discounted or ignored. Advocacy, whether organized around grassroots movements, networks of urban-based NGOs or transnational coalitions, is the path often chosen by those who suffer immediate losses of livelihoods and are left to their own devices (Young, 2001).

Advocacy coalitions in the Mekong region have been fast to move at challenging the alleged benefits of dams. The Pak Mun Dam story (see Chapter 3) is exemplary of several dimensions of political struggles around water development projects: this case study shows that compensations have been secured after repeated, protracted, costly and painful demonstrations and initiatives. Compensations and dam management adjustments were repeatedly promised and then denied, and sustained mobilization was necessary to ensure these mitigation measures. In other cases, including the Theun-Hinboun Dam Project (see Chapter 2) and the NT2 Dam Project (see Chapter 4), where the money spent on impact mitigation and the effort at ensuring transparency have arguably notably exceeded those of earlier projects, it is apparent that the constant pressure and scrutiny of outsiders have helped to raise the degree of adherence to standards or decent practices.

Yet, advocacy coalitions have their weaknesses. Grassroots movements, such as the Assembly of the Poor, that formed around the Pak Mun struggle may be undermined when the state organizes local opposition groups, engages in hostile media discourse or compensates free-riders. NGOs also have different priorities, with some more focused on conservation or biodiversity, and others more livelihood or human rights oriented. The IUCN, for example, supported the NT2 dam because it saw the revenue it would create as a means of establishing and maintaining protected areas in the Nakai Plateau around the proposed reservoir (Bakker, 1999). Other organizations, such as International Rivers or TERRA, opposed it on grounds of the expected destruction that it would bring to the local environment and to the impacts upon the livelihoods of local villagers around the dam site (see Chapter 4). While NGOs often accurately represent marginalized and vulnerable people, their advocacy narratives can drastically simplify complex development conflicts (see Chapter 3).

Perhaps the greatest hurdle that these actors face lies in promoting their messages of conservation, preservation and socio-environmental responsibility in areas where conditions of extreme poverty frequently prevail, without being seen as opposed to 'development' (see Chapter 4). Where the 'balance point' precisely lies between projects that clearly benefit private interests rather than collective ones, on the one hand, and total paralysis, on the other, is hard to establish; in many cases, debates seem to pit developers unprepared to admit that a particular project may be unsound against activists who take expected impacts as a reason for opposing any project. While the scope for energy savings or greener energy generation is substantial, meeting long-term projections in energy demand is likely to include projects that do have impacts.

Transboundary governance

While knowledge production, negotiations or political struggle often unfold at the national level, the linkages between the Mekong region countries through their dependence upon the same river system also opens up opportunities for improving water governance at a regional or basin level. There are now several overlapping institutions that have potential to contribute to improving transboundary governance.

The 1995 Mekong Agreement established a Mekong River Commission Secretariat and basic common principles and procedures. In the years since, member countries have struggled to negotiate specific and meaningful rules for water utilization, project notification and the coordination of development plans (see Chapter 14). For the most part, members have successfully maintained a situation where they can pursue their own interests unfettered as much as possible by concerns of other states. But it is also a situation in which individual and collective influence over decisions and activities by upstream China is modest. The MRC and its secretariat, in particular, have often had to tread a thin line between strong competing interests of member countries and those of donor countries and multilateral agencies. They have also had to commit to promoting participation while not threatening the long-term agenda of member states (Sneddon and Fox, 2007).

It is apparent[10] that countries are reluctant to give up sovereignty and that national interests prevail over transboundary interests (Hirsch and Mørck-Jensen, 2006). The 1995 Mekong Agreement, largely weakened to accommodate Thai interests and prerequisites (Ratner, 2003; Goh, 2004), is lacking 'legal teeth' to enforce its provisions (Dore, 2003). The representation of the Mekong River as a legal structure, as implied in the 1995 agreement, privileges the state and practices of sovereignty and confines transboundary management to an issue of allocation rules limited to the main stem of the river (Fox, 2000). In any case, there are no easy or consensual metrics to assess the effectiveness of the MRC. Sneddon and Fox (2006) caution that successful 'cooperation' might well result in ecological alterations and resource degradation for local people who depend upon river basins for their livelihoods. The MRC, ultimately, is 'owned' by its member states and cannot be expected to act against their agendas. It was first weakened by a post-Cold War context that provided fewer incentives for states to cooperate (Ratner, 2003) and is now at risk of being increasingly sidelined because of the irruption of private banks and investors making direct deals with governments in the region.

Other intergovernmental frameworks for cooperation, such as the Greater Mekong Sub-Region (GMS) initiatives established and facilitated by the ADB, have not yet played a central role on water, although they have become increasingly important in related energy and transport sectors. Hydropower-related initiatives illustrate the potential of multilateral actors to support development and acceptance of standards – for example, for investment projects.

Civil society networks have also made some effort to go beyond national boundaries and to tackle regional governance problems. One of their advantages is that they are often less intimidated by dominant actors or beholden to prevailing options and agendas. Thai-based and international organizations, however, still dominate many of these initiatives. These are also limitations in terms of continuity of effort as such cooperation is often not strongly institutionalized. Even so, the contributions of non-state actors and the networks that they drive and support are becoming an emergent feature of water governance in the Mekong region.

The five interconnected pathways towards shifting Mekong water governance deserve the attention of both practitioners and scholars. Both can help to shed light on possible approaches in specific Mekong contexts, as well as to develop a more fundamental understanding of how these pathways are activated, subverted or sustained (Foran, 2007).

Conclusions

The future of the waterscapes of the Mekong region has been, and will continue to be, contested. However, the recent history of water governance gives grounds for both concern and hope.

On the one hand, vested, powerful interests continue to dominate decision-making around major hydropower and irrigation infrastructure projects. They do so by keeping key information about plans secret or hard to access, project procedures closed, and by labelling queries, debate and opposition as 'anti-development' and undermining legitimate concerns on impacts by reference to uncertainties. Sophisticated technologies of mapping, modelling and assessment, and even stakeholder consultation, are often turned around and made to serve project sponsors. With little transparency, much of the debate is reduced to ideological rhetoric and positioning.

The promises of benefits from flood protection, dams or irrigation schemes are often not realized. The devil that dwells in development projects' details usually writes the next phase of the story: 'alternatives jobs' do not materialize as expected; the markets for cash crops and aquaculture products proposed as alternatives are nowhere to be found; resettlement takes longer than planned due to delays in new house construction or to villagers refusing to budge; flow alteration incurs severe impacts upon fisheries, recession agriculture or embankment stability; etc. In other words, the social and environmental complexity that is glossed over at the planning stage suddenly erupts and strikes back: the state 'tunnel vision' that had oversimplified the real world (Scott, 1998) is laid bare; the time- and cost-cutting logic of investors works against identification and compensation of impacts.

This, perhaps, explains why the repeated assurance that development actors have 'learned from past mistakes', or that all necessary safeguards are being enforced, sounds hollow. Even when standards have been established and are supposed to be adhered to, capital-intensive projects inserted in contexts where affected populations have limited political clout, environmental values are not well recognized, and poor governance and corruption are pervasive and tend to generate costs and risks that are unequally distributed. Claims of processes that screen poor investments or generate 'good dams' end up being hard to uphold and involve a degree of wishful thinking.[11] Experience tells us that irrigation projects that promise hundreds of thousands of hectares in Cambodia or Thailand, even if eventually not developed on the scale announced, must be considered with much circumspection.

On the one hand, all well-wishing stakeholders may feel compelled to adhere to a vision whereby 'a river of promises is to be transformed into a river of prosperity', new vast paddy irrigation schemes convert water into 'white gold', and hydropower dams are 'powering progress' and 'kick-start[ing] development'[12] in order to 'lift people from poverty and promote sustainable development for all'.[13] On the other, no comfort is offered by recalling sorrowful episodes of the recent past, including the loss of lives and destruction of livelihoods in the Sesan Valley in Cambodia after the construction of the Yali Falls Dam in Vietnam; the late recognition – under public pressure – of the impacts generated by the Theun-Hinboun Dam; the disruptions and mayhem wrought by the Pak Mun Dam for the production of around 0.2 per cent of Thai electricity generation; and the flurry of dam and irrigation projects under consideration and that are being planned again with insufficient mechanisms to assess impacts, crowd out unsound projects or come up with just compensations. Indeed, it makes one uneasy to compare the US$1 million first reserved for mitigation and compensation by the Theun-Hinboun Power Company, with annual revenues of around US$60 million, and the US$2 billion in revenue to the Government of Laos (let alone the return to investors) over 25 years expected from the NT2 dam with the US$90 million earmarked for all social and environmental compensations and mitigations (see Chapter 4).

'Local' issues or problems are downplayed by picturing them against national strategies and interests and then 'scaled out' by framing regional development and cooperation as an overriding goal and irresistible transformation towards prosperity (Mitchell, 1998; Sneddon and Fox, 2006). Basin hydrologic models depict macro-level changes in the flow regime, but not local impacts and ecosystem productivity. Regional cooperation agreements focus on the main stem of the Mekong River and leave wider systemic relationships with tributaries, as well as land and water use, to the responsibility of individual states.

In a recent interview, reflecting on the hydropower explosion, a regional analyst and campaigner for International Rivers acknowledged that pleasing everyone is just not possible: many projects will be undertaken and impacts will have to be dealt with. Some impacts are amenable to mitigation, but not all. To avoid the 'race to the bottom' suggested by the new deals made between governments in the region and banks or construction companies with poor or no social/environmental commitment, it is necessary to constantly and tirelessly reopen and challenge the 'black box' of decision-making, redress power imbalances, and contest the production and mobilization of particular registers of knowledge.

Examples exist of governments, business and communities pursuing, with varying degrees of enthusiasm, iterative and fair approaches to evaluating projects and alternatives. Such examples underline the diverse knowledge sources and understandings that need to be brought together to comprehend livelihoods, ecosystem services, burdens and risks at multiple levels. They also underline the importance of maintaining arenas for deliberation in which people can challenge and express dissent about projects in both their grand conceptions and specific details. As schemes become more elaborate, the needs for public scrutiny and contestation correspondingly increase. Before the waterscapes of the Mekong region are irreversibly transformed, it is crucial that a diverse range of alternatives are fully explored by those who must continue to live within them.

NOTES

1 An ADB official at the Exploring Water Futures Together dialogue in Vientiane, 2006.
2 As illustrated in the film *Tongpan* directed by Paijong Lai-sakul (1977). See also Sluiter (1992).
3 'More often than not the ADB has forced the government to undertake an EIA' (King, 2006).
4 Pers comm at the Vientiane Mekong Dialogue in July 2006.
5 NT2 promoters argued that the project's preparation was a model for future hydropower development and could be used to strengthen the Lao government's capacity to manage new hydropower projects (see Chapter 4).
6 The Raymond Morrison Knudsen-Brown Rootes Jones (RMK-BRJ) company did 97 per cent of the works undertaken by the American army in Vietnam.
7 See, for example, Wheeler (1970), MRC (1995) and Le-Huu and Nguyen-Duc (2003) for unsullied views of Mekong cooperation efforts.
8 The chairman of the Lao National Economic Committee, in 1995, quoted in Goh (2004).
9 Regulation begets bypass strategies, as shown, for example, by the logging bans in China and Thailand that have merely displaced logging activities to poorer neighbour states with looser control (Lang, 2002).
10 Although this is the dominant view of analysts, several accounts stick to the image of the success story mentioned earlier. Le-Huu and Nguyen-Duc (2003), for example, consider that 'the Mekong Committee and current MRC have provided a forum for the four member countries to work out the best solution so that no development is missed or unnecessarily delayed'.

11 The World Bank's December 2004 *Country Economic Memorandum* pointed to the weak governance environment in Laos and noted that without significant governance improvements upfront, hydropower revenues will not result in good development outcomes (see Chapter 4). Likewise, the ADB noted in its technical assistance paper for NT2 that 'the government's capacity to implement large-scale complex hydropower projects still remains a major concern'. According to one diplomat based in Vientiane, the Laotian government is 'pretty good at starting then stopping' its promised reforms, and passing but not implementing regulations to get more foreign aid (Richardson, 2002).

12 'So that we can compete with other countries': an official with the Prime Minister's Office, quoted in Richardson (2002).

13 The primary goal of the GMS programme, as stated at the GMS Summit Meeting in 2002.

References

AOP (Assembly of the Poor) (2000) *Siphok panha sammacha kon chon* [*Sixteen Problems of the Assembly of the Poor*], AOP, Bangkok

Baird, I. G. and Shoemaker, B. (2007) 'Unsettling experiences: Internal resettlement and international aid agencies in Laos', *Development and Change*, vol 38, no 5, pp865–888

Bakker, K. (1999) 'The politics of hydropower: Developing the Mekong', *Political Geography*, vol 18, pp209–232

Bangkok Post (1991) 'Bank chiefs hear both sides of dam issue', *Bangkok Post*, 19 October

Bangkok Post (2004a) '"Plodprasop proposes new pilot project". Bid to end row over Water-Grid project', *Bangkok Post*,14 April

Bangkok Post (2004b) 'Water crisis looms, says grid study: Diversion from across border will be needed', *Bangkok Post*, 13 June

Bangkok Post (2008) 'Samak to revive old dam project: PM on collision course with environmentalists', *Bangkok Post*, 6 June

Barney, K. (2007) *Power, Progress and Impoverishment: Plantations, Hydropower, Ecological Change and Community Transformation in Hinboun District, Lao PDR*, YCAR Papers no 1, Political Ecology Series, York,Toronto

Beck, U. (1992) *Risk Society: Towards a New Modernity*, Sage, London

Bell, P. F. (1969) 'Thailand's northeast: Regional underdevelopment, "insurgency", and official response', *Pacific Affairs*, vol 42, no 1, pp47–54

Biggs, D. A. (2006) 'Reclamation nations: The US Bureau of Reclamation's role in water management and nation building in the Mekong Valley, 1945–1975', *Comparative Technology Transfer and Society*, vol 4, no 3, pp225–246

Blake, D. J. H. (2008) 'The multi-dimensional commons of the lower Nam Songkhram River wetlands, Thailand', Paper submitted to The 12th Biennial Conference of the International Association of Study of the Commons, 14–18 July 2008, University of Gloucester, Cheltenham, UK

Blake, D., Carson, B. and Tubtim, N. (2005) *Review of the Environmental Management Division*, Theun-Hinboun Power Company, 2 March, 2005

Breukers, S. (1999) 'Who defines the "problem", who defines "development"? The case of the Songkhram Irrigation Project', *Watershed*, vol 4, no 2, pp87–97

Bruns, B. (1991) *The Stream the Tiger Leaped: A Study of Intervention and Innovation in Small Scale Irrigation Development in Northeast Thailand*, PhD thesis, Cornell University, NY

CCILMB (Committee for Coordination of Investigations of the Lower Mekong Basin) (1970) *Rapport sur le Plan Indicatif du Bassin*, Bangkok

CCILMB (Committee for Coordination of Investigations of the Lower Mekong Basin) (1972) *The Mekong Project 1972*, Bangkok

Chomchai, P. (1994) *The United States, the Mekong Committee and Thailand: A Study of American Multilateral and Bilateral Assistance to Northeast Thailand since the 1950s*, Asian Studies Monograph no 051, Institute of Asian Studies, Chulalongkorn University, Bangkok

CIFOR (Centre for International Forestry Research) (2004) *Forests and Floods: Drowning in Fiction or Thriving on Facts*, CIFOR, Bogor, Indonesia and FAO, Bangkok, Thailand

Cornford, J. and Matthews, N. (2007) *Hidden Costs: The Underside of Economic Transformation in the Greater Mekong Subregion*, Oxfam Australia

Darling, F. C. (1962) 'American Policy in Thailand', *Political Research Quarterly*, vol 15, no 93, pp93–110

de Koninck, R. (2005) *L'Asie du Sud-Est*, Armand Colin, Paris

Dore, J. (2003) 'The governance of increasing Mekong regionalism', in Mingsarn Kaosa-ard and John Dore (eds) *Social Challenges for the Mekong Region*, White Lotus, Bangkok, Thailand, pp405–440

Dore, J. (2007) 'Multi-stakeholder platforms (MSPS): Unfulfilled potential', in Lebel, L., Dore, J., Daniel, R. and Koma, Y. (eds) *Democratizing Water Governance in the Mekong Rregion*, Mekong Press, Chiang Mai

Drooj, M. (2006) Oral presentation, 6–7 July, Vientiane, Laos

Dryzek, J. S. (2000) *Deliberative Democracy and Beyond: Liberals, Critics, Contestations*, Oxford University Press, Oxford

Dugan, P. (2008) *Examining the Barrier Effect of Mainstream Dams to Fish Migration in the Mekong, with an Integrated Perspective to the Design of Mitigation Measures: Conclusions from an Independent Expert Group Meeting*, 22–23 September 2008, Vientiane, Laos

du Pont, P. (2005) *Nam Theun 2 Hydropower Project (NT2) Impact of Energy Conservation, DSM, and Renewable Energy Generation on EGAT's Power Development Plan*, World Bank, accessed 24 March, http://siteresources.worldbank.org/INTLAOPRD/Resources/DSMmarch2005.pdf

EarthRights International (2005) *Flooding the Future: Hydropower and Cultural Survival in the Salween River Basin*, 28 April, www.earthrights.org/burma

The Economist (2005) 'Dams back in fashion. But have the right lessons been learned?', *The Economist*, 7 April

Ekachai, S. (2008) 'Destructive mega dreams', *Bangkok Post*, 12 June

Feldhoff, T. (2002) 'Japan's construction lobby activities – systemic stability and sustainable regional development', *ASIEN*, vol 84, pp34–42

FIVAS (2007) *Ruined Rivers, Damaged Lives – The Impacts of the Theun-Hinboun Hydropower Project on Downstream Communities in Lao PDR*, FIVAS, Oslo, Norway

Floch, P., Molle, F. and Loiskandl, W. (2007) *Marshalling Water Resources: A Chronology of Irrigation Development in the Chi-Mun River Basin, Northeast Thailand*, Mekong Programme on Water, Environment and Resilience, Chiang Mai University, Thailand International Water Management Institute, Colombo

Foran, T. (2006) *Rivers of Contention: Pak Mun Dam, Electricity Planning, and State–Society Relations in Thailand, 1932–2004*, PhD thesis, Division of Geography, Department of Geosciences, University of Sydney, Australia

Foran, T. (2007) *Advocacy Pathways towards Decision Making in Thailand*, Working paper, Unit for Social and Environmental Research, Chiang Mai University, Chiang Mai, Thailand

Forsyth, T. and Walker, A. (2008) *Forest Guardians, Forest Destroyers: The Politics of Environmental Knowledge in Northern Thailand*, University of Washington Press, WA

Fox, C. A. (2000) *Flexible Sovereignty and the Politics of Hydropower Development in the Mekong River Basin*, PhD thesis, University of Oregon, OR

Friesen, K. M. (1999) *Damming the Mekong: Plans and Paradigms for Developing the River Basin from 1951 to 1995*, PhD thesis, Faculty of the School of International Service, American University, Washington DC

Goh, E. (2004) 'The hydro-politics of the Mekong river basin: Regional cooperation and environmental security', in Thakur, R. and Newman, E. (eds) *Broadening Asia's Security Discourse and Agenda: Political, Social, and Environmental Perspectives*, United Nations University Press, Tokyo

Greacen, C. and Palettu, A. (2007) 'Electricity sector planning and hydropower', in Lebel, L., Dore, J., Daniel, R. and Saing Koma, Y. (eds) *Democratizing Water Governance in the Mekong region*, Mekong Press, Chiang Mai

Hirsch, P. (1998) 'Dams, resources and the politics of environment in mainland Southeast Asia', in Hirsch, P. and Warren, C. (eds) *The Politics of Environment in Southeast Asia: Resources and Resistance*, Routledge, London, Chapter 3, pp55–70

Hirsch, P. and Mørck-Jensen, K. (2006) *National Interests and Transboundary Governance in the Mekong*, Australian Mekong Resource Centre, University of Sydney, Australia

Hirsch, P. and Wyatt, A. (2004) 'Negotiating local livelihoods: Scales of conflict in the Se San River Basin', *Asia Pacific Viewpoint*, vol 45, no 1, April, pp51–68

Hoanh, C. T., Tuong, T. P., Gallop, K. M., Gowing J. W., Kam, S. P., Khiem, N. T. and Phong, N. D. (2003) 'Livelihood impacts of water policy changes: Evidence from a coastal area of the Mekong river delta', *Water Policy*, vol 5, pp475–488

Hori, H. (2000) *The Mekong: Environment and Development*, United Nations University, United Nations University Press, Tokyo, Japan

Hudson-Rodd, N. and Shaw, B. J. (2003) 'Mekong river development: Whose dreams? Which visions?', *Water International*, vol 28, no 2, pp268–275

Ingersol, J. (1968) 'Mekong river basin development: Anthropology in a new setting', *Anthropological Quarterly*, vol 41, no 3, pp147–167

IUCN, TEI, IWMI and M-POWER (2007) *Exploring Water Futures Together: Mekong Region Waters Dialogue: Resource Papers from Regional Dialogue*, IUCN, Thailand Environment Institute, International Water Management Institute and the Mekong Programme on Water, Environment and Resilience, Bangkok, Colombo and Chiang Mai, p132

Jenkins, D. (1968) 'The Lower Mekong Scheme', *Asian Survey*, vol 8, no 6, pp456–464

Karl, H. A., Susskind, L. E. and Wallace, K. H. (2007) 'A dialogue not a diatribe: Effective integration of science and policy through joint fact finding', *Environment*, vol 49, pp20–34

Keeley, J. and Scoones, I. (1999) *Understanding Environmental Policy Processes: A Review*, IDS Working Paper 89, Environment Group, Institute of Development Studies, University of Sussex, Brighton, UK

King, P. (2006) Oral presentation, 6–7 July, Vientiane, Laos

Kirmani, S. S. (1990) 'Water, peace and conflict management: The experience of the Indus and Mekong river basins', *Water International*, vol 15, pp200–205

Lamberts, D. and Koponen, J. (2008) 'Flood pulse alterations and productivity of the Tonle Sap ecosystem: A model for impact assessment', *Ambio*, vol 37, no 3, pp178–184

Lang, G. (2002) *Deforestation, Floods, and State Reactions in China and Thailand*, Working Papers Series no 21, Southeast Asia Research Centre, Hong Kong

Le-Huu, T. and Nguyen-Duc, L. (2003) *Mekong Case Study*, Technical Documents in Hydrology, PC-CP series, No 10, UNESCO, Paris

Manorom, K. (2007) 'People's EIA: A mechanism for grassroot participation in environmental decision-making', *Watershed*, vol 12, no 1, pp26–30

McCool, D. (1987) *Command of the Waters: Iron Triangles, Federal Water Development, and Indian Water*, University of California Press, Berkeley, CA

McDowell, D., Scudder, T. and Talbott, L. (2008) *Lao PDR Nam Theun 2 Multipurpose Project*, 13th and 14th Reports of the International Environmental and Social Panel of Experts, February–April

Menon, P. K. (1972), 'Financing the Lower Mekong River Basin Development', *Pacific Affairs*, vol 44, no 4, pp566–579

Missingham, B. D. (2003) *The Assembly of the Poor in Thailand: From Local Struggles to National Social Movement*, Silkworm Books, Chiang Mai

Mitchell, M. (1998) 'The political economy of Mekong basin development', in Hirsch, P. and Warren, C. (eds) *The Politics of Environment in Southeast Asia: Resources and Resistance*, Routledge, London, pp71–89

Molle, F. (2006) *Planning and Managing Water Resources at the River Basin Level: Emergence and Evolution of a Concept*, Comprehensive Assessment Research Report, no 16, IWMI, Colombo, Sri Lanka

Molle, F. (2008a) 'Nirvana concepts, narratives and policy models: Insights from the water sector', *Water Alternatives*, vol 1, no 1, pp131–156

Molle, F. (2008b) 'Why enough is never enough: The societal determinants of river basin closure', *International Journal of Water Resource Development*, vol 24, no 2, pp247–256

MRC (Mekong River Commission) (1995) *Mekong River Commission towards Sustainable Development*, MRC, Bangkok

MRC (2005) *Public Participation in the Lower Mekong Subregion*, MRC, Vientiane, Laos

MRC (2008) *Flood Situation Report, August 2008*, MRC Technical Paper No. 21, 1 September, Mekong River Commission Secretariat, Vientiane

MRCS/WUP-FIN (2007) *Final Report – Part 2: Research Findings and Way Forward. WUP-FIN Phase 2: Hydrological, Environmental and Socio-Economic Modelling Tools for the Lower Mekong Basin Impact Assessment*, Mekong River Commission and Finnish Environment Institute Consultancy Consortium, Vientiane, Laos, www.eia.fi/wup-fin/wup-fin2/publications.htm.

NEDECO/TEAM (1983) *Pre-Feasibility Study of the Nam Songkhram Basin Irrigation and Flood Control Development*, vol 1, Main report, March 1983, Code 4.56.056, Prepared by NEDECO, Arnhem, The Netherlands, in association with TEAM, Bangkok, Thailand, Interim Committee for Coordination of Investigations of the Lower Mekong Basin, Bangkok

Öjendal, J., Mathur, V. et al (2002) *Environmental Governance in the Mekong: Hydropower Site Selection Processes in the Se San and Sre Pok Basins*, SEI/RESPSI Report Series no 4, Stockholm Environment Institute, Bangkok

Parnwell, M. and Bryant, R. (eds) (1996) *Environmental Change in South-East Asia: People, Politics and Sustainable Development*, Routledge, London

Phillips, D., Daoudy, M., McCaffrey, S., Öjendal, J. and Turton, A. R. (2006) *Transboundary Water Cooperation as a Tool for Conflict Prevention and Broader Benefit-Sharing*, Ministry for Foreign Affairs Expert Group on Development Issues (EGDI), Stockholm

Phongpaichit, P. and Piriyarangsan S. (1996) *Corruption and Democracy in Thailand*, Silkworm Books, Chang Mai, Thailand

Porter, I. and Shivakumar, J. (2008) *Doing Dams Right: The Challenges of Lao Nam Theun 2 (Directions in Development)*, World Bank Publications, World Bank, Washington, DC

Ratner, B. (2003) 'The politics of regional governance in the Mekong River Basin', *Global Change*, vol 15, no 1, pp59–76

Reisner, M. (1986) *Cadillac Desert: The American West and Its Disappearing Water*, Penguin, New York, NY

Repetto, R. (1986) *Skimming the Water: Rent Seeking and the Performance of Public Irrigation Systems*, Research Report no 4, World Resources Institute, Washington, DC, p47

Richardson, M. (2002) 'In its water, Laos sees power to cut poverty', *International Herald Tribune*, 11 March

Rigg, J. (1997) *Southeast Asia: The Human Landscape of Modernisation and Development* Routledge, London

Roe, E. (1991) 'Development narratives, or making the best of blueprint development', *World Development*, vol 19, no 4, pp287–300

Roongrueng, C. (1999) 'There is no other alternative', *Watershed: People's Forum on Ecology*, vol 4, no 2, pp10–16

Samudavanija, C. (1995) *Economic Development and Democracy in Thailand's Industrialisation and Its Consequences*, St Martin's Press, Canberra, Australia

Scholtens, B. and Dam, L. (2007) 'Banking on the Equator: Are banks that adopted the Equator Principles different from non-adopters?', *World Development*, vol 35, no 8, pp1307–1328

Scott, J. C. (1998) *Seeing like a State: How Certain Schemes to Improve the Human Condition Have Failed*, Yale University Press, New Haven and London

Sewell, W. R. and White, G. (1966) *The Lower Mekong: An Experiment in International River Development*, Carnegie Endowment for International Peace, New York, NY, cited in Ingerssol (1968)

Sinh, B. T. (2003) 'The cultural politics of development and environment in Vietnam', in Kaosa-Ard, K. and Dore, J. (eds) *Social Challenges for the Mekong Region*, White Lotus Bangkok, Thailand, pp375–405

Sluiter, L. (1992) *The Mekong Currency*, Project for Ecological Recovery/TERRA, Bangkok

Sneddon, C. and Fox, C. (2006) 'Rethinking transboundary waters: A critical hydropolitics of the Mekong Basin', *Political Geography*, vol 25, pp181–202

Sneddon, C. and Fox, C. (2007) 'Power, development, and institutional change: participatory governance in the lower Mekong basin', *World Development*, vol 35, no 12, pp2161–2181

Sretthachau, C. (1999) 'People movement against dams in Thailand', Paper presented at the Environmental NGOs' International Symposium on Dams, 29 December 1999, Seoul, Korea

SSEO (Shan Sapawa Environmental Organization) (2006) *Warning Signs: An Update on Plans to Dam the Salween in Burma's Shan State*, Chiang Mai, Thailand

Sutton, R. (1999) *The Policy Process: An Overview*, Working paper no 118, ODI, London

Tubb, J. A. (1966) 'A consideration of the fisheries problem of the Lower Mekong Basin', *Indian Journal of Power and River Valley Development*, Mekong Project Number, vol 16, pp62–64

Vaaler, P. M., James, B. E. and Aguilera, R. V. (2008) 'Risk and capital structure in Asian project finance', *Asia Pacific Journal of Management*, vol 25, no 1, pp25–50

Warner, J. F. (2006) 'More sustainable participation? Multi-stakeholder platforms for integrated catchment management', *Water Resources Development*, vol 22, pp15–35

Wheeler, R. A. (1958) *Programme of Studies and Investigations for Comprehensive Development: Lower Mekong Basin*, United Nations Technical Assistance Mission, Headed by Lt General R. A. Wheeler (TAA/AFE 3), UNESCAP, Bangkok

Wheeler, V. M. (1970) 'Co-operation for development in the Lower Mekong Basin', *The American Journal of International Law*, vol 64, no 3, pp594–609

White, G., F., de Vries, E., Dunkerley, H. B. and Krutilla, J. V. (1962) *Economic and Social Aspects of Lower Mekong Development*, UNESCAP, Bangkok

Wipatayotin, A. (2008) 'MRC defends China over Thai floods', *Bangkok Post*, 16 August

Woodall, B. (1993) 'The logic of collusive action: The political roots of Japan's Dango System', *Comparative Politics*, vol 25, no 3, pp297–312

World Bank/ADB (Asian Development Bank) (2006) *Future Directions for Water Resources Management in the Mekong River Basin*, Mekong Water Resources Assistance Strategy, World Bank/ADB

Young, I. M. (2001) 'Activist challenges to deliberative democracy', *Political Theory*, vol 29, pp670–690

Index

Note: *Italic* page numbers refer to figures and tables.

For Product Safety Concerns and Information please contact our EU
representative GPSR@taylorandfrancis.com
Taylor & Francis Verlag GmbH, Kaufingerstraße 24, 80331 München, Germany

www.ingramcontent.com/pod-product-compliance
Ingram Content Group UK Ltd.
Pitfield, Milton Keynes, MK11 3LW, UK
UKHW022006180926
475701UK00003B/229

9781138021181